OXFORD GRADUATE TEXTS IN MATHEMATICS

Series Editors

OXFORD GRADUATE TEXTS IN MATHEMATICS

Books in the series

1. Keith Hannabuss: *An Introduction to Quantum Theory*
2. Reinhold Meise and Dietmar Vogt: *Introduction to Functional Analysis*
3. James G. Oxley: *Matroid Theory*
4. N. J. Hitchin, G. B. Segal, and R. S. Ward: *Integrable Systems: Twistors, Loop Groups, and Riemann Surfaces*
5. Wulf Rossmann: *Lie Groups: An Introduction through Linear Groups*
6. Qing Liu: *Algebraic Geometry and Arithmetic Curves*
7. Martin R. Bridson and Simon M. Salamon (eds): *Invitations to Geometry and Topology*
8. Shmuel Kantorovitz: *Introduction to Modern Analysis*
9. Terry Lawson: *Topology: A Geometric Approach*
10. Meinolf Geck: *An Introduction to Algebraic Geometry and Algebraic Groups*
11. Alastair Fletcher and Vladimir Markovic: *Quasiconformal Maps and Teichmüller Theory*
12. Dominic Joyce: *Riemannian Holonomy Groups and Calibrated Geometry*
13. Fernando Villegas: *Experimental Number Theory*
14. Péter Medvegyev: *Stochastic Integration Theory*
15. Martin A. Guest: *From Quantum Cohomology to Integrable Systems*
16. Alan D. Rendall: *Partial Differential Equations in General Relativity*
17. Yves Félix, John Oprea, and Daniel Tanré: *Algebraic Models in Geometry*
18. Jie Xiong: *Introduction to Stochastic Filtering Theory*
19. Maciej Dunajski: *Solitons, Instantons, and Twistors*
20. Graham R. Allan: *Introduction to Banach Spaces and Algebras*
21. James Oxley: *Matroid Theory, Second Edition*
22. Simon Donaldson: *Riemann Surfaces*
23. Clifford Henry Taubes: *Differential Geometry: Bundles, Connections, Metrics and Curvature*
24. Gopinath Kallianpur and P. Sundar: *Stochastic Analysis and Diffusion Processes*

Stochastic Analysis and Diffusion Processes

GOPINATH KALLIANPUR AND P. SUNDAR

OXFORD
UNIVERSITY PRESS

OXFORD
UNIVERSITY PRESS

Great Clarendon Street, Oxford, OX2 6DP,
United Kingdom

Oxford University Press is a department of the University of Oxford. It furthers the University's objective of excellence in research, scholarship, and education by publishing worldwide. Oxford is a registered trade mark of Oxford University Press in the UK and in certain other countries

First Edition published in 2014

Published in the United States of America by Oxford University Press
198 Madison Avenue, New York, NY 10016, United States of America

British Library Cataloguing in Publication Data
Data available

Library of Congress Control Number: 2013943837

ISBN 978–0–19–965707–0

Dedicated to

Krishna Kallianpur

and

Mrs. Sundara Padmanabhan

Preface

The idea of writing a book on stochastic analysis arose from a suggestion that we write an enjoyable book on Brownian motion and stochastic analysis. After starting our work, we felt that the word "enjoyable" should not be construed as a synonym for "heuristic and non-rigorous"; rather, it should mean "readable and concise with full details". Writing such a book is a tall order, and at best, only partial success is all that one can hope for.

The origin of the book owes itself to our lecture notes formed over a number of years and drawn from many inspiring sources. In the present book, we build the basic theory of stochastic calculus in a self-contained manner in the first six chapters. Starting with Kolmogorov's construction of stochastic processes, Brownian motion and martingales are presented with a view to build stochastic integration theory and study stochastic differential equations. This would constitute the first part of the book.

The next six chapters deal with the probabilistic behavior of diffusion processes and certain finer aspects, applications, and extensions of the theory. One can view it as the second part of this book. The selection of material for the second part of the book reflects our own tastes and provides only a glimpse of some of the active areas of research. Stochastic analysis being so vast, important topics such as Malliavin calculus, stochastic control, and filtering theory, though of interest to us, had to be left out. Instead, we start with martingale problems, a method unique to stochastic analysis, and proceed to discuss the connection between stochastic analysis and partial differential equations and then study Gaussian solutions of stochastic equations. Jump Markov processes, invariant measures, and large deviations principle for diffusions are presented in successive chapters, though each of these can easily form the subject matter for a whole book.

The book is written for graduate students, applied mathematicians, and anyone interested in learning stochastic calculus. The reader is assumed to be knowledgeable in probability theory at a graduate level. A course on stochastic analysis can be designed using the first part of this book along with parts of Chapter 8. A selection of the last six chapters can be used for a second course on stochastic analysis.

Regarding interdependence of chapters, the first part of the book is connected naturally as a sequence with one exception: Chapter 4 is not needed to read Chapters 5 and 6. A good knowledge of the first part is required to read any of the following chapters. However, the second part affords more flexibility in reading. For instance, each of the Chapters 9, 10, and 12 can be read independently of the others.

We are quite indebted to several of our friends and colleagues who helped and inspired us to write this book. We thank several of our colleagues, especially G. Ferreyra, H.-H. Kuo, U. Manna, P. E. Protter, B. Rüdiger, A. Sengupta, S. S. Sritharan, W. Woyczynski, and H. Yin. Several of our graduate students read parts of the book and spotted numerous typos. We appreciate their efforts and thank them for their careful reading. We thank our families, especially Krishna, Kathy, and Vijay for being patient with us and helping us cheerfully while the book was written. Thanks to Ms. Elizabeth Farrell for a thorough proofreading of the entire manuscript.

Contents

1 Introduction to Stochastic Processes

The main result that guarantees the existence of a wide class of stochastic processes is the Kolmogorov consistency theorem. Though the Kolmogorov construction of stochastic processes is set on a very large space equipped with a small σ-field, it is canonical and its applicability is wide. First, we give a proof of this result, followed by important examples that illustrate its usefulness. Next, we introduce basic terminology and notation that are useful in stochastic analysis. The chapter ends with a brief overview of stopping times, associated σ-fields, and progressive measurability.

1.1 The Kolmogorov Consistency Theorem

Throughout the book, Ω will denote an abstract space which, in probability theory, is called the sample space or the space of all outcomes. Let $\mathcal{F}$ denote a σ-field of subsets of Ω, known as the class of events or measurable sets in Ω. A measure P on $(\Omega, \mathcal{F})$ is said to be a probability measure if it is a nonnegative, countably additive set function with $P(\Omega) = 1$. The triplet $(\Omega, \mathcal{F}, P)$ is called a probability space.

In several applications, it is more natural to encounter a finitely additive probability measure, P_0, on a field $\mathcal{G}$ of sets rather than a measure on a σ-field $\mathcal{F}$. The first question that arises is to find conditions under which P_0 can be extended to a probability measure on the σ-field generated by $\mathcal{G}$. The answer is provided by a well-known theorem on extension of measures.

The following proposition is quite useful and easy to prove.

Proposition 1.1.1 *Suppose P_0 is a finitely additive probability measure defined on a field $\mathcal{G}$ of subsets of a space Ω. Let P_0 be continuous at $\emptyset$, that is, if $E_n \in \mathcal{G}$ for all n and $E_n \downarrow \emptyset$, then $P_0(E_n) \downarrow 0$. Then P_0 is a probability measure on $\mathcal{G}$.*

Proof Take any sequence $\{E_n\}$ of pairwise disjoint sets from $\mathcal{G}$ such that $\cup_{j=1}^{\infty} E_j \in \mathcal{G}$. Then for any finite n, $\cup_{j=n+1}^{\infty} E_j \in \mathcal{G}$. Also, one can easily check that $\cup_{j=n+1}^{\infty} E_j \downarrow \emptyset$ as

$n \to \infty$. Therefore $P_0\left(\cup_{j=n+1}^{\infty} E_j\right) \downarrow 0$ as $n \to \infty$. By finite additivity,

$$P_0\left(\cup_{j=1}^{\infty} E_j\right) = P_0\left(\cup_{j=1}^{n} E_j\right) + P_0\left(\cup_{j=n+1}^{\infty} E_j\right)$$

$$= \sum_{j=1}^{n} P_0\left(E_j\right) + P_0\left(\cup_{j=n+1}^{\infty} E_j\right).$$

The partial sums $\sum_{j=1}^{n} P_0\left(E_j\right)$ form a bounded increasing sequence indexed by n, and hence as $n \to \infty$, we have

$$P_0\left(\cup_{j=1}^{\infty} E_j\right) = \sum_{j=1}^{\infty} P_0\left(E_j\right). \qquad \blacksquare$$

The theorem stated below is a standard result in measure theory. A proof of it can be found in many texts (for e.g., [56], pp. 19-23).

Theorem 1.1.2 (The Kolmogorov extension theorem) *If P_0 is a probability measure on a field $\mathcal{G}$ of subsets of Ω, and if $\mathcal{F}$ denotes the σ-field generated by $\mathcal{G}$, then P_0 can be uniquely extended to a probability measure on $\mathcal{F}$.*

A function $X : \Omega \to \mathbb{R}^1$ is called a random variable provided X is an $\mathcal{F}$-measurable function. In probability theory, what we are about to observe is a random variable. The probabilistic information on X is fully contained in the distribution of X. Recall that given a random variable X on $(\Omega, \mathcal{F}, P)$, the distribution of X is the measure μ defined on $(\mathbb{R}^1, \mathcal{B})$ by $\mu(B) = P\{X \in B\}$ for all Borel sets $B \in \mathcal{B}$.

A natural question that arises is the following converse: Given a probability measure μ on $(\mathbb{R}^1, \mathcal{B})$, can one construct a random variable X on some probability space $(\Omega, \mathcal{F}, P)$ such that the distribution of X coincides with the given measure μ? The answer to this question is quite simple. One can take $\Omega = \mathbb{R}^1$, $\mathcal{F} = \mathcal{B}$, and $P = \mu$. Define $X(\omega) = \omega$ for all $\omega \in \Omega$. Thus one can go back and forth between a random variable and its distribution.

If the range of X is $\mathbb{R}^n$, then X is said to be an n-dimensional random vector. The discussion in the above paragraph carries over for probability measures on $\mathbb{R}^n$.

In applications, one assumes the existence of a family of random variables defined on a probability space. When a finite number, n, of random variables out of the given family is considered, it is a random vector and hence gives rise to a finite-dimensional distribution which is a probability measure on $\mathbb{R}^n$. As we vary the selections of the random variables from the given family, we obtain a family of finite-dimensional measures. The assumption mentioned above can be removed, had one started with a family of finite-dimensional distributions. In this situation, it is unclear if there exists a common probability space on which a corresponding family of random variables can be defined. To get an affirmative answer, one needs a certain consistency property in the above family of finite-dimensional measures. This is the content of a famous result known as the

Kolmogorov consistency theorem. We start with a lemma on regularity of probability measures that holds in general for measures defined on the Borel σ-field of complete, separable metric spaces (see [33]). However, we prove the lemma for measures on $(\mathbb{R}^n, \mathcal{B}^n)$.

Lemma 1.1.3 *Let μ be any given probability measure on $(\mathbb{R}^n, \mathcal{B}^n)$. Given any $B \in \mathcal{B}^n$, and $\epsilon > 0$, there exist a compact set K and an open set G such that $K \subseteq B \subseteq G$ with $\mu(G \setminus K) < \epsilon$.*

Proof Let $\mathcal{A}$ denote the class of all Borel sets B that satisfy the stated property. It suffices to show that $\mathcal{A}$ contains all closed sets and is a σ-field.

Step 1 Clearly, the empty set $\emptyset \in \mathcal{A}$. Let $B_N = [-N, N]^{\times n}$, the n-dimensional closed box with origin as the center and side-length $2N$. Since $\mu(B_N) \uparrow 1$ as N increases to ∞, one can choose N large enough so that $\mu(\mathbb{R}^n \setminus B_N) < \epsilon$. Thus, the full space $\mathbb{R}^n \in \mathcal{A}$.

Step 2 We will show that $\mathcal{A}$ is closed under complements and countable unions. If $A \in \mathcal{A}$, then by the definition of $\mathcal{A}$, it is possible to choose $K \subseteq A \subseteq G$ such that $\mu(G \setminus K) < \epsilon/2$. As in Step 1, one can choose N large so that $\mu(B_N^c) < \epsilon/2$.

Clearly, $G^c \subseteq A^c \subseteq K^c$.

$$\begin{aligned}\mu\left(K^c \setminus (G^c \cap B_N)\right) &= \mu\left(K^c \cap (G \cup B_N^c)\right) \\ &\leq \mu(G \setminus K) + \mu(B_N^c) \\ &< \epsilon.\end{aligned}$$

Thus, $A^c \in \mathcal{A}$.

If $\{A_j\}$ is a sequence of sets in $\mathcal{A}$, one can choose $K_j \subseteq A_j \subseteq G_j$ such that $\mu(G_j \setminus K_j) < \epsilon/2^{j+1}$. Therefore,

$$\mu\left((\cup G_j) \setminus (\cup K_j)\right) \leq \sum_j \mu\left(G_j \setminus K_j\right) < \epsilon/2.$$

Let G denote $\cup_j G_j$. Then, $\lim_{N\to\infty} \mu(G \setminus (\cup_{j=1}^N K_j)) < \epsilon/2$. Therefore, for large enough N, one gets $\mu(G \setminus (\cup_{j=1}^N K_j)) < \epsilon$. Thus $\cup_{j=1}^\infty A_j \in \mathcal{A}$. We have thus shown that $\mathcal{A}$ is a σ-field.

Step 3 The class $\mathcal{A}$ contains all closed sets. For, if F is a closed set, then let F_δ denote the delta neighborhood of F. That is,

$$F_\delta = \{x : |x - a| < \delta \text{ for some } a \in F\}.$$

The set F_δ is an open set and decreases to F as $\delta \downarrow 0$. Therefore, one can choose δ small enough so that $\mu(F_\delta \setminus F) < \epsilon/2$.

As in Step 1, choose N large so that $\mu(B_N^c) < \epsilon/2$. Then $\mu(F_\delta \setminus (F \cap B_N)) < \epsilon$ which completes the proof. ∎

Let T denote an index set such as $[0, \infty)$.

Definition 1.1.1 *For any $n \in \mathbb{N}$, let $t_1 < \cdots < t_n$ be any selection (that is, finite sequence) of distinct elements in T. Let $\mu_{t_1,\ldots,t_n}$ be a probability measure on the Borel σ-field of $\mathbb{R}^n$ that corresponds to the selection. The family of probability measures $\{\mu_{t_1,\ldots,t_n} : t_1 < \cdots < t_n, t_i \in T \; \forall \; i, n \in \mathbb{N}\}$ is said to be a consistent family of probability measures if the following condition holds:*

Let $n \in \mathbb{N}$, and $t_1 < \cdots < t_{n+1}$ be any choice of $n+1$ distinct elements from T. Let $B_1, \ldots, B_n$ be one-dimensional Borel sets, arbitrarily chosen. Then, for any k, we have

$$\mu_{t_1,\ldots,t_{n+1}} (B_1 \times \cdots \times B_k \times \mathbb{R} \times B_{k+1} \times \cdots \times B_n) = \mu_{t_1,\ldots,t_k,t_{k+2},\ldots,t_{n+1}} (B_1 \times \cdots \times B_n).$$

Definition 1.1.2 *Let $\mathbb{R}^T$ denote the space of all real-valued functions defined on T. A subset of $\mathbb{R}^T$ is said to be a finite-dimensional* cylinder set *if it is of the form $\{x \in \mathbb{R}^T : (x_{t_1}, \ldots, x_{t_n}) \in B\}$ for any $n \in \mathbb{N}$ and distinct $t_1 < \cdots < t_n \in T$ and B, Borel in $\mathbb{R}^n$.*

The class of all cylinder sets in $\mathbb{R}^T$ will be denoted $\mathcal{C}$. The σ-field generated by $\mathcal{C}$ is denoted by $\mathcal{F}$. For $x \in \mathbb{R}^T$, the projection $x \to (x_{t_1}, \ldots, x_{t_n})$ is denoted by $\pi_{t_1,\ldots,t_n}$.

Theorem 1.1.4 (Kolmogorov consistency theorem) *Suppose the family of probability measures $\{\mu_{t_1,\ldots,t_n} : t_1 < \cdots < t_n, n \in \mathbb{N}\}$ is consistent.*

(i) Then there exists a probability measure μ on $(\mathbb{R}^T, \mathcal{F})$ such that $\mu\pi^{-1}_{t_1,\ldots,t_n} = \mu_{t_1,\ldots,t_n}$ for any $n \in \mathbb{N}$ and distinct $t_1 < \cdots < t_n$.

(ii) If $\Omega := \mathbb{R}^T$ and $P := \mu$, then on $(\Omega, \mathcal{F}, P)$, there exist random variables $\{X_t : t \in T\}$ such that for any distinct $t_1 < \cdots < t_n$, the joint distribution of $X_{t_1}, \ldots, X_{t_n}$ coincides with $\mu_{t_1,\ldots,t_n}$.

Proof

Step 1 The class $\mathcal{C}$ is a field of subsets of $\mathbb{R}^T$. For each $C \in \mathcal{C}$, define a set function

$$P_0(C) = \mu_{t_1,\cdots,t_n}(B)$$

if $C = \{x \in \mathbb{R}^T : (x_{t_1}, \ldots, x_{t_n}) \in B\}$. By using the consistency condition, it follows that P_0 is well defined. It is easy to verify that P_0 is a finitely additive probability measure on $\mathcal{C}$. If P_0 were countably additive on $\mathcal{C}$, we can use the extension theorem to obtain a unique extension of P_0 to a measure on $\mathcal{F}$.

Step 2 By the proposition, it suffices to show that P_0 is continuous at $\emptyset$. We will use the method of contradiction. Let $A_n \in \mathcal{C} \;\; \forall \;\; n$, and $A_n \downarrow \emptyset$. Suppose there exists $\delta > 0$ such that $\lim_{n\to\infty} P_0(A_n) \geq \delta$.

Let $(\{t_1, \ldots, t_{k_n}\}, B_n)$ be a representation in finite dimensions for A_n for each n. Then,

$$P_0(A_n) = \mu_{t_1,\ldots,t_{k_n}}(B_n).$$

By inserting as many sets as needed between each A_j and A_{j+1}, we can assume without loss of generality that $k_n = n$ for all n.

Given any $0 < \epsilon < \delta$, by Lemma 1.1.3, there exists a compact set F_n such that $F_n \subseteq B_n$ and

$$\mu_{t_1,\ldots,t_n}(B_n \setminus F_n) < \epsilon/2^n.$$

Let us denote by E_n the cylinder set represented by $(\{t_1, \ldots, t_n\}, F_n)$.

Then, $P_0(A_n \setminus E_n) < \epsilon/2^n$. Let Λ_n denote $\cap_{j=1}^n E_j$. Then,

$$P_0(A_n - \Lambda_n) = P_0\left(\cup_{j=1}^n (A_n - E_j)\right) \le \sum_{j=1}^n P_0(A_j \setminus E_j) < \epsilon$$

since $A_n \subseteq A_j$ for $j \le n$. Using $\Lambda_n \subseteq E_n \subseteq A_n$, we get

$$P_0(\Lambda_n) \ge P_0(A_n) - \epsilon \ge \delta - \epsilon > 0.$$

Each Λ_n is nonempty, since $P_0(\Lambda_n) > 0$. Therefore, for each n, there exists a $\mathbf{x}^{(n)} \in \Lambda_n$. By using the definition of Λ_n, it follows that for each k, $(x_{t_1}^{(n)}, \ldots, x_{t_k}^{(n)}) \in F_k$ for all $n \ge k$.

Therefore, as a sequence in n, $\{x_{t_1}^{(n)}\}$ is in a closed bounded set F_1 of the real line so that there exists a convergent subsequence $\{x_{t_1}^{(1,n)}\}$ with limit denoted by $\alpha_1 \in F_1$.

The sequence $\{x_{t_1}^{(1,n)}, x_{t_2}^{(1,n)}\}$ is a sequence in F_2 and has a convergent subsequence whose limit is denoted by $(\alpha_1, \alpha_2) \in F_2$. This procedure can be continued to obtain an element $(\alpha_1, \ldots, \alpha_k) \in F_k$ for any finite k.

It is clear that there exist elements y of $\mathbb{R}^T$ such that $y_{t_k} = \alpha_k$ for all k. Any such element y has the property that $(y_{t_1}, \ldots, y_{t_k}) \in F_k$ for any finite k. Therefore, $y \in E_k \subseteq A_k$ for all k, which implies that $\cap_{n=1}^\infty A_n$ is non-empty. This contradicts the assumption that $\cap_{n=1}^\infty A_n$ is empty. We have thus proved that P_0 on $\mathcal{C}$ is countably additive. Invoking the Kolmogorov extension theorem, there exists a unique probability measure μ on $(\mathbb{R}^T, \mathcal{F})$ that extends P_0. The proof of part (i) is over.

Step 3 Set $\Omega = \mathbb{R}^T$ and $P = \mu$ on $(\Omega, \mathcal{F})$. Let ω denote a generic element of Ω. Define the functions $X_t(\omega) = \omega_t$ for all $t \in T$ and $\omega \in \Omega$.

Clearly, $\{X_t \le x\} = \{\omega : \omega_t \le x\} \in \mathcal{C} \subseteq \mathcal{F}$, so that X_t is a random variable. Further,

$$P\left\{\cap_{j=1}^n (X_{t_j} \le x_j)\right\} = \mu\left\{\omega : \cap_{j=1}^n (\omega_{t_j} \le x_j)\right\} = \mu_{t_1,\ldots,t_n}\left\{\times_{j=1}^n (-\infty, x_j]\right\},$$

so that the finite-dimensional distribution of the vector random variable $X_{t_1}, \ldots, X_{t_n}$ is given by $\mu_{t_1,\ldots,t_n}$. ∎

It is possible to reformulate the Kolmogorov consistency theorem in terms of characteristic functions. Toward building such a statement, consider a probability measure P

on $(\mathbb{R}^T, \mathcal{F})$. Let F^T denote the collection of all points λ in $\mathbb{R}^T$ such that all but a finite number of coordinates of λ are zero. Define for any $\lambda \in F^T$,

$$\phi(\lambda) = \int_{\mathbb{R}^T} e^{i(\lambda,x)} P(dx) \tag{1.1.1}$$

where $(\lambda, x) = \sum_j \lambda_{t_j} x_{t_j}$ is a finite sum.

Let $(t_1 < \cdots < t_n)$ be a fixed finite number of distinct elements of T, and let λ be any element of F^T with $\lambda_t = 0$ if $t \neq t_j$ for any $j = 1, \ldots, n$. The function $\phi(\lambda)$ restricted to such λ is known as the *section* of ϕ determined by $(t_1, \ldots, t_n)$.

Theorem 1.1.5 (A reformulation of the Kolmogorov consistency theorem) *Let ϕ be a given complex-valued function on F^T. If any arbitrary section of a function ϕ is a characteristic function, then there exists a probability measure P on $(\mathbb{R}^T, \mathcal{F})$ such that* (1.1.1) *holds for all $\lambda \in F^T$.*

We call the family of random variables $\{X_t : t \in T\}$ obtained in the consistency theorem the canonical process. It should be observed that the space Ω in part (ii) of the above theorem is too large while the σ-field $\mathcal{F}$ is too small. In fact, Ω is the space of all real-valued functions defined on T. The smallness of $\mathcal{F}$ can be inferred from the following result and examples. We will assume that T is nondenumerable. In what follows, the symbols $\omega(t_j)$ and ω_{t_j} are synonymous. Let A be a subset in Ω. If there exist a countable set $\{t_1, t_2, \ldots\}$ in T and a Borel set $B \in \mathcal{B}(\mathbb{R}^T)$ such that $A = \{\omega : (\omega(t_1), \omega(t_2), \ldots) \in B\}$, then A will be called as a **set with a countable base.**

The following theorem is due to Dynkin, who extracted a particularly useful part of an otherwise general, result due to Sierpinski. The theorem is given without a proof. Before stating it, let us recall that a collection $\mathcal{A}$ of subsets of a space E is known as a **Dynkin class** or a λ-system if $E \in \mathcal{A}$ and if $\mathcal{A}$ is closed under proper differences and increasing unions.

Theorem 1.1.6 (The Dynkin class theorem) *Let $\mathcal{S}$ be a π system, that is, a collection of subsets of E that is closed under finite intersections. If $\mathcal{A}$ is a λ-system with $\mathcal{S} \subseteq \mathcal{A}$, then $\sigma(\mathcal{S}) \subseteq \mathcal{A}$.*

Lemma 1.1.7 *Let $\mathcal{D}$ denote the class of all sets with a countable base. Then $\mathcal{F} := \sigma(\mathcal{C}) \subseteq \mathcal{D}$.*

Proof It is clear that $\mathcal{C} \subseteq \mathcal{D}$. The class $\mathcal{C}$, being a field, is a π-system.

By taking any countable set $\{t_1, t_2, \ldots\}$ in T and $B = \mathbb{R}^T$, we get the full space $\Omega \in \mathcal{D}$. Let $A_1, A_2 \in \mathcal{D}$ and $A_1 \subseteq A_2$. If A_1 is represented as $A_1 = \{\omega : (\omega(t_1), \omega(t_2), \ldots) \in B_1\}$, then there exists a Borel set B_2 such that $A_2 = \{\omega : (\omega(t_1), \omega(t_2), \ldots) \in B_2\}$ with $B_1 \subseteq B_2$. Therefore, $A_2 \backslash A_1 = \{\omega : (\omega(t_1), \omega(t_2), \ldots) \in B_2 \backslash B_1\}$, so that $\mathcal{D}$ is closed under proper differences.

If $\{A_j\}$ is an increasing sequence in $\mathcal{D}$ and if $A_1 = \{\omega : (\omega_{t_1}, \omega_{t_2}, \ldots) \in B_1\}$, then each A_j can be represented by using the same set $\{t_1, t_2, \ldots\}$. In fact, $A_j = \{\omega : (\omega_{t_1}, \omega_{t_2}, \ldots) \in B_j\}$ with $B_i \subseteq B_j$ for all $i \leq j$. Thus, $\cup A_j = \{\omega : (\omega_{t_1}, \omega_{t_2}, \ldots) \in \cup B_j\}$, which shows that $\mathcal{D}$ is closed under increasing unions.

The class $\mathcal{D}$ is thus a Dynkin class, and hence, $\mathcal{F} = \sigma(\mathcal{C}) \subseteq \mathcal{D}$. ■

Example 1.1.8 *Let Ω_c be the set of all continuous functions in Ω. Then Ω_c is not in $\mathcal{F}\sigma(\mathcal{C})$.*

Proof If Ω_c were in $\mathcal{F}$, then by the above lemma, Ω_c is a set with a countable base so that

$$\Omega_c = \{\omega : (\omega_{t_1}, \omega_{t_2}, \ldots) \in B\}.$$

However, this is not possible since discontinuous functions are also included in the set on the right side. ■

Next, we show that for any probability measure P on $(\Omega, \mathcal{F})$, the inner measure $\underline{P}$ of Ω_c is zero. Recall that if A is any subset of Ω, then the outer measure of A is defined by

$$\bar{P}(A) = \inf\left\{P(E) : A \subseteq E = \cup_j E_j \quad \text{where} \quad E_j \in \mathcal{C}\right\}.$$

The inner measure $\underline{P}(A)$ is defined by $\underline{P}(A) = 1 - \bar{P}(A^c)$.

Proposition 1.1.9 $\underline{P}(\Omega_c) = 0$.

Proof We will show that $\bar{P}(\Omega_c^c) = 1$. Let $(\Omega_c^c) \subseteq E = \cup_j E_j$ where $E_j \in \mathcal{C}$ for all j. Let $S = \{s_1, s_2, \ldots\} \subseteq T$ be a countable base for E so that there exists a Borel $B \in \mathcal{B}(\mathbb{R}^\infty)$ such that

$$E = \{\omega : (\omega_{s_1}, \omega_{s_2}, \ldots) \in B\}. \tag{1.1.2}$$

If $\omega* \in \Omega$ is any continuous, real-valued function on T, then let ω_0 be a *discontinuous* function on T whose values coincide with the values of $\omega*$ on S. It follows that $\omega* \in E$, since the right side of (1.1.2) depends only on the function values on the countable set S. Thus, $E = \Omega$, and hence $P(E) = 1$, which yields that, $\bar{P}(\Omega_c^c) = 1$. ■

Another example of a set with zero inner measure is given by

$$\Omega_m = \{\omega \in \Omega : \omega(t) \quad \text{is a Lebesgue measurable function of } t\}.$$

The proof is similar to the one given above and consists in showing that $\bar{P}(\Omega_m^c) = 1$.

We will now construct a rather important measure P on $(\Omega, \mathcal{F})$ when the index set $T = [0, \infty)$ by specifying the family of finite-dimensional measures as follows:

$$\mu_{t_1, t_2, \ldots, t_n}(B_1 \times B_2 \cdots \times B_n) = \int_{B_1} \cdots \int_{B_n} \Pi_{j=1}^n p\left(t_j - t_{j-1}; x_j - x_{j-1}\right) dx_n \cdots dx_1 \tag{1.1.3}$$

where $0 < t_1 < t_2 \cdots < t_n$, B_j are any Borel sets in $\mathbb{R}$, $t_0 = 0$, $x_0 = 0$, and

$$p(t; x) = \frac{1}{\sqrt{2\pi t}} exp\left\{-\frac{x^2}{2t}\right\}$$

for all $t > 0$ and any real number x. When $t = 0$, $\mu_0(B) = \delta_0(B)$ for any Borel set B.

It is easy to show that the above family of probability measures satisfies the Kolmogorov consistency condition so that there exists a family of random variables $\{X_t\}$ on the probability space $(\Omega, \mathcal{F}, P)$ such that their finite-dimensional distributions are given by $\{\mu_{t_1,t_2,\ldots,t_n}\}$. Recall that the process was defined by $X_t(\omega) = \omega(t)$ for all $\omega \in \Omega$. By the specification of measures (1.1.3), the following properties of $\{X_t\}$ follow easily:

(i) The random variable X_0 is identically equal to 0.
(ii) For any $t > 0$, X_t is an $N(0,t)$ random variable.
(iii) For any $0 < s < t$, $X_t - X_s$ is an $N(0, t-s)$ random variable.
(iv) If $0 < s_1 < t_1 \le s_2 < t_2 \cdots \le s_n < t_n$, the random variables $\{X_{t_j} - X_{s_j} : j = 1, \ldots n\}$ are independent.

We will proceed to show that under this measure P, $\bar{P}(\Omega_c) = 1$.

In terms of the canonical process $\{X_t\}$, this assertion would translate to $P(\omega : \{X_t(\omega)\} \in E) = 1$ for any set $E \in \mathcal{F}$ that contains Ω_c. Such sets E have a countable base and yet must successfully diagnose the property of continuity in t of ω. In other words, E must detect the uniform continuity of ω in the interval $[0, T]$ for all $T > 0$. Toward this goal, we prove below the following propositions.

Proposition 1.1.10 *Let $T := \{t_j : j = 0, 1, \ldots, n\}$ arranged in increasing order with $t_0 = 0$ and $t_n = 1$. Then, for any x,*

(i) $P\{\max_{t\in T} X_t > x\} \le 2P\{X_1 > x\}$, and
(ii) $P\{\max_{t\in T} |X_t| > x\} \le 2P\{|X_1| > x\}$.

Proof Define $Y_j = X_{t_j} - X_{t_{j-1}}$, and $\tau = \inf\{j : X_{t_j} > x\}$. Then $P\{\max_{tinT} X_t > x\} = \cup_{j=1}^{n} p\{\tau = j\}$.

By the symmetry of the distribution of $X_1 - X_{t_j}$, we can write

$$\frac{1}{2} P\left\{\max_{t\in T} X_t > x\right\} = \sum_{j=1}^{n} P\{\tau = j\} P\left\{X_1 - X_{t_j} \ge 0\right\}.$$

The event $\{\tau = j\}$ belongs to the σ-field generated by $\{X_{t_1}, \ldots, X_{t_j}\}$, namely $\sigma\left(X_{t_1}, \ldots, X_{t_j}\right)$ which is the same as

$$\sigma\left(X_{t_1}, X_{t_2} - X_{t_1}, \ldots, X_{t_j} - X_{t_{j-1}}\right).$$

Therefore, the events $\{\tau = j\}$ and $\{X_1 - X_{t_j} > 0\}$ are independent so that

$$\begin{aligned}\frac{1}{2}P\left\{\max_{t\in T} X_t > x\right\} &= \sum_{j=1}^{n} P\left\{\tau = j, X_1 - X_{t_j} \geq 0\right\}\\ &\leq \sum_{j=1}^{n} P\{\tau = j, X_1 > x\}\\ &\leq P\{X_1 > x\}.\end{aligned}$$

The first inequality in the proposition is thus proved. To prove the second inequality, take any $x > 0$ and note that

$$\begin{aligned}\left\{\max_{t\in T_n} |X_t| > x\right\} &= \left\{\max_{t\in T_n} X_t > x\right\} \cup \left\{\min_{t\in T_n} X_t < -x\right\}\\ &= \left\{\max_{t\in T_n} X_t > x\right\} \cup \left\{\max_{t\in T_n} (-X_t) > x\right\}.\end{aligned}$$

The finite-dimensional distribution of $\{-X_t\}$ is the same as that of $\{X_t\}$. Using the inequality (i),

$$\begin{aligned}P\left\{\max_{t\in T_n} |X_t| > x\right\} &= P\left\{\max_{t\in T_n} X_t > x\right\} + P\left\{\max_{t\in T_n} (-X_t) > x\right\}\\ &\leq 2\left(P\{X_1 > x\} + P\{-X_1 > x\}\right)\\ &= 2P\left\{|X_1| > x\right\}.\end{aligned}$$

■

Remark

1. The inequalities in the above proposition are valid even if T is a countable set in $[0, 1]$. Indeed, write T as $\cup T_n$ where T_n is an increasing sequence of finite ordered subsets of T such that $0, 1 \in T_n$ for each n. For each T_n, the inequalities hold. Let $n \to \infty$ to complete the proof.
2. For simplicity, we have taken $t_0 = 0$ and $t_n = 1$. For general $t_0 = a$ and $t_n = b$ the inequalities would appear as follows:

$$P\left\{\max_{t\in T} X_t - X_a > x\right\} \leq 2P\{X_b - X_a > x\}$$

$$P\left\{\max_{t\in T} |X_t - X_a| > x\right\} \leq 2P\left\{|X_b - X_a| > x\right\}.$$

Proposition 1.1.11 *On any countable set D in $[0, \infty)$ and any for any finite $T > 0$, the process $\{X_t\}$ restricted to $D \cap [0, T]$ is uniformly continuous with probability one.*

Proof Without loss of generality, let $T = 1$. Let us enlarge D by including the set of points $\{k/2^n : k = 1, \ldots, 2^n\}$ for all $n \geq 1$, and still denote it as D. Define

$$M_n = \sup_{\{t_i, t_j \in D : |t_j - t_i| \leq 1/2^n\}} |X_{t_j} - X_{t_i}|.$$

We need to show that $\lim_{n \to \infty} M_n = 0$ almost surely (a.s.). Since M_n is a decreasing sequence, it suffices to show that for any $\epsilon > 0$,

$$\lim_{n \to \infty} P\{M_n > \epsilon\} = 0.$$

Let $I_k = [k/2^n, (k+1)/2^n]$. Define

$$Z_k = \sup_{t \in I_k \cap D} |X_t - X_{k/2^n}| \quad \forall \ k = 0, 1, \ldots, 2^n - 1.$$

By the triangle inequality, $M_n \leq 3 \max_k Z_k$. It is enough to show that

$$P\left\{\max_k Z_k > \epsilon\right\} \to 0$$

as $n \to \infty$. Clearly,

$$P\left\{\max_k Z_k > \epsilon\right\} = P\left(\cup \{Z_k > \epsilon\}\right) \leq \sum_{k=0}^{2^n - 1} P\{Z_k > \epsilon\}. \tag{1.1.4}$$

By the remark following the above proposition,

$$\begin{aligned} P\left\{\max_{t \in I_k \cap D} |X_t - X_{k/2^n}| > \epsilon\right\} &\leq 2P\left\{|X_{(k+1)/2^n} - X_{k/2^n}| > \epsilon\right\} \\ &= 2P\left\{|X_{1/2^n}| > \epsilon\right\}. \end{aligned}$$

By using this estimate in (1.1.4),

$$P\left\{\max_k Z_k > \epsilon\right\} \leq 2^{n+1} P\left\{|X_{1/2^n}| > \epsilon\right\}$$

which tends to zero as $n \to \infty$. In fact, as n tends to ∞, it is easily seen that $2^n P\{|X_{1/2^n}| > \epsilon\} \to 0$ by the Markov inequality. The proof is thus completed. ■

Theorem 1.1.12 *The outer measure $\bar{P}$ of the set of continuous functions in Ω is one.*

Proof If E is any set in $\mathcal{F}$, then E is countably based, and we can augment the countable base by all the dyadic rationals. Let us call the enlarged countable base D.

Set $D_n = D \cap [0, n]$ for all integers $n \geq 1$.

Define $U = \cap_n \{\omega : \omega|_{D_n}$ is uniformly continuous$\}$. If $\Omega_c \subseteq E$, then $U \subseteq E$. By the above proposition, for each n,

$$P\left\{\omega : \omega|_{D_n} \text{ is uniformly continuous}\right\} = 1.$$

Therefore, $P(E) = 1$. We conclude that $\bar{P}(\Omega_c) = 1$. ■

Remark To sum up, we have proved in Proposition 1.1.9 and in Theorem 1.1.12 that $\bar{P}\{(\Omega_c)^c\} = 1$ and $\bar{P}(\Omega_c) = 1$. This is possible since the outer measure is not an additive set function.

The following proposition allows us to localize P to Ω_c and construct a probability measure on $(\Omega_c, \mathcal{F} \cap \Omega_c)$.

Proposition 1.1.13 *Let $(\Omega, \mathcal{F}, P)$ be any probability space, and let $A \subseteq \Omega$ be a subset having P-outer measure one. Then, there exists a unique probability measure Q on $(A, \mathcal{F} \cap A)$ with the property that $Q(E \cap A) = P(E)$ for all $E \in \mathcal{F}$.*

Proof We have to show that the measure Q is well defined. If E_1 and E_2 are in $\mathcal{F}$, and $E_1 \cap A = E_2 \cap A$, we need to prove that $P(E_1) = P(E_2)$. The symmetric difference $E_1 \Delta E_2$ satisfies

$$\begin{aligned} I_{E_1 \Delta E_2} &= I_{E_1} + I_{E_2} \quad (\text{mod } 2) \\ &= I_{\{E_1 \setminus A\}} + I_{\{E_2 \setminus A\}} \quad (\text{mod } 2) \\ &= I_{\{E_1 \setminus A\} \Delta \{E_2 \setminus A\}} \\ &= I_{\{(E_1 \Delta E_2) \setminus A\}} \\ &\leq I_{\{\Omega \setminus A\}} \end{aligned}$$

where the second equality is due to $E_1 \cap A = E_2 \cap A$.

Since A has outer measure one, $P(E_1 \Delta E_2) = 0$. Therefore, $P(E_1) = P(E_2)$.

Thus Q is well defined. The uniqueness of Q is easily seen. ■

The measure P that we constructed on $(\Omega, \mathcal{F})$ can thus be restricted to Ω_c, and named as Q. The canonical process $\{Y_t\}$ on $(\Omega_c, \mathcal{F} \cap \Omega_c, Q)$ has the same finite-dimensional distributions as the canonical process $\{X_t\}$ on $(\Omega, \mathcal{F}, P)$. In addition, the process $\{Y_t\}$ has continuous paths with Q measure one.

1.2 The Language of Stochastic Processes

Let C denote the space of all continuous, real-valued functions defined on $[0, \infty)$. Let C be equipped with the metric

$$d(f,g) = \sum_{n=1}^{\infty} \frac{1}{2^n} \left(1 \wedge \sup_{t \in [0,n]} |f(t) - g(t)|\right)$$

for any $f, g \in C$. Under this topology of uniform convergence on compact subsets of $[0, \infty)$, the space C is a complete separable metric space. Let the Borel σ-field of C be denoted by $\mathcal{B}$. Let $\mathcal{G}$ denote the σ-field in C generated by the class of all finite-dimensional cylinder sets.

Lemma 1.2.1 *The σ-field $\mathcal{G}$ is equal to the Borel σ-field $\mathcal{B}$.*

Proof Let G be an open subset of C. Then, G is a countable union of open balls. Consider $B_\epsilon(g)$, the ϵ ball in C centered at g. Clearly,

$$B_\epsilon(g) = \left\{ f \in C : \sum_{n=1}^{\infty} \frac{1}{2^n} 1 \left(\wedge \sup_{t \in [0,n] \cap \mathbb{Q}} |f(t) - g(t)| < \epsilon \right) \right\}.$$

The set on the right side is in $\mathcal{G}$ by Lemma (1.1.7). Hence we can conclude that, $\mathcal{B} \subseteq \mathcal{G}$.

To prove the converse, consider the set $A = \{f \in C : f(t) \in O\}$ for any fixed $t > 0$ and any open set O in $\mathbb{R}^1$. If h is any element in A, then there exists $\epsilon > 0$ such that the interval $(h(t) - \epsilon, h(t) + \epsilon) \in O$. It follows that $B_{\epsilon/2}(h) \subseteq A$ so that A is an open set in C. Hence, $A \in \mathcal{B}$. The proof is completed upon noting that sets of the form A generate $\mathcal{G}$. ∎

It is clear from the above lemma that the probability measure Q constructed in the previous section can indeed be considered as a measure on $(C, \mathcal{B})$ since $\Omega_c = C$ and $\mathcal{F} \cap \Omega_c = \mathcal{B}$. The measure Q is known as the **Wiener measure**.

Thus far, the family of random variables that we have encountered has been the canonical (or the coordinate) process, which is just one example of what are known as stochastic processes. Before we proceed further, a few basic definitions are needed and given below.

Definition 1.2.1 *Let $(\Omega, \mathcal{F}, P)$ be any complete probability space, and S, a complete separable metric space. Let $\mathcal{B}$ denote the σ-field of Borel sets of S. Then, $X = \{X_t : t \in T\}$ is called a stochastic process if, for each $t \in T$, X_t is an S-valued random variable. The space S is called the state space of X.*

The index set T, from now on, is either a finite interval, $[0, \infty)$, or $(-\infty, \infty)$. It usually represents time. For each fixed $\omega \in \Omega$, the function $t \to X_t(\omega)$ is known as the **sample path** or **trajectory** associated with ω. A process $X = \{X_t\}$ is said to be continuous, right-continuous, or left-continuous if for P-almost all ω, the trajectory of ω has this property.

Definition 1.2.2 *A stochastic process $X = \{X_t\}, t \in T$, is a measurable (or jointly measurable) process if the map $(t, \omega) \to X_t(\omega)$ is measurable with respect to $\mathcal{B}(T) \times \mathcal{F}$.*

The basic regularity that one would expect in a stochastic process is its measurability. *We will assume that all processes considered in this book are measurable.*

Definition 1.2.3 *Let $X = \{X_t\}$ and $Y = \{Y_t\}$ be two stochastic processes defined on $(\Omega, \mathcal{F}, P)$. They are said to be* **versions** *or* **modifications of** *each other if*

$$P\{\omega : X_t(\omega) = Y_t(\omega)\} = 1 \text{ for each } t.$$

If X and Y are continuous (or right-continuous, or left-continuous) processes, and if X is a version of Y, then the following stronger statement holds:

$$P\{\omega : X_t(\omega) = Y_t(\omega) \ \ \forall t \in \mathbb{R}^+\}\, 1.$$

The above equality is called **indistinguishability** of the processes X and Y. A weaker notion of modification is the concept of **equivalence** of processes given below. It makes sense even when two processes are defined on different probability spaces.

Definition 1.2.4 *Let $X = \{X_t\}$ and $Y = \{Y_t\}$ be two stochastic processes defined on $(\Omega, \mathcal{F}, P)$ and $(\Omega', \mathcal{F}', P')$ respectively. The processes are said to be* **equivalent** *if for every $\{t_1, \ldots, t_n\} \subseteq T$ and $B_j \in \mathcal{B}$ for $j = 1, \ldots, n$*

$$P\{\omega : X_{t_1}(\omega) \in B_1, \ldots, X_{t_n}(\omega) \in B_n\} = P'\left\{\omega' : X'_{t_1}(\omega') \in B_1, \ldots, X'_{t_n}(\omega') \in B_n\right\}.$$

Using the concept of equivalence of processes, we have:

Definition 1.2.5 *Let $X = \{X_t\}$ be a given stochastic process on $\left(\mathbb{R}^T, \mathcal{F}, P\right)$. Let $A \subseteq \mathbb{R}^T$ and $\mathcal{A}$ be the σ-field generated by the finite-dimensional Borel cylinder subsets of A. Then, X is said to have a* **realization** *in A if there exists a probability measure P' on $(A, \mathcal{A})$ such that the coordinate process $Y_t(\omega) = \omega(t)$ for $\omega \in A$ is equivalent to X.*

For instance, the process $\{X_t\}$ (with finite-dimensional distributions specified by equation (1.1.3)) defined on $\left(\mathbb{R}^T, \mathcal{F}, P\right)$ is equivalent to the canonical process $\{Y_t\}$ on $(C, \mathcal{B}, Q)$ and hence has a realization in the space of continuous functions. It is worthwhile to note that such a property does not hold even for the simplest process $X = \{X_t\}$ of mutually independent and identically distributed (iid) random variables as shown in [39]:

Example 1.2.2 *Let $T = [0, 1]$. Let $\{X_t\}$ be iid random variables with range of X_t containing at least two distinct values. Then, the process X doesn't have a realization in $C[0, 1]$.*

Proof Assume the contrary. Suppose μ is the probability measure on $C = C[0, 1]$ induced by the distribution of $\{X_t\}$. Let a be chosen such that $\mu\{x \in C : x(1) > a\}$ is strictly between 0 and 1. Call it δ. Define

$$F_k = \{x \in C : x(1 - 1/n) > a \text{ for } n = k + 1, \ldots, 2k\}.$$

Since $\mu(F_k) = \delta^k$, we get $\sum_{k=1}^{\infty} \mu(F_k) < \infty$. By the Borel-Cantelli lemma, $\mu\{\limsup_{k\to\infty} F_k\} = 0$. This implies that $\mu\{x \in C : x(1) > a\} = 0$, a contradiction since $\delta > 0$. ■

In fact, there is no measurable process equivalent to X as the following example illustrates. However, such processes arise as models for white noise, and are treated in a different framework.

Example 1.2.3 *Let $X = \{X_t\}$ be iid random variables as in the previous example, where $t \in [0, 1]$. Then, there is no measurable process equivalent to X.*

Proof Assume the contrary. Choose a large enough K, and define the process $Y_t = X_t 1_{(X_t \le K)}$ where Y_t takes at least two distinct values. Let $E(Y_t) = m$ and variance $V(Y_t) = \sigma^2$ so that $Z_t := \dfrac{Y_t - m}{\sigma}$ is a process consisting of iid random variables with zero expectation and unit variance. Our supposition would imply that $\{Z_t\}$ is a measurable process. We will establish a contradiction.

If I is any subinterval of $[0, 1]$, then $E \int_I \int_I |Z_t Z_s| \, dt \, ds < \infty$.

By Fubini's theorem,

$$E\left[\left(\int_I Z_t \, dt\right)^2\right] = E\left(\int_I \int_I Z_t Z_s \, dt \, ds\right) = \int_I \int_I E\left(Z_t Z_s\right) \, dt \, ds = 0 \tag{1.2.1}$$

since $E(Z_t Z_s) = 0$ if $t \neq s$, and is $= 1$ if $t = s$. Hence, $\int_I Z_t(\omega) \, dt = 0$ P-a.a. ω. We thus obtain a P-null set N_I such that $\int_I Z_t(\omega) \, dt = 0$ if ω is not in N_I. Let $\mathcal{I}$ be the class of all intervals in $[0, 1]$ with rational endpoints. Define $N = \cup_{I \in \mathcal{I}} N_I$. Clearly, $P(N) = 0$, and for $\omega \in N^c$, $\int_a^b Z_t(\omega) \, dt = 0$ for any $[a, b] \subseteq [0, 1]$.

Thus, for any $\omega \in N^c$, $Z_t(\omega) = 0$, except possibly for a set of Lebesgue measure zero. Therefore, by Fubini,

$$\int_0^1 \int_\Omega Z_t^2(\omega) \, P(d\omega) \, dt = 0,$$

a contradiction since $E(Z_t^2) = 1$, and hence the left side in the above equation is equal to 1. ■

1.3 Sigma Fields, Measurability, and Stopping Times

Let $(\Omega, \mathcal{F}, P)$ be a given probability space. An increasing family of sub σ-fields $(\mathcal{F}_t : t \in \mathbb{R}^+)$ of $\mathcal{F}$, that is, $\mathcal{F}_s \subseteq \mathcal{F}_t \subseteq \mathcal{F}$ for any $s \le t$, is known as a **filtration**. A stochastic process $X = \{X_t\}$ is said to be **$\mathcal{F}_t$-adapted** if each X_t is measurable with respect to $\mathcal{F}_t$.

The sub σ-field $\mathcal{F}_t$ provides us information about the way in which the full space Ω is partitioned at time t. As time increases, the partitions become finer. One refers to $\mathcal{F}_t$ as the information available up to time t.

If $X = \{X_t\}$ is a given process, let $\sigma(X_s : 0 \le s \le t)$ be the smallest σ-field with respect to which the random variables X_s are measurable for all $s \le t$. Let us denote this σ-field by $\mathcal{F}_t^{X,0}$. The family $\left\{\mathcal{F}_t^{X,0} : t \ge 0\right\}$ is known as the **natural filtration** for the process X. Let us also define

$$\mathcal{F}_t^X = \mathcal{F}_t^{X,0} \vee \{P - \text{null sets of } \mathcal{F}\}$$

that is, the augmentation of $\mathcal{F}_t^{X,0}$ by the P-null sets of $\mathcal{F}$.

We will *assume* in this book that

(i) *The space* $(\Omega, \mathcal{F}, P)$ *is a complete probability space,* and
(ii) *The* σ *field* $\mathcal{F}_0$ *contains all P-null sets in* $\mathcal{F}$.

Let $(\Omega, \mathcal{F}, (\mathcal{F}_t), P)$ be the given probability space with a **filtration**.

Definition 1.3.1 *A random variable* $\tau : \Omega \to [0, \infty]$ *is called a* **stopping time** *if the event* $\{\tau \leq t\} \in \mathcal{F}_t$ *for any* $t \in \mathbb{R}^+$.

Note that a stopping time is allowed to take the value ∞.

Example 1.3.1 *Let X be an adapted continuous process. Let F be a closed set. Then,* $\tau = \{t > 0 : X_t(\omega) \in F\}$ *is a stopping time.*

Proof Define $G_n = \{x : d(x, F) < 1/n\}$, where $d(x, F)$ denotes the distance from the point x to the set F. Then, G_n is an open set. By continuity of the process,

$$\{\tau \leq t\} = \{X_t \in F\} \cup \left\{\cap_n \cup_{s \in \mathbb{Q} \cap [0,t)} \{X_s \in G_n\}\right\}.$$

The expression on the right side belongs to $\mathcal{F}_t$. ■

The stopping time τ given in the above example is known as the **hitting time** of F for X.

If τ_1 and τ_2 are stopping times, then $\tau_1 \wedge \tau_2 := \min\{\tau_1, \tau_2\}$, $\tau_1 \vee \tau_2 := \max\{\tau_1, \tau_2\}$, and $\tau_1 + \tau_2$ are all stopping times. If $\{\tau_n\}$ is a sequence of stopping times, then $\sup \tau_n$ is a stopping time. These statements are easy to prove, and hence left as exercises.

Definition 1.3.2 *Let* τ *be a stopping time. Then, the* σ*-field* $\mathcal{F}_\tau$ *is defined to be*

$$\{A \in \mathcal{F} : A \cap \{\tau \leq t\} \in \mathcal{F}_t \text{ for all } t \geq 0\}.$$

It is easy to verify that $\mathcal{F}_\tau$ defined above is indeed a σ-field and that τ is $\mathcal{F}_\tau$ measurable. If τ_1 and τ_2 are two stopping times with $\tau_1 \leq \tau_2$, then $\mathcal{F}_{\tau_1} \subseteq \mathcal{F}_{\tau_2}$.

Advantages of a Right Continuous Filtration

A filtration $\{\mathcal{F}_t\}$ is said to be *right-continuous* if $\mathcal{F}_{t^+} = \mathcal{F}_t$ for each t where $\mathcal{F}_{t^+} = \cap_{u>t} \mathcal{F}_u$.

When a given filtration satisfies the additional requirement of right-continuity, the space $(\Omega, \mathcal{F}, (\mathcal{F}_t), P)$ is said to satisfy the **usual conditions**. Some of the niceties that accrue as a result are listed below:

(i) If $\{\tau_n\}$ is a sequence of stopping times, $\inf_n \tau_n$, $\liminf \tau_n$, and $\limsup \tau_n$ are stopping times.
(ii) If $\tau = \inf_n \tau_n$, then $\mathcal{F}_\tau = \cap_n \mathcal{F}_{\tau_n}$.
(iii) Let X be an adapted, right-continuous process. The hitting time of an open set is a stopping time.

The statements above are quite easy to prove and are left to the reader. Whenever right-continuity of a filtration is used, it would be expressly mentioned.

Progressive Measurability

Definition 1.3.3 *Let $X = \{X_t\}$ be a stochastic process on $(\Omega, \mathcal{F}, (\mathcal{F}_t), P)$. We call X* **progressively measurable** *with respect to $\mathcal{F}_t$ if for each $t \in \mathbb{R}^+$, the map*

$$(s, \omega) \to X_s(\omega)$$

from $[0, t] \times \Omega$ to $\mathbb{R}$ is measurable with respect to $\mathcal{B}[0, t] \times \mathcal{F}_t$.

The following proposition shows that the class of progressively measurable processes is huge and contains several processes of interest. If a filtration $(\mathcal{F}_t)$ is given and doesn't change, we will simply write "progressively measurable" instead of "progressively measurable with respect to $\mathcal{F}_t$".

Proposition 1.3.2 *Let $X = \{X_t\}$ be an $\mathcal{F}_t$-adapted, right-continuous (or left-continuous) process. Then, X is progressively measurable.*

Proof Suppose X has right-continuous sample paths. Fix any $t > 0$. For each n, define the stepwise approximation process X^n on $[0, t]$ by

$$X_s^n := \sum_{j=0}^{n-1} X_{(j+1)t/n} 1_{(jt/n,(j+1)t/n]}(s)$$

with $X_0^n = X_0$. Clearly, X^n is $\mathcal{B}([0, t]) \times \mathcal{F}_t$-measurable. By right-continuity of X, we get $\lim_{n\to\infty} X_s^n(\omega) = X_s(\omega)$ for all $(s, \omega) \in [0, t] \times \Omega$. Therefore, X as a function on $[0, t] \times \Omega$ is also $\mathcal{B}([0, t]) \times \mathcal{F}_t$-measurable.

One can likewise prove it when X has left-continuous paths. ■

A nice feature of progressively measurable processes is given in the next proposition.

Proposition 1.3.3 *Let $X = \{X_t\}$ be progressively measurable, and let τ be a finite stopping time. Then, the random variable X_τ is $\mathcal{F}_\tau$-measurable.*

Proof We will first show that the process $\{X_{\tau\wedge t}\}$ is progressively measurable.

Indeed, the process $\{\tau \wedge t\}$ is continuous and adapted to $(\mathcal{F}_t)$ and hence is progressively measurable by the previous proposition. Therefore, the map

$$(s, \omega) \to (\tau(\omega) \wedge s, \omega)$$

is measurable from $([0, t] \times \Omega, \mathcal{B}([0, t]) \times \mathcal{F}_t)$ into itself. By progressive measurability of X, we thus obtain that the map

$$(s, \omega) \to X_{\tau(\omega)\wedge s}(\omega)$$

is $\mathcal{B}([0, t]) \times \mathcal{F}_t$-measurable as a map from $[0, t] \times \Omega$ to $\mathbb{R}$.

Thus, $\{X_{\tau\wedge t}\}$ is a progressively measurable process.

$X_{\tau\wedge t}$ is $\mathcal{F}_t$-measurable for any fixed t. For any Borel set B in $\mathbb{R}$, it follows that for any t,

$$\{X_\tau \in B\} \cap \{\tau \le t\} = \{X_{\tau\wedge t} \in B\} \cap \{\tau \le t\} \in \mathcal{F}_t.$$

In other words, X_τ is $\mathcal{F}_\tau$-measurable. ■

Some Deep Results for Stochastic Processes

We list below a few results that are powerful, and are very difficult to prove. It is important to know them, though we do not prove these theorems in this book. The first result is due to Dellacherie [12].

Theorem 1.3.4 *Let $X = \{X_t\}$ be any progressively measurable process on $(\Omega, \mathcal{F}, (\mathcal{F}_t), P)$ where $(\mathcal{F}_t)$ satisfies the usual conditions. For any Borel set B in $\mathbb{R}$, the hitting time τ_B is a stopping time.*

The next result is due to Chung and Doob [5] (also see [53] p. 68). It is quite clear that a progressively measurable process is measurable and adapted. The converse is almost (but not quite!) true, and is given by this famous result.

Theorem 1.3.5 *If a stochastic process $X = \{X_t\}$ is measurable and adapted to a filtration $(\mathcal{F}_t)$, then it has a progressively measurable modification.*

Many processes of interest happen to be right-continuous and adapted and are therefore progressively measurable by Proposition 1.3.2. It is for this reason that we do not need Theorem 1.3.5 in many instances.

Exercises

1. Let $\Omega = (0, 1]$ equipped with its Borel σ-field. For any $n \in \mathbb{N}$, define the random variables

$$X_n = 1_{(0,\frac{1}{n})}.$$

   Let $\mathcal{A} = \cup_{n=1}^{\infty} \sigma(X_1, \ldots, X_n)$, and $\mathcal{F} = \sigma(\mathcal{A})$.

   (i) Characterize the sets in $\mathcal{A}$ and in $\mathcal{F}$.

   (ii) For any Borel set $B \in \mathcal{B}(\mathbb{R}^n)$, assign

$$P\{(X_1, \ldots, X_n) \in B\} = 1 \ \text{ if } \ (1, \ldots, 1) \in B$$

   and equal to zero otherwise. Prove that P is additive on $\mathcal{A}$ but there is no extension of P to a probability measure on $\mathcal{F}$.

2. Let P and Q be two probability measures on $(\Omega, \mathcal{F})$. Let $\mathcal{S}$ be a π system such that $\mathcal{F} = \sigma(\mathcal{S})$. If $P = Q$ on $\mathcal{S}$, show that $P = Q$ on $\mathcal{F}$.
3. Let $\mathcal{S}$ be as in the previous problem. Suppose L is a space of $\mathcal{F}$-measurable functions such that

   (i) $1 \in L$; $I_A \in L \ \forall A \in \mathcal{S}$,

   (ii) $f, g \in L$, then $af + bg \in L$ for all nonnegative constants a, b, and

   (iii) If f_n is a non-decreasing sequence of nonnegative functions in L such that $\lim_{n\to\infty} f_n = f$, then $f \in L$.

   Show that L contains all nonnegative $\mathcal{F}$-measurable functions. This is the *monotone class* theorem for measurable functions.
4. Let $(\mathcal{F}_t : 0 \le t \le \infty)$ be a filtration on $(\Omega, \mathcal{F})$. Let τ be a stopping time with respect to $(\mathcal{F}_t)$.

   (i) Show that $\mathcal{F}_\tau$ is a σ-field.

   (ii) Prove that τ is $\mathcal{F}_\tau$ measurable.
5. Let $(\mathcal{F}_t : 0 \le t \le \infty)$ be a filtration on $(\Omega, \mathcal{F})$. Let τ_1, τ_2 be two stopping times with respect to $(\mathcal{F}_t)$. Show that $\tau_1 \wedge \tau_2$, $\tau_1 \vee \tau_2$, and $\tau_1 + \tau_2$ are stopping times.
6. Let S, T be two stopping times with $S \le T$ a.s. Show that

   (i) $\mathcal{F}_S \subset \mathcal{F}_T$, and

   (ii) $\mathcal{F}_S \cap \mathcal{F}_T = \mathcal{F}_{S\wedge T}$.
7. Suppose that a filtration $(\mathcal{F}_t)$ is right-continuous. If $\{\tau_n\}$ is a sequence of stopping times with respect to $(\mathcal{F}_t)$, prove that $\inf \tau_n$, $\liminf_{n\to\infty} \tau_n$, and $\limsup_{n\to\infty} \tau_n$ are stopping times. Note that $\sup \tau_n$ is a stopping time even without right-continuity of the filtration.
8. Given any filtration $(\mathcal{F}_t)$, show that $(\mathcal{F}_{t+})$ is a right-continuous filtration.
9. Consider a probability space $(\Omega, \mathcal{F}, P)$ with a filtration $(\mathcal{F}_t)$. Let $\mathcal{P}$ be the σ-field on $\mathbb{R}^+ \times \Omega$ generated by all continuous, adapted, real-valued processes. Show that $\mathcal{P}$ is generated by the family of sets

   $$\mathcal{R} = \{(s,t] \times F : s \le t, F \in \mathcal{F}_s\} \cup \{0 \times F : F \in \mathcal{F}_0\}.$$

   Conclude that $\mathcal{P}$ is generated by all adapted, left-continuous real-valued processes. $\mathcal{P}$ is known as the *predictable* σ-field.
10. Suppose that X and Y are equivalent processes and X is right-continuous. Show that there exists a modification of Y which is right-continuous.

2 Brownian Motion

Brownian motion was first discovered in 1828 by Robert Brown, an English botanist, when he observed the random motion of pollen suspended in liquid. The motion was later understood as the effect of molecular bombardment of particles. In 1900, Louis Bachelier observed a similar phenomenon in his study on the fluctuation of stock prices in financial markets. Indeed, he introduced a process known as geometric Brownian motion since stock prices have to be positive. In 1902, Albert Einstein, in his thesis, studied Brownian motion to understand the movement of particles. Brownian motion, as understood by physicists such as Einstein and S. Chandrasekhar, differs from the modern mathematical theory of it.

In this chapter, we start with the definition of a Brownian motion which plays a central role in the development of stochastic analysis. Brownian motion enjoys several pleasant features—it is a process which is Gaussian, Markov, self-similar, a martingale, and has stationary, independent increments. Brownian motion is also known as a Wiener process in honor of Norbert Wiener who constructed a measure on the space of continuous functions under which the canonical process $X_t(\omega) = \omega(t)$ is a Brownian motion. Wiener's work appeared in a series of papers in the early 1920s, a decade before Kolmogorov's monograph that set probability theory on a rigorous mathematical foundation.

After the definition, we present a construction of Brownian motion using Haar functions and the Kosambi-Karhunen-Loéve (KKL) expansion [42], [50] for a standard Wiener process. Next, basic path properties of the process are proved in full. It is shown that almost every Brownian path is

(i) locally Hölder continuous of any order $\alpha < 1/2$,

(ii) nowhere differentiable, and

(iii) a function with a finite quadratic variation for suitable partitions.

We give a full proof of the reflection principle of André, and some of its consequences. Several other applications of the principle can be found in Revuz and Yor [62] and Varadhan [73].

2.1 Definition and Construction of Brownian Motion

Let the index set $\mathbf{T}$ denote either a finite interval $[0, T]$ or $\mathbb{R}^+$.

Definition 2.1.1 *A real-valued stochastic process $\{B_t\}$ defined on a complete probability space $(\Omega, \mathcal{F}, P)$ is called a* **Brownian motion** *with variance parameter σ^2 if it satisfies the following conditions:*

(1) *Initial condition: $B_0(\omega) = 0$ P – a.s.*

(2) *Stationary increments property: For any s and t with $0 \leq s < t$, the random variable $B_t - B_s$ has a normal (i.e., Gaussian) distribution $N(0, \sigma^2(t-s))$. That is, for any real number a,*

$$P\{B_t - B_s \leq a\} = \frac{1}{\sigma\sqrt{2\pi(t-s)}} \int_{-\infty}^{a} \exp\left\{-\frac{x^2}{2\sigma^2(t-s)}\right\} dx$$

(3) *Independent increments property: For any n time points $\{t_j\}$ satisfying $0 \leq t_1 < t_2 < \cdots < t_n$, the random variables $B_{t_1}, B_{t_2} - B_{t_1}, \ldots, B_{t_n} - B_{t_{n-1}}$ are mutually independent.*

(4) *Continuity of paths: For almost all ω, the sample paths $t \to B_t(\omega)$ are continuous.*

When $\sigma^2 = 1$, the process is known as the *standard* one-dimensional Brownian motion. From now on, we will consider the standard Brownian motion unless stated otherwise.

The requirement (2) in the definition implies that the distribution of $B_t - B_s$ coincides with that of $B_{t+h} - B_{s+h}$ for any shift h such that $s + h \geq 0$.

The joint probability distribution of the random vector $(B_{t_1}, B_{t_2}, \ldots, B_{t_n})$, for any distinct t_j with $0 \leq t_1 < t_2 < \cdots < t_n$ is multivariate normal, and its joint density function is quite easy to write:

$$f(x_1, \ldots, x_n) = \prod_{j=1}^{n} \frac{1}{\sqrt{2\pi(t_j - t_{j-1})}} \exp\left\{-\frac{(x_j - x_{j-1})^2}{2(t_j - t_{j-1})}\right\}$$

by using the property of independent increments.

The construction of a Brownian Motion on $[0, 1]$ is given below. The use of Haar functions in the construction is due to Ciesielski [6].

We first recall the definition of the Haar family of functions on $[0, 1]$ denoted by $\{g_{00}, g_{nj}$, for $n = 1, 2, \ldots$ and $j =$ any positive odd integer $< 2^n\}$ where $g_{00} \equiv 1$ and

$$g_{nj}(s) = \begin{cases} 2^{(n-1)/2} & \text{if } \frac{j-1}{2^n} \leq s < \frac{j}{2^n}, \\ -2^{(n-1)/2} & \text{if } \frac{j}{2^n} \leq s < \frac{j+1}{2^n}, \\ 0 & \text{otherwise.} \end{cases}$$

It is well known that the Haar family forms a complete orthonormal system (CONS) in $L^2[0,1]$. Define

$$G_{nj}(t) = \int_0^t g_{nj}(s)\,ds.$$

The graph of the function G_{nj} is a *tent* that forms a triangle over the interval $\left[\frac{j-1}{2^n}, \frac{j+1}{2^n}\right]$ with a maximum height of $2^{-(n+1)/2}$ attained at the midpoint of the interval. Clearly, for any fixed n, the triangles formed by the graphs of G_{nj} are non-overlapping for different j's. This property of G_{nj} will be used repeatedly in what follows. In fact, sums over j that involve $G_{nj}(t)$ for a fixed n and t reduce to a single term.

Let I_n denote the set $\{j : j \text{ is positive, odd and } < 2^n\}$ for $n \geq 1$. Let $I_0 = \{0\}$. Let $\{X_{nj} : j \in I_n,\ n \geq 1\}$ be a countable family of mutually independent standard normal variables defined on a common probability space $(\Omega, \mathcal{A}, P)$. Indeed, by the Kolmogorov consistency theorem, we can take $\Omega = \mathbb{R}^\infty$, a countable product of real lines, and $\mathcal{A}$ as the σ-field $\mathcal{B}(\mathbb{R}^\infty)$ generated by the Borel cylinder sets, and P as the countable product of $N(0,1)$ measures on $\mathbb{R}$. However, such a probability space is not a complete measure space. Therefore, let us redefine $\mathcal{A}$ as the completion of $\mathcal{B}(\mathbb{R}^\infty)$ with respect to P.

Define the process $Y_n(t,\omega) = \sum_{j\in I_n} X_{nj}(\omega)G_{nj}(t)$. Note that the effect of each X_{nj} is localized to a time-interval of length $\frac{1}{2^{n-1}}$.

Set $M_n = \max_{j\in I_n} |X_{nj}|$ and $L_n = \max_{j\in I_n} X_{nj}$. For any positive number a, and $n \geq 1$, we have

$$\begin{aligned}
P\{M_n > a\} &\leq 2P\{L_n > a\} && \text{by symmetry of } X_{nj}\\
&= 2P\left\{e^{L_n} > e^a\right\}\\
&\leq \frac{2}{e^a} Ee^{L_n}\\
&\leq \frac{2}{e^a} 2^{n-1} e^{1/2} && \text{since } e^{L_n} \leq \sum_{j\in I_n} e^{X_{nj}}\\
&\leq \frac{2^{n+1}}{e^a}.
\end{aligned}$$

Choosing $a = 2(n+1)\log 2$, we get $P\{M_n > a\} \leq 2^{-(n+1)}$ and hence,

$$\sum_{n=1}^{\infty} P\{M_n > a\} < \infty.$$

By the first part of the Borel-Cantelli lemma,

$$P\left\{M_n \leq 2(n+1)\log 2 \ \text{for all large enough } n\right\} = 1. \tag{2.1.1}$$

Let us call the almost sure set obtained in (2.1.1) as Ω_0. For $\omega \in \Omega_0$, it then follows that

$$\max_{0 \le t \le 1} |Y_n(t,\omega)| \le 2^{-(n+1)/2}\, 2(n+1)\log 2$$

if n is large enough. Therefore, $P\{\omega : \sum_{n=0}^{\infty} \max_{0\le t\le 1} |Y_n(t,\omega)| < \infty\} = 1$. Thus $\sum_{n=0}^{\infty} Y_n(t,\omega)$ converges uniformly in t for all $\omega \in \Omega_0$. Define for all t

$$W_t(\omega) = \begin{cases} \sum_{n=0}^{\infty} Y_n(t,\omega) & \text{if } \omega \in \Omega_0, \\ 0 & \text{otherwise.} \end{cases}$$

We will show that $\{W_t\}$ thus defined is a standard Brownian motion on $(\Omega, \mathcal{A}, P)$ for $t \in [0,1]$. Indeed, if $\omega \in \Omega_0$, $W_t(\omega)$ is the uniform limit of continuous functions, and hence, the sample path $W_t(\omega)$ is a continuous function of t. Thus condition 4 of our definition is satisfied. The finite-dimensional distributions of W are Gaussian. Indeed, if a sequence of independent Gaussian random variables converges almost surely, then the limit is also Gaussian as seen by considering the convergence of their characteristic functions. In case the limit is a degenerate random variable, we view it as a normal variable with zero variance.

Since the Haar family is a CONS in $L^2[0,1]$, by the Parseval theorem,

$$\|1_{[0,t]}\|^2_{L^2[0,1]} = \sum_{n=0}^{\infty} \sum_{j \in I_n} (1_{[0,t]}, g_{nj})^2 = \sum_{n=0}^{\infty} \sum_{j \in I_n} G^2_{nj}(t).$$

Hence for each t, W_t is also the $L^2(P)$-limit of the partial sums:

$$E\left(W_t - \sum_{n=0}^{m} \sum_{j \in I_n} X_{nj}\, G_{nj}(t)\right)^2 = \sum_{n=m+1}^{\infty} \sum_{j \in I_n} G^2_{nj}(t) \to 0$$

as $m \to \infty$, so that

$$\begin{aligned} E[W_t\, W_s] &= \lim_{m\to\infty} E\left[\left(\sum_{n=0}^{m} \sum_{j \in I_n} X_{nj}\, G_{nj}(t)\right)\left(\sum_{n=0}^{m} \sum_{j \in I_n} X_{nj}\, G_{nj}(s)\right)\right] \\ &= \sum_{n=0}^{\infty} \sum_{j \in I_n} (1_{[0,t]}, g_{nj})(1_{[0,s]}, g_{nj}) \\ &= (1_{[0,t]}, 1_{[0,s]}) = \min(t,s). \end{aligned}$$

Also, $EW_t = 0$ for all t, and $W_0 \equiv 0$. Using the above calculation, it is easy to verify that the conditions of a Brownian motion are satisfied by $W = \{W_t\}$. In fact, the verification of independent increments property is quite simplified by the following two basic observations:

(i) pairwise independence of a finite set of Gaussian random variables is equivalent to their mutual independence, and

(ii) two Gaussian random variables are independent if and only if they are uncorrelated.

Remark 2.1.2 The canonical process $\{Y_t\}$ that was constructed on $(C, \mathcal{B}, Q)$ in Proposition 1.1.13 from Chapter 1 is in fact a Brownian motion. Let $\mathcal{A}$ be the completion of $\mathcal{B}$ with respect to Q. The measure space $(C, \mathcal{A}, Q)$ is known as the **Wiener space**, and Q is called the **Wiener measure.**

Next, we proceed to extend the above method to construct a Brownian motion on $\mathbb{R}^+$.

The Brownian motion on $\mathbb{R}^+$

Using the Haar family $\{g_{nj}\}$, we define the functions $\{h_{nj}\}$ on $\mathbb{R}^+$ as follows:

$$h_{nj}(t) = \left(\frac{2}{\pi}\right)^{1/2} (1+t^2)^{-1/2}\, g_{nj}\left(\frac{2}{\pi}\arctan t\right) \quad \text{for } 0 \le t < \infty$$

where $j \in I_n$ if $n \ge 1$, and $j = 0$ if $n = 0$.

Lemma 2.1.1 *The sequence $\{h_{nj}\}$ is a CONS in $L^2(\mathbb{R}^+)$.*

Proof Let $(.\,,\,.)$ denote the $L^2(\mathbb{R}^+)$-inner product. By a change of variables, it is easy to obtain

$$\begin{aligned}(h_{nj}\,,\,h_{mk}) &= \frac{2}{\pi}\int_0^\infty (1+t^2)^{-1}\, g_{nj}\left(\frac{2}{\pi}\arctan t\right) g_{mk}\left(\frac{2}{\pi}\arctan t\right) \\ &= \int_0^1 g_{nj}(t)\, g_{mk}(t)\, dt \;=\; \delta_{mn}\,\delta_{jk}.\end{aligned}$$

where δ denotes the Kronecker delta. To show the completeness of $\{h_{nj}\}$, let us suppose that $(f, h_{nj}) = 0$ for all n and $j \in I_n$. This yields, upon a change of variables,

$$\left(\frac{\pi}{2}\right)^{1/2}\int_0^1 f\left(\tan\left(\frac{\pi}{2}u\right)\right)\sqrt{\sec^2\left(\frac{\pi}{2}u\right)}\, g_{nj}(u)\, du = 0$$

for all n and $j \in I_n$, so that

$$f\left(\tan\left(\frac{\pi}{2}u\right)\right)\sqrt{\sec^2\left(\frac{\pi}{2}u\right)} = 0 \text{ for almost all } u \in [0,1].$$

Therefore, $f(t) = 0$ for almost all $t \in \mathbb{R}^+$ with respect to the Lebesgue measure. ■

Let $H_{nj}(t) = \int_0^t h_{nj}(u)\,du$.

Lemma 2.1.2

$$\sum_{j\in I_n} |H_{nj}(t)| \le \sqrt{(1+a^2)}\left(\frac{\pi}{2}\right)^{1/2} 2^{-(n-1)/2} \quad \textit{for any} \quad t \in [0,a],\ a < \infty.$$

Proof For $t \in [0,a]$,

$$\begin{aligned} H_{nj}(t) &\le \sqrt{(1+a^2)}\left(\frac{2}{\pi}\right)^{1/2} \int_0^t (1+u^2)^{-1} \left| g_{nj}\left(\frac{2}{\pi}\arctan u\right)\right| du \\ &\le \sqrt{(1+a^2)}\left(\frac{\pi}{2}\right)^{1/2} 2^{-(n-1)/2} \end{aligned}$$

by using a change of variables and the definition of g_{nj} in finding an upper bound for the above integral. ■

Let $Y_n(t,\omega) = \sum_{j\in I_n} X_{nj}(\omega)\, H_{nj}(t)$. We make the following estimate:

$$\begin{aligned} \max_{t\in[0,a]} |Y_n(t,\omega)| &\le M_n(\omega) \sum_{j\in I_n} |H_{nj}(t)| \\ &\le M_n(\omega)\sqrt{(1+a^2)}\left(\frac{\pi}{2}\right)^{1/2} 2^{-(n-1)/2}. \end{aligned}$$

As before, a use of the Borel-Cantelli lemma yields an a.s. set Ω_0 such that for any $\omega \in \Omega_0$, there exists an $n_0(\omega)$ such that for all $n \ge n_0(\omega)$, we have

$$\max_{[0,a]} |Y_n(t,\omega)| \le \sqrt{(1+a^2)}\left(\frac{\pi}{2}\right)^{1/2} 2^{-(n-3)/2}(n+1)\log 2.$$

Hence, $\sum_{n=0}^{\infty} \max_{[0,a]} |Y_n(t,\omega)| < \infty$ $P-$ a.s. The series

$$\sum_{n=0}^{\infty}\sum_{j\in I_n} X_{nj}(\omega) H_{nj}(t) \tag{2.1.2}$$

converges uniformly on *each* interval $[0,a]$ P-a.s.

Define $W_t(\omega)$, for any $t \in \mathbb{R}^+$, to be the sum of the series (2.1.2) if $\omega \in \Omega_0$, and to be zero, otherwise. The process $\{W_t\}$ thus constructed is a Brownian motion on $(\Omega, \mathcal{A}, P)$.

The Wiener measure on $C = C(\mathbb{R}^+)$ is defined the same way as in the case of $C[0,1]$ (see Remark 2.1.2). In the notation of Proposition 1.2.1, if $\mathcal{A}$ denotes the completion of $\mathcal{B}$ (the Borel σ-field on C) with respect to Q, the coordinate process on C is a Wiener process on $(C, \mathcal{A}, Q)$. The Wiener measure is given by Q on $\mathcal{A}$.

Next, we show the KKL expansion of a Brownian motion. The KKL expansion is quite useful in simulating the paths of a Brownian motion. In fluid dynamics, turbulence theory, and numerical analysis, KKL-type expansion is known as proper orthogonal decomposition (POD).

Let $R(t,s)$ for any $0 \leq s,t \leq 1$ be a continuous covariance. By the theory of integral equations (cf. [63]), the self-adjoint operator R defined on $L^2[0,1]$ by

$$(Rf)(t) = \int_0^1 R(t,s)f(s)ds \text{ for } f \in L^2[0,1]$$

has eigenvalues λ_n with corresponding eigenfunctions ϕ_n. By Mercer's theorem (cf. [63]),

$$R(t,s) = \sum_{n=1}^{\infty} \lambda_n \phi_n(t)\phi_n(s) \tag{2.1.3}$$

where the series converges uniformly.

Let $X := \{X_t\}$ be a mean-square continuous Gaussian process with mean 0 and covariance $R(t,s)$. Clearly, $R(t,s)$ is continuous. Define the Gaussian random variables

$$Z_n = \int_0^1 X_s \phi_n(s)\, ds.$$

Then, we have

$$E(Z_n Z_m) = \int_0^1 \int_0^1 R(t,s)\phi_n(t)\phi_m(s)\, dt\, ds = \lambda_n \delta_{nm}$$

where δ_{nm} is the Kronecker delta. Thus, Z_n are independent random variables with

$$EZ_n = 0 \text{ and } E(Z_n^2) = \lambda_n.$$

KKL Expansion

Consider

$$E\left[\left|X_t - \sum_{n=1}^{N} Z_n\phi_n(t)\right|^2\right] = E\left(X_t^2\right) - 2\sum_{n=1}^{N} \phi_n(t)E(Z_nX_t) + \sum_{n=1}^{N} \phi_n^2(t)E(Z_n^2).$$

$$E(Z_nX_t) = E\left(X_t \int_0^1 X_s\phi_n(s)\, ds\right) = \int_0^1 R(t,s)\phi_n(s)\, ds = \lambda_n\phi_n(t).$$

Therefore,

$$\begin{aligned} E\left[\left|X_t - \sum_{n=1}^{N} Z_n\phi_n(t)\right|^2\right] &= R(t,t) - 2\sum_{n=1}^{N} \lambda_n\phi_n^2(t) + \sum_{n=1}^{N} \lambda_n\phi_n^2(t) \\ &= R(t,t) - \sum_{n=1}^{N} \lambda_n\phi_n^2(t) \\ &\to 0 \quad \text{from (2.1.3) as } N \to \infty. \end{aligned}$$

We then have

$$X_t = \sum_{n=1}^{\infty} Z_n \phi_n(t) \tag{2.1.4}$$

where convergence is in $L^2(P)$ sense. Writing $\xi_n = \frac{Z_n}{\sqrt{\lambda_n}}$,

$$X_t = \sum_{n=1}^{\infty} \sqrt{\lambda_n} \xi_n \phi_n(t) \tag{2.1.5}$$

with convergence in $L^2(P)$. The ξ_n are iid $N(0, 1)$ random variables. The series (2.1.4) or (2.1.5) is called the KKL expansion for X_t.

KKL Expansion for Brownian Motion

Since $R(t, s) = \min\{t, s\}$, $0 \leq t, s \leq 1$, we first find the eigenfunctions and eigenvalues of R. To do this, consider the integral equation

$$\int_0^1 \min\{t, s\} \phi(s) ds = \lambda \phi(t). \tag{2.1.6}$$

That is,

$$\lambda \phi(t) = \int_0^t s\phi(s)\, ds + t \int_t^1 \phi(s)\, ds. \tag{2.1.7}$$

The right side shows that ϕ is differentiable and

$$\lambda \phi'(t) = t\phi(t) + \int_t^1 \phi(s) ds - t\phi(t) = \int_t^1 \phi(s)\, ds. \tag{2.1.8}$$

From this, we obtain

$$\lambda \phi''(t) = -\phi(t).$$

For convenience, write $\mu = \frac{1}{\lambda}$. Then ϕ satisfies the equation

$$y''(t) + \mu y(t) = 0. \tag{2.1.9}$$

From (2.1.7) and (2.1.8), we have $\phi(0) = 0$ and $\phi'(1) = 0$. The solution

$$\phi(t) = A \cos \sqrt{\mu} t + B \sin \sqrt{\mu} t$$

gives $A = \phi(0) = 0$. Since $\phi'(t) = B\sqrt{\mu} \cos \sqrt{\mu} t$, we get $\phi'(1) = 0$. Hence $\sqrt{\mu_n} = (2n + 1)\frac{\pi}{2}$ so that

$$\lambda_n = \frac{1}{(n + \frac{1}{2})^2 \pi^2},$$

and $\phi_n(t) = c_n \sin \frac{1}{\sqrt{\lambda_n}} t$. The normalizing constant c_n is $\sqrt{2}$. Thus, we get

$$Z_n = \sqrt{2} \int_0^1 B(t) \sin(n + \frac{1}{2})\pi t \, dt.$$

The Brownian motion can thus be expanded into

$$B_t = \sqrt{2} \sum_{n=0}^{\infty} Z_n \sin(n + \frac{1}{2})\pi t, \tag{2.1.10}$$

known as the KKL expansion with the series converging a.s. since Z_n are independent random variables and

$$\sum_{n=0}^{\infty} E(Z_n^2) = \sum_{n=0}^{\infty} \frac{1}{(n + \frac{1}{2})^2 \pi^2} < \infty.$$

2.2 Essential Features of a Brownian Motion

In this section we present the basic properties of a Brownian motion. First, we collect certain transformations of a Brownian motion that lead us to a Brownian motion, making the observation that Proposition 1.1.10 in Chapter 1 and the remark following it can be cast as statements about a Brownian motion. We collect them in the lemma given below.

Lemma 2.2.1 *Let $\{B_t\}$ be a Brownian motion, and let $M_1 = \max_{t \in [0,1]} |B_t|$. Then, for any x,*

$$P\left\{ \max_{t \in [0,1]} |B_t| > x \right\} \leq 2P\left\{ |B_1| > x \right\},$$

and hence $E(\max_{t \in [0,1]} |B_t|) \leq 2E|B_1|$.

Theorem 2.2.2 *Let $\{B_t : t \geq 0\}$ be a Brownian motion. Then, each of the following processes is a Brownian motion:*

(i) The process $-B_t$.

(ii) For any fixed $s > 0$, $X_t = B_{t+s} - B_s$.

(iii) (Brownian scaling) For any $c > 0$, $Y_t = \frac{1}{\sqrt{c}} B_{ct}$.

(iv) (time inversion) Define $Z_t = \begin{cases} tB_{1/t} & \text{if } t > 0, \\ 0 & \text{if } t = 0. \end{cases}$

Proof Each of the processes in (i), (ii) and (iii) has continuous paths and take the value zero at time 0. The finite-dimensional distributions of each process is Gaussian. We will compute the covariances in (iii) and (iv): For any $t, s > 0$,

$$E(Y_tY_s) = E\left\{\frac{1}{\sqrt{c}}B_{ct}\frac{1}{\sqrt{c}}B_{cs}\right\} = \frac{1}{c}\min\{ct, cs\} = \min\{t, s\},$$
$$E(Z_tZ_s) = E\{tB_{1/t}sB_{1/s}\} = ts\min\{1/t, 1/s\} = \min\{t, s\}.$$

Thus the covariances are as required. The proof will be over if we show that $\{Z_t\}$ is continuous at time 0. Setting $u = 1/t$, we need to show that $\lim_{u\to\infty}\frac{B_u}{u} = 0$ a.s. Indeed, if $n \le u < n+1$, then

$$\begin{aligned}\left|\frac{B_u}{u}\right| &\le \left|\frac{B_n}{u}\right| + \left|\frac{B_u}{u} - \frac{B_n}{u}\right| \\ &\le \left|\frac{B_n}{n}\right| + \left|\frac{B_u - B_n}{u}\right| \\ &\le \left|\frac{B_n}{n}\right| + \frac{1}{n}\max_{u\in[n,n+1]}|B_u - B_n|.\end{aligned}$$

The first term on the left side converges to 0 P-a.s. by the strong law of large numbers. It remains to show that the second term also has limit 0.

Let M_n denote $\max_{u\in[n,n+1]}|B_u - B_n|$. Then, $\{M_n\}$ are iid. random variables since $\{B_t\}$ has stationary independent increments. By part (ii) of the above lemma, we get $EM_1 < 2E|B_1| < \infty$. It is well known that if a random variable $X \ge 0$ with $EX < \infty$, then $\sum_{n=1}^{\infty} P\{X > n\} \le EX$. For any $\epsilon > 0$,

$$\sum_{n=1}^{\infty} P\{M_n/n > \epsilon\} = \sum_{n=1}^{\infty} P\{M_1 > n\epsilon\} \le E(M_1/\epsilon) < \infty.$$

By the Borel-Cantelli lemma, $P\{M_n/n > \epsilon \text{ infinitely often}\} = 0$. By the arbitrariness of ϵ, $M_n/n \to 0$ a.s., which finishes the proof. ■

Remark 2.2.1 One can infer from the above proof that P-a.s., we have $\lim_{u\to\infty}\dfrac{B_u}{u} = 0$.

Proposition 2.2.3 *Let $B := \{B_t\}$ be a standard one-dimensional Brownian motion. Then*

$$P\left\{\limsup_{t\to\infty}|B_t| = \infty\right\} = 1.$$

Proof Consider the set

$$\left\{\limsup_{t\to\infty}|B_t| = \infty\right\} = \cap_{n=1}^{\infty}\left\{\limsup_{t\to\infty}|B_t| > n\right\}.$$

By path continuity of Brownian motion,

$$\left\{\limsup_{t\to\infty} |B_t| > n\right\} = \left\{\limsup_{t\in\mathbb{Q},t\to\infty} |B_t| > n\right\}$$
$$= \cap_{k=1}^{\infty} \cup_{t\in\mathbb{Q},t\geq k} \left\{|B_t| > n\right\}.$$

Therefore,

$$P\left\{\limsup_{t\to\infty} |B_t| > n\right\} \geq \limsup_{t\in\mathbb{Q},t\to\infty} P\left\{|B_t| > n\right\}$$
$$= \limsup_{t\in\mathbb{Q},t\to\infty} \sqrt{\frac{1}{2\pi t}} \int_n^{\infty} \exp\left\{-\frac{x^2}{2t}\right\} dx$$
$$=1.$$

■

We turn next to study the Hölder-continuity of Brownian paths. Recall that a function f defined on $\mathbb{R}$ is said to be locally **Hölder-continuous** of order $\alpha > 0$ if for every $L > 0$, there exists a finite constant c such that $|f(t) - f(s)| \leq c|t-s|^{\alpha}$ for all real $t, s \in [-L, L]$. In other words, for every $L > 0$,

$$\sup\left\{|f(t) - f(s)|/|t-s|^{\alpha} : |t|, |s| \leq L,\ t \neq s\right\} < \infty.$$

The result given below uses the fact that for any $0 \leq s < t$, and any positive integer k,

$$E(B_t - B_s)^{2k} = (2k-1)(2k-3)\cdots 3.1\,(t-s)^k.$$

Theorem 2.2.4 *Let $\{B_t\}$ be a Brownian motion defined on a probability space $(\Omega, \mathcal{F}, P)$. For P-almost every $\omega \in \Omega$, the Brownian path $B_{\cdot}(\omega)$ is Hölder-continuous of any order $< 1/2$.*

Proof We will work with $T = [0, 1]$. It suffices to show that for any $\alpha < 1/2$,

$$E\left[\sup\left\{|B_t - B_s|/|t-s|^{\alpha} : s, t \in T, s \neq t\right\}\right] < \infty.$$

First, note that for any positive integer k, there exits a constant C_k such that

$$E(|B_t - B_s|^{2k}) \leq C_k|t-s|^k. \tag{2.2.1}$$

Fix a positive integer n, and let D_n denote the set $\{j/2^n : j = 0, 1, \ldots, 2^n\}$. Let

$$M_n = \max_{j=1,\ldots,2^n} |B_{j/2^n} - B_{(j-1)/2^n}|.$$

Using equation (2.2.1), we get

$$E\left(M_n^{2k}\right) \leq \sum_{j=1}^{2^n} E|B_{j/2^n} - B_{(j-1)/2^n}|^{2k}$$
$$\leq 2^n C_k \frac{1}{2^{nk}} = C_k \frac{1}{2^{(n-1)k}}.$$

Let D denote $\cup_n D_n$, the set of all dyadic rationals in $[0, 1]$. Let s, t be two members of D such that $0 < |s - t| < 1/2^m$ for some n. Then we can find a finite sequence $\{s_j : j \geq 1\}$ in D that increases to s where each $s_j \in D_{m+j}$ and the last term of the sequence is equal to s. Likewise, a finite increasing sequence $\{t_i\}$ can be constructed for t. Then,

$$B_t - B_s = \sum_{i \geq m+1} (B_{t_{i+1}} - B_{t_i}) + B_{t_1} - B_{s_1} + \sum_{j \geq m+1} (B_{s_{j+1}} - B_{s_j}).$$

Therefore,

$$|B_t - B_s| \leq 2 \sum_{i \geq m+1} M_i + M_m \leq 2 \sum_{i \geq m} M_i.$$

For any $\alpha > 0$, define

$$H_\alpha = \sup\left\{|B_t - B_s|/|t - s|^\alpha : s, t \in D, s \neq t\right\}.$$

Then,

$$J_\alpha \leq \sup_m \left\{2^{(m+1)\alpha} \sup_{|t-s| \leq 1/2^m} |B_t - B_s| : s, t \in D, s \neq t\right\}$$
$$\leq \sup_m \left\{2^{(m+1)\alpha}\, 2 \sum_{i \geq m} M_i\right\}$$
$$\leq 2^{\alpha+1} \sum_{i \geq m+1} 2^{i\alpha} M_i.$$

Using the notation $\|.\|_p$ for the $L^p(P)$ norm, and the Minkowski inequality,

$$\|H_\alpha\|_{2k} \leq 2^{\alpha+1} C_k \sum_{i-0}^{\infty} 2^{i\alpha} \|M_i\|_{2k}$$
$$\leq 2^{\alpha+1} C_k \sum_{j=0}^{\infty} 2^{i(\alpha-(k-1)/2k)},$$

which is finite if $\alpha < (k-1)/2k = 1/2 - 1/2k$. Since k can be any large integer, we have shown that P-a.s., $H_\alpha < \infty$ for all $\alpha < 1/2$. ■

The proof given above has the essential arguments for establishing a famous result known as the *Kolmogorov continuity criterion.*

We will later show that the Brownian paths are not Hölder continuous of any order $\geq 1/2$. Though the sample paths of a Brownian motion are continuous, they are quite exceptional. In fact, Paley, Wiener, and Zygmund [58] proved that almost every Brownian path is nowhere differentiable. It should be recalled that the construction by Wierstrass of continuous nowhere differentiable functions is quite involved. The following elegant proof is due to Dvoretzky, Erdös, and Kakutani [19].

Theorem 2.2.5 *Let $\{B_t\}$ be a Brownian motion defined on a probability space $(\Omega, \mathcal{F}, P)$. For P-almost every $\omega \in \Omega$, the Brownian path $B_.(\omega)$ is nowhere differentiable.*

Proof

Step 1 In this step, we consider only deterministic functions. We will work with $T = [0, 1]$. Let α be any fixed positive number. If a function f has a derivative $f'(t)$ at $t \in T$ with $|f'(t)| < \alpha$, then there exists a large N such that for all $n \geq N$,

$$|f(t) - f(s)| < \alpha|t - s| \quad \text{if } s \text{ satisfies } |t - s| \leq 2/n. \tag{2.2.2}$$

This suggests that we define the set

$$A_n = \{f \in C : \exists\, t \text{ such that } |f(t) - f(s)| \leq \alpha|t - s| \;\; \forall \;\; |t - s| < 2/n\}.$$

The sequence $\{A_n\}$ is increasing in n with limit

$$A = \{f \in C : |f'(t)| \leq \alpha \;\text{ for some }\; t \in T\}.$$

Note that if (2.2.2) holds and $k/n \leq t \leq (k+1)/n$, then the following inequalities hold:

$$\begin{aligned} |f((k+1)/n) - f(k/n)| &\leq \alpha/n, \\ |f((k+2)/n) - f((k+1)/n)| &\leq 3\alpha/n, \\ |f(k/n) - f((k-1)/n)| &\leq 3\alpha/n. \end{aligned}$$

In order to get three such inequalities without troublesome notation, we chose the time-width in (2.2.2) as $2/n$. It is clear that A_n is contained in the union of the sets

$$\left\{\max\left(\left|f\left(\frac{k}{n}\right) - f\left(\frac{k-1}{n}\right)\right|, \left|f\left(\frac{k+1}{n}\right) - f\left(\frac{k}{n}\right)\right|, \left|f\left(\frac{k+2}{n}\right) - f\left(\frac{k+1}{n}\right)\right|\right) \leq \frac{3\alpha}{n}\right\}$$

as k varies from 1 to $n-2$.

Step 2 From the above containment,

$$P\{B_{.} \in A_n\} \leq \sum_{k=1}^{n-2} P\left\{\max_{j=0,1,2}\left(\left|B_{(k+j)/n} - B_{(k+j-1)/n}\right|\right) \leq 3\alpha/n\right\}$$
$$\leq \sum_{k=1}^{n-2} \left(P\left\{\left|B_{k/n} - B_{k-1/n}\right| \leq 3\alpha/n\right\}\right)^3$$

by using the stationary, independent increment property of Brownian motion. Equivalently,

$$P(A_n) \leq (n-2)\left(P\left\{\sqrt{n}|B_{1/n}| \leq 3\alpha/\sqrt{n}\right\}\right)^3. \tag{2.2.3}$$

Note that $\sqrt{n}\left|B_{k/n} - B_{(k-1)/n}\right|$ is a standard normal variable, so that

$$P\left\{\sqrt{n}\left|B_{k/n} - B_{k-1/n}\right| \leq 3\alpha/\sqrt{n}\right\} = \sqrt{2/\pi}\int_0^{3\alpha/\sqrt{n}} e^{-x^2/2}dx \leq 6\alpha/\sqrt{2\pi n}. \tag{2.2.4}$$

Using (2.2.4) in the inequality (2.2.3), we get

$$P(A_n) \leq (n-2)\left(\frac{6\alpha}{\sqrt{2\pi n}}\right)^3.$$

Thus $P\left\{\omega : B_{.}(\omega) \in A_n\right\}$ approaches zero as $n \to \infty$. The proof is over by the arbitrariness of α. ■

Remark 2.2.2 With slight changes to the above proof, one can show that almost every Brownian path is nowhere Hölder-continuous of order β for any $\beta > 1/2$. In fact, the only additional observation that is required is the basic inequality $a^\beta + b^\beta \leq 2(a+b)^\beta$ for any positive a, b.

Corollary 2.2.6 *For P-almost every $\omega \in \Omega$, the Brownian path $B_{.}(\omega)$ has infinite variation on every finite interval.*

Proof Assume the contrary. Then, there exists an interval I such that

$$P\left\{\omega : B_{.}(\omega) \text{ on } I \text{ have bounded variation}\right\} > 0.$$

According to a standard result in real analysis (see [64], p. 100), if f is a function of bounded variation on an interval I, then it has a derivative almost everywhere in I. Thus, with positive probability, Brownian paths on I have a derivative somewhere in I. This contradicts Theorem 2.2.5. ■

Let P_n denote the partition $t_0^{(n)} = S < t_1^{(n)} < \ldots < t_{k_n}^{(n)} = T$ of the interval $[S, T]$. We will drop the superscripts for notational simplicity. Define

$\|P_n\| = \max\{t_j - t_{j-1} : j = 1, \ldots, k_n\}$. Let $\|P_n\| \to 0$ as $n \to \infty$. Then the **quadratic variation** of a Wiener process over $[S, T]$ is defined as the limit (in a suitable sense) of $Q_n := \sum_{t_j \in P_n} (B_{t_j} - B_{t_{j-1}})^2$ as $\|P_n\| \to 0$.

Though a typical Brownian trajectory has paths of unbounded variation, the observation that $E(B_t - B_s)^2 = t - s$ for any $0 \le s < t$ gives us an inkling that the quadratic variation exists.

Theorem 2.2.7 *Let $\{P_n\}$ be partitions of $[S, T]$ with $\|P_n\| \to 0$. Then, as $n \to \infty$, $Q_n \to T - S$ in $L^2(P)$.*

If $\sum_{n=1}^{\infty} \|P_n\| < \infty$, then $Q_n \to T - S$ P-a.s.

Proof Without loss of generality, let $S = 0,\ T = 1$. We can write

$$Q_n - 1 = \sum_{j=1}^{k_n} \left((B_{t_j} - B_{t_{j-1}})^2 - (t_j - t_{j-1})\right).$$

The terms of the above sum have zero means and are mutually independent. Therefore,

$$E(Q_n - 1)^2 = \sum_{j=1}^{k_n} E\left((B_{t_j} - B_{t_{j-1}})^2 - (t_j - t_{j-1})\right)^2.$$

By factoring $(t_j - t_{j-1})$ from the jth term, and noting that each $\frac{(B_{t_j} - B_{t_{j-1}})^2}{(t_j - t_{j-1})}$ is a chi-squared random variable X with one degree of freedom, we get

$$\begin{aligned} E(Q_n - 1)^2 &= E(X - 1)^2 \left(\sum_{j=1}^{k_n} (t_j - t_{j-1})^2\right) \\ &\le E(X - 1)^2 \, \|P_n\|, \end{aligned}$$

which proves the $L^2(P)$-convergence.

To prove the second statement, consider for any $\epsilon > 0$,

$$\begin{aligned} P\{|Q_n - 1| > \epsilon\} &\le \frac{1}{\epsilon^2} E(Q_n - 1)^2 \\ &\le \frac{1}{\epsilon^2} E(X - 1)^2 \, \|P_n\|. \end{aligned}$$

Using the hypothesis $\sum_{n=1}^{\infty} \|P_n\| < \infty$, and the Borel Cantelli lemma, the proof is completed. ■

Remark 2.2.3 The almost-sure convergence of Q_n holds even under the weaker condition $\|P_n\| \to 0$ provided that $\{P_n\}$ is a refining sequence of partitions, that is, $P_n \subset P_{n+1}$ for all n. The proof of this result is due to Doob [14].

2.3 The Reflection Principle

André's reflection principle states that, for a standard one-dimensional Brownian motion and any $T > 0$,

$$P\left\{\sup_{0\le s\le T} B_s \ge a\right\} = 2P\{B_T \ge a\} \quad \text{for all} \quad a > 0.$$

We already know the inequality

$$P\left\{\sup_{0\le s\le T} B_s > a\right\} \le 2P\{B_T > a\}.$$

from Proposition 1.1.10.

The reflection principle gives us the probability that a Brownian motion crosses the line $y = a$ by time T. An intuitive proof of this principle relies on the observation that if a Brownian path crosses the level $y = a$, then there are as many paths that end up above a at time T as there are that end up below a at time T. Indeed, to see this, simply draw a path that rises to level a for the first time at $s < T$ and ends up below a at time T.

Reflect this curve from time s onwards about the line $y = a$ to get a path that ends up above level a at time T.

Hence, we can conclude that

$$P\left\{\sup_{0\le s\le T} B_s \ge a, B_T > a\right\} = P\left\{\sup_{0\le s\le T} B_s \ge a, B_T < a\right\}.$$

Since $P\{B_T = a\} = 0$, we obtain

$$\begin{aligned} P\left\{\sup_{0\le s\le T} B_s \ge a\right\} &= P\left\{\sup_{0\le s\le T} B_s \ge a, B_T > a\right\} + P\left\{\sup_{0\le s\le T} B_s \ge a, B_T \le a\right\} \\ &= P\left\{\sup_{0\le s\le T} B_s \ge a, B_T > a\right\} + P\left\{\sup_{0\le s\le T} B_s \ge a, B_T < a\right\} \\ &= 2P\left\{\sup_{0\le s\le T} B_s \ge a, B_T > a\right\} \\ &= 2P\{B_T > a\} = 2P\{B_T \ge a\}. \end{aligned}$$

We will now prove the result rigorously. For any fixed $n \in \mathbb{N}$, define the random variables

$$X_j = B_{\frac{jT}{n}} - B_{\frac{(j-1)T}{n}} \quad \text{for} \quad j = 1, \ldots, n.$$

Then X_j are independent $N(0, \frac{T}{n})$ random variables, and

$$B_{\frac{kT}{n}} = \sum_{j=1}^{k} X_j.$$

We will denote $B_{\frac{kT}{n}}$ as Y_k.

Proposition 2.3.1 *For any given $\epsilon > 0$ and $a > 0$, we have*

$$2P\{B_T > a + 2\epsilon\} - 2\sum_{j=1}^{n} P\{X_j > \epsilon\} \leq P\left\{\max_{1\leq k\leq n} Y_k > a\right\}. \tag{2.3.1}$$

Proof

Step 1 Define the stopping time

$$\tau = \inf\{k : Y_k > a\}$$

with the convention that infimum of the empty set is ∞. We have

$$\{\tau = n,\ B_T > a + 2\epsilon\} \subseteq \{X_n > 2\epsilon\}$$

since $Y_n = B_T$. Therefore,

$$P\{B_T > a + 2\epsilon\} \leq P\{\tau < n,\ B_T > a + 2\epsilon\} + P\{X_n > 2\epsilon\}. \tag{2.3.2}$$

We will now break up the set $A := \{\tau < n,\ B_T > a + 2\epsilon\}$ into $B \cup C$ where

$$B = \{\tau < n,\ B_T > a + 2\epsilon,\ B_T - Y_\tau \leq \epsilon\}$$

and

$$C = \{\tau < n,\ B_T > a + 2\epsilon,\ B_T - Y_\tau > \epsilon\}.$$

If $\omega \in B$, then $Y_{\tau(\omega)}(\omega) > a + \epsilon$ so that $X_{\tau(\omega)}(\omega) > \epsilon$. Hence,

$$\begin{aligned} P(A) &\leq P\{\tau < n, X_\tau > \epsilon\} + P(C) \\ &\leq \sum_{j=1}^{n-1} P\{X_j > \epsilon\} + P\{\tau < n,\ B_T - Y_\tau > \epsilon\}. \end{aligned} \tag{2.3.3}$$

From (2.3.2) and (2.3.3), it follows that

$$P\{B_T > a + 2\epsilon\} \leq \sum_{j=1}^{n} P\{X_j > \epsilon\} + P\{\tau < n,\ B_T - Y_\tau > \epsilon\}. \tag{2.3.4}$$

Step 2 Now, we will estimate $P\{\tau < n,\ B_T - Y_\tau > \epsilon\}$. Let us consider

$$
\begin{aligned}
P\{\tau < n,\ B_T - Y_\tau > \epsilon\} &= \sum_{k=1}^{n-1} P\{\tau = k,\ B_T - Y_k > \epsilon\} \\
&= \sum_{k=1}^{n-1} P\{\tau = k\}\, P\{B_T - Y_k > \epsilon\} \quad \text{by independence;} \\
&= \sum_{k=1}^{n-1} P\{\tau = k\}\, P\{B_T - Y_k < -\epsilon\} \quad \text{by symmetry;} \\
&= P\{\tau < n,\ B_T - Y_\tau < -\epsilon\}.
\end{aligned}
$$

Let us denote the latter event as D. Then,

$$P(D) \leq P\{\tau < n, B_T - Y_\tau \leq -\epsilon, B_T > a\} + P\{\tau < n, B_T \leq a\}.$$

If ω lies in the set $\{\tau < n, B_T - Y_\tau \leq -\epsilon, B_T > a\}$, then $\tau(\omega) < n$ and $Y_{\tau(\omega)}(\omega) > a + \epsilon$. Since τ is the first time that the random variables X_k exceed a, we can conclude that $\tau(\omega) < n$ and $X_{\tau(\omega)}(\omega) > \epsilon$. Hence,

$$P(D) \leq P\{\tau < n, X_\tau > \epsilon\} + P\{\tau < n, B_T \leq a\},$$

which implies that

$$P\{\tau < n,\ B_T - Y_\tau > \epsilon\} \leq \sum_{k=1}^{n-1} P\{X_k > \epsilon\} + P\{\tau < n, B_T \leq a\}. \tag{2.3.5}$$

Step 3 Combining (2.3.4) and (2.3.5), we have

$$P\{B_T > a + 2\epsilon\} \leq 2\sum_{j=1}^{n} P\left\{X_j > \epsilon\right\} + P\{\tau < n, B_T \leq a\}. \tag{2.3.6}$$

Note that

$$
\begin{aligned}
P\left\{\max_{1\leq k\leq n} Y_k > a\right\} &= P\{\tau \leq n, B_T \leq a\} + P\{\tau \leq n, B_T > a\} \\
&= P\{\tau < n, B_T \leq a\} + P\{B_T > a\}
\end{aligned} \tag{2.3.7}
$$

since $\{B_T > a\} \subseteq \{\tau \leq n\}$. Using the inequality

$$P\{B_T > a + 2\epsilon\} \leq P\{B_T > a\}$$

and the bound (2.3.6) in (2.3.7),

$$2P\{B_T > a + 2\epsilon\} - 2\sum_{j=1}^{n} P\left\{X_j > \epsilon\right\} \leq P\left\{\max_{1\leq k\leq n} Y_k > a\right\}. \qquad \blacksquare$$

Theorem 2.3.2 *Let $\{B_t\}$ be a a standard one-dimensional Brownian motion. For all $T > 0$ and $a > 0$, we have*

$$P\left\{\sup_{0\le s\le T} B_s \ge a\right\} = 2P\{B_T \ge a\}.$$

Proof Consider the random variables $Y_k = B_{\frac{kT}{n}}$. By Proposition 2.3.1,

$$2P\{B_T > a + 2\epsilon\} - 2\sum_{j=1}^{n} P\{X_j > \epsilon\} \le P\left\{\max_{1\le k\le n} Y_k > a\right\}.$$

Since X_j are independent $N(0, \frac{T}{n})$ random variables, let us call $P\{X_j > \epsilon\}$ as p_n. We can then write $2\sum_{j=1}^{n} P\{X_j > \epsilon\}$ as $2np_n$. By a direct calculation, one can show that $2np_n \to 0$ as $n \to \infty$.

The path continuity of Brownian motion implies that as $n \to \infty$,

$$P\left\{\max_{1\le k\le n} Y_k > a\right\} \to P\left\{\sup_{0\le s\le T} B_s > a\right\}.$$

Thus we obtain

$$2P\{B_T > a + 2\epsilon\} \le P\left\{\sup_{0\le s\le T} B_s > a\right\}.$$

Allowing $\epsilon \to 0$,

$$2P\{B_T > a\} \le P\left\{\sup_{0\le s\le T} B_s > a\right\}. \tag{2.3.8}$$

On the other hand, from Proposition 1.1.10, we know that

$$P\left\{\sup_{1\le k\le n} Y_k > a\right\} \le 2P\{B_T > a\}$$

so that by letting $n \to \infty$,

$$P\left\{\sup_{0\le s\le T} B_s > a\right\} < 2P\{B_T > a\}. \tag{2.3.9}$$

The bounds (2.3.8) and (2.3.9) yield the desired result. ∎

We discuss two applications of the reflection principle. Several others are left as exercises. Fix any $a > 0$. Let τ_a denote the first time that a Brownian motion reaches level a.

We will find the distribution of τ_a from the reflection principle. Consider, for any $t > 0$,

$$\begin{aligned} P\{\tau_a \le t\} &= P\left\{\max_{0\le s\le t} B_s \ge a\right\} \\ &= 2P\{B_t \ge a\} \\ &= \frac{2}{\sqrt{2\pi t}} \int_a^\infty \exp\left\{-\frac{x^2}{2t}\right\} dx \\ &= \sqrt{\frac{2}{\pi}} \int_{\frac{a}{\sqrt{t}}}^\infty \exp\left\{-\frac{x^2}{2}\right\} dx \end{aligned}$$

by a change of variables. Differentiating with respect to t, we obtain the probability density function of τ_a:

$$f_{\tau_a}(t) = \frac{a}{\sqrt{2\pi}} t^{-3/2} \exp\left\{-\frac{a^2}{2t}\right\}. \tag{2.3.10}$$

As a second application, we show that the Brownian paths oscillate about the initial state zero in any small (non-empty) time interval $[0, \delta)$. Let us define the sets

$$A_n = \left\{\omega : \text{there exists a } \ t \in \left[0, \frac{1}{n}\right) \ \text{ such that } \ B_t(\omega) > 0\right\} \quad \text{and}$$

$$B_n = \left\{\omega : \text{there exists a } \ t \in \left[0, \frac{1}{n}\right) \ \text{ such that } \ B_t(\omega) < 0\right\}.$$

Let $A := \cap_{n=1}^\infty A_n$, and $B := \cap_{n=1}^\infty B_n$. Then $A_n = \left\{\sup_{0\le t\le 1/n} B_t > 0\right\}$ for each n. Hence,

$$\begin{aligned} P(A) &= \lim_{n\to\infty} P\left\{\sup_{0\le t\le 1/n} B_t > 0\right\} \\ &= \lim_{n\to\infty} P\left\{\cup_{m=1}^\infty \sup_{0\le t\le 1/n} B_t > \frac{1}{m}\right\} \\ &= \lim_{n\to\infty} \lim_{m\to\infty} P\left\{\sup_{0\le t\le 1/n} B_t > \frac{1}{m}\right\} \\ &=2 \lim_{n\to\infty} \lim_{m\to\infty} P\left\{B_{\frac{1}{n}} > \frac{1}{m}\right\} \end{aligned}$$

by the reflection principle. Hence,

$$P(A) = 2 \lim_{n\to\infty} P\left\{B_{\frac{1}{n}} > 0\right\}$$
$$= 1.$$

Likewise, $P(B) = 1$. Therefore, $P(A \cap B) = 1$. Thus almost every Brownian path oscillates around the starting value 0 in any small time interval.

Exercises

1. Prove that the collection

$$\{g_{00}, g_{jn} : j, \text{ any positive odd integer } < 2^n, \ \ n \geq 1\}$$

of Haar functions is complete in $L^2[0, 1]$.

2. Prove that for any $t > 0$ and any positive integer k,

$$E\left[(B_t^{2k}\right] = t^k\,(2k-1)(2k-3)\cdots 3.1.$$

3. Let X and Y be two independent random variables having the same distribution such that

$$EX = 0 \ \ \text{and} E(X^2) = 1.$$

Suppose that $Z_1 = \frac{X+Y}{\sqrt{2}}$ and $Z_2 = \frac{X-Y}{\sqrt{2}}$ have the same distribution as X.

(i) Let $\phi(t) = E(e^{itX})$. Show that $\phi'(0) = 0$, and $\phi''(0) = -1$.
Also show that $\phi^2(\frac{t}{\sqrt{2}}) = \phi(t)$.

(ii) Show that ϕ is a real-valued, and $\phi(t) > 0$ for all t.

4. As a continuation of Problem 2,

(i) prove that

$$\frac{\log \phi(t)}{t^2} = \frac{\log \phi\left(\frac{t}{(\sqrt{2})^n}\right)}{\left(\frac{t}{(\sqrt{2})^n}\right)^2}.$$

(ii) Conclude that $\dfrac{\log \phi(t)}{t^2} = -\dfrac{1}{2}$, and hence X is a standard normal random variable.

5. Show that the independent increment property in the definition of a Brownian motion can be replaced by the following statement:
For any $0 \leq s < t \leq u < v$, $B_t - B_s$ and $B_v - B_u$ are independent.

6. Let $0 < t_1 < t < t_2$. Prove that the conditional density of $B(t)$ given $B(t_1) = a$ and $B(t_2) = b$ is a normal density with mean

$$a + (b - a)\frac{(t - t_1)}{(t_2 - t_1)} \quad \text{and variance} \quad (t_2 - t)\frac{(t - t_1)}{(t_2 - t_1)}.$$

7. Let T_a denote the time at which $\{B_t\}$ first attains the value $a > 0$.
   (i) Using the reflection principle, find the density of T_a.
   (ii) Using your answer for part (i), find

$$P\left\{\min_{0 \le u \le t} B(u) \le 0 \,|\, B(0) = a\right\}.$$

8. Recall that a function f defined on the real line is said to be locally Hölder-continuous of order $\alpha > 0$ if for every $L > 0$, there exists a finite constant c such that $|f(t) - f(s)| \le c|t - s|^\alpha$ for all real $t, s \in [-L, L]$. In other words, for every $L > 0$,

$$\sup\left\{|f(t) - f(s)|/|t - s|^\alpha : |t|, |s| \le L,\ t \neq s\right\} < \infty.$$

   Show that almost every Brownian path is nowhere Hölder-continuous of order β for any $\beta > 1/2$.

9. Show that $\dfrac{B(t)}{t^p} \to 0$ a.s. as $t \to \infty$ where $\{B(t)\}$ is a Wiener process and $p > 1/2$.

10. Show that $P\{B(t) > \epsilon\} \le \frac{1}{\epsilon}\sqrt{\frac{t}{2\pi}}e^{-\frac{\epsilon^2}{2t}}$ for any $t, \epsilon > 0$.

3 Elements of Martingale Theory

The theory of martingales is quite extensive and useful, and was developed primarily by Doob and Meyer. In this chapter, a concise treatment of martingales is presented. We start the chapter with definitions and various examples of martingales. This is followed by a discussion of Wiener martingales and their essential features. The strong Markov property of Wiener martingales is established, and various applications of it are shown. The Doob-Meyer decomposition plays a major role in the development of stochastic analysis. It is presented with complete details in Section 3.4. The Meyer process for L^2-martingales is given in Section 3.5. Local martingales are introduced and discussed in the final section of this chapter.

3.1 Definition and Examples of Martingales

We will adopt the convention that the index set is $\mathbb{R}^+ = [0, \infty)$ unless stated otherwise, and the letters s, t will denote instants of time with $s \leq t$. Let $(\Omega, \mathcal{F}, P)$ be a complete probability space with a given family of σ-fields $\{\mathcal{F}_t\}$ such that $\mathcal{F}_s \subseteq \mathcal{F}_t \subseteq \mathcal{F}$.

Definition 3.1.1 *A stochastic process $X = \{X_t\}$ defined on $(\Omega, \mathcal{F}, P)$ is said to be an $\mathcal{F}_t$-martingale if*

(i) each random variable X_t is measurable with respect to $\mathcal{F}_t$ and is in $L^1(P)$, and

(ii) the conditional expectation $E(X_t|\mathcal{F}_s) = X_s$ a.s.

Remark 3.1.1 In the above definition, if condition (ii) is changed to the inequality $E(X_t|\mathcal{F}_s) \geq X_s$ a.s., then X is called a **submartingale**. On the other hand, if the inequality is reversed, the resulting process is known as a **supermartingale**. The prefixes "sub" and "super" are as in the terminology used in the theory of harmonic functions. A martingale is both a submartingale as well as a supermartingale.

When the index set T is discrete, say, $\mathbb{N}$ instead of $\mathbb{R}^+$, we call the process a discrete parameter martingale. Martingales arose as models for fair games of chance. We illustrate the notion with several examples.

Example 3.1.1 *Suppose a fair coin is tossed repeatedly. Let the tosses be independent. Define*

$$\xi_n = \begin{cases} 1 & \text{if the } n^{th} \text{ toss shows heads,} \\ -1 & \text{otherwise.} \end{cases}$$

Let $S_n = \sum_{j=1}^{n} \xi_j$, *and* $\mathcal{F}_n = \sigma(\xi_1, \ldots, \xi_n)$. *Then* $\{S_n\}$ *is a* $\mathcal{F}_n$*-martingale, since the* $\{\xi_j\}$ *are independent with zero means. In fact, for any* $m < n$,

$$E(S_n|\mathcal{F}_m) = E(S_m + (S_n - S_m)|\mathcal{F}_m) = S_m + E(S_n - S_m) = S_m.$$

The process $\{S_n\}$ *is an important process known as the symmetric one-dimensional* **random walk**. *Likewise, the partial sums* S_n *of any sequence of centered, independent random variables form a martingale.*

Example 3.1.2 *Let* $\Omega = [0, 1)$, $\mathcal{F}$ *be the* σ*-field of Lebesgue measurable sets in* $[0, 1)$, *and* P, *the Lebesgue measure on* Ω. *Take any* $f \in L^1(P)$. *Let* $\mathcal{D}_n$ *denote the* n^{th} *dyadic* σ*-field in* $[0, 1)$, *that is,*

$$\mathcal{D}_n = \sigma([0, 1/2^n), [1/2^n, 2/2^n), \ldots, [(2^n - 1)/2^n, 1)).$$

Then $X_n := E(f|\mathcal{D}_n)$ *is a* $\mathcal{D}_n$*-martingale, since for any* $m < n$,

$$E(X_n|\mathcal{D}_m) = E\left(E(f|\mathcal{D}_n)|\mathcal{D}_m\right) = E(f|\mathcal{D}_m)$$

since $\mathcal{D}_m \subseteq \mathcal{D}_n$.

This martingale is an example of what are known as **Doob martingales** in discrete time. The value of X_n on any of the intervals $[k/2^n, (k+1)/2^n)$ that generates $\mathcal{D}_n$ is simply the average value of f over that interval.

Example 3.1.3 *Let* $\{X_n\}$ *be a given* $\mathcal{F}_n$*-martingale, and* $\{V_n\}$ *be a sequence of random variables with* V_n *being* $\mathcal{F}_{n-1}$ *measurable for all* $n \geq 1$, *and* V_0, *an* $\mathcal{F}_0$*-random variable. Define*

$$(V.X)_n = V_0 + V_1(X_1 - X_0) + V_2(X_2 - X_1) + \cdots + V_n(X_n - X_{n-1}).$$

If $(V.X)_n \in L^1(P)$ *for all* n, *then* $V.X$ *is a* $\mathcal{F}_n$ *martingale. Indeed, for any* $m < n$,

$$E\left((V.X)_n|\mathcal{F}_m\right) = (V.X)_m + E\left(V_{m+1}(X_{m+1} - X_m) + \cdots + V_n(X_n - X_{n-1})|\mathcal{F}_m\right).$$

Observe that each summand in the second term on the right is zero since

$$\begin{aligned} E\left(V_{m+j}(X_{m+j} - X_{m+j-1})|\mathcal{F}_m\right) &= E\left(E(V_{m+j}(X_{m+j} - X_{m+j-1})|\mathcal{F}_{m+j-1})|\mathcal{F}_m\right) \\ &= E(V_{m+j}E(X_{m+j} - X_{m+j-1}|\mathcal{F}_{m+j-1})|\mathcal{F}_m) \\ &= 0. \end{aligned}$$

Thus $V.X$ is a $\mathcal{F}_n$-martingale. The process $V.X$ is known as the **martingale transform** *of V by X.*

Example 3.1.4 *Let $(X_n, \mathcal{F}_n)$ be a martingale, and τ be an $\mathcal{F}_n$-stopping time. Define the process X_n^τ by*

$$X_n^\tau = \begin{cases} X_n & \text{if } n \leq \tau, \\ X_\tau & \text{otherwise.} \end{cases}$$

The process $\{X_n^\tau\}$, also denoted $\{X_{\tau\wedge n}\}$ is an $\mathcal{F}_n$-martingale, known as the **stopped martingale.** *To prove this, take*

$$V_n = \begin{cases} 1 & \text{if } n \leq \tau, \\ 0 & \text{otherwise} \end{cases}$$

in the previous example. Then

$$|(V.X)_n| \leq |X_{\tau\wedge n}| \leq \sum_{j=0}^{n} |X_j| \in L^1(P)$$

and $(V.X)_n = X_n^\tau$.

Example 3.1.5 *If $W = \{W_t\}$ is a Wiener process and $\mathcal{F}_t^{W,0} = \sigma(W_s : 0 \leq s \leq t)$ for any $t \geq 0$, then W is a $\mathcal{F}_t^{W,0}$-martingale by using the independent increments property of a Brownian motion:*

$$E(W_t|\mathcal{F}_s) = E(W_t - W_s|\mathcal{F}_s) + W_s = E(W_t - W_s) + W_s = W_s \text{ a.s.} \tag{3.1.1}$$

Example 3.1.6 *Define $M_t = e^{\alpha W_t - \alpha^2 t/2}$ for all $t \geq 0$ where α is a real number. Then $(M_t, \mathcal{F}_t^{W,0})$ is a martingale. For,*

$$\begin{aligned} E(M_t \mid \mathcal{F}_s^{W,0}) &= e^{\alpha W_s - \alpha^2 s/2} E\left(e^{\alpha(W_t - W_s) - \alpha^2(t-s)/2} \mid \mathcal{F}_s^{W,0}\right) \\ &= e^{\alpha W_s - \alpha^2 s/2} E\left(e^{\alpha(W_t - W_s) - \alpha^2(t-s)/2}\right) \\ &= e^{\alpha W_s - \alpha^2 s/2} = M_s. \end{aligned}$$

For $\alpha = 1$, we will call M_t as the **stochastic exponential** *of W_t, and denote it by $\mathcal{E}(W)_t$.*

Example 3.1.7 *Let $(X_t, \mathcal{F}_t)$ be any given martingale. Let $f \geq 0$ be a convex function on $\mathbb{R}^1$ such that $Ef(X_t) < \infty$ for all $t \geq 0$. Then, $(f(X_t), \mathcal{F}_t)$ is a submartingale by the Jensen inequality for conditional expectations.*
In particular, the processes $(|X_t|, \mathcal{F}_t)$ and $(X_t^+, \mathcal{F}_t)$ are submartingales.

We end this section with a discrete-time version of Doob's optional sampling theorem for bounded stopping times.

Theorem 3.1.8 *Let $(X_n, \mathcal{F}_n)$ be a martingale, and S and T be two bounded stopping times such that $S \leq T$. Then X_S and X_T are integrable, and $E(X_T \mid X_S) = X_S$ a.s.*

Proof Since T is bounded, there exists a $k \in \mathbb{N}$ such that $T \leq k$. Let $\{Y_n\}$ be any $\mathcal{F}_n$-martingale. Clearly, Y_S is in $L^1(P)$. Let $A \in \mathcal{F}_S$. For all $j \leq k$, $A \cap \{S = j\} \in \mathcal{F}_j$ so that

$$\int_{A\cap\{S=j\}} (Y_k - Y_S)\, dP = \int_{A\cap\{S=j\}} (Y_k - Y_j)\, dP = 0;$$

summing over j, one obtains $\int_A (Y_k - Y_S)\, dP = 0$, so that $E(Y_k - Y_S \mid \mathcal{F}_S) = 0$. Take Y_n to be the stopped martingale X_n^T to complete the proof. ∎

3.2 Wiener Martingales and the Markov Property

We start this section with a discussion on filtrations. Many results on martingales require the filtration $(\mathcal{F}_t)_{t\geq 0}$ to satisfy the usual conditions of completeness and right-continuity. The natural filtration $\mathcal{F}_t^{W,0} = \sigma(W_s : 0 \leq s \leq t)$ for the Wiener process W *does not* satisfy the usual conditions.

To see that $\mathcal{F}_t^{W,0}$ is not right-continuous, consider the event that W has a local maximum at time t. It belongs to $\mathcal{F}_{t+}^{W,0}$. However, the event cannot be a member of $\mathcal{F}_t^{W,0}$ since one needs to peek into the Brownian path for an infinitesimally short time after t to conclude that t is a point of local maximum.

However, $\mathcal{F}_t^{W,0}$ is a left-continuous filtration. Indeed,

$$\mathcal{F}_{t-}^{W,0} = \sigma\left\{\cup_{s<t}\mathcal{F}_s^{W,0}\right\}$$

and $W_t = \lim_{n\to\infty} W_{t_n}$ for any sequence $\{t_n\}$ strictly increasing to t. Therefore, W_t is measurable with respect to $\mathcal{F}_{t-}^{W,0}$ since each W_{t_n} has this property. Thus $\mathcal{F}_{t-}^{W,0} = \mathcal{F}_t^{W,0}$. Here, we have used only the left-continuity of the paths of W in this argument.

There are various advantages to having a right-continuous filtration, as the following theorem of Doob illustrates:

Theorem 3.2.1 *Let $X = \{X_t\}$ be an $\mathcal{F}_t$-martingale. If $\mathcal{F}_t$ satisfies the usual conditions, then X has a version with cadlag paths.*

To construct a right-continuous filtration in the context of a Brownian motion, define

$$\mathcal{N} = \left\{E \subset \Omega : \exists G \in \mathcal{F} \text{ such that } E \subset G,\ P(G) = 0\right\}.$$

Consider the filtration

$$\mathcal{F}_t^W = \sigma\left\{(W_s : 0 \leq s \leq t) \cup \mathcal{N}\right\}$$

known as the *augmented* Brownian filtration. We will prove later that $\{\mathcal{F}_t^W\}$ satisfies the usual conditions as a consequence of the strong Markov property.

The Brownian motion $\{W_t\}$ is clearly adapted to the filtration $\mathcal{F}_t^W$. The distributional properties of the increments of the process do not hinge on the underlying filtration. Also, $\{W_t\}$ is an $\mathcal{F}_t^W$-martingale as in Example 3.1.5 since the calculation in (3.1.1) holds for the filtration $\mathcal{F}_t^W$.

We start with the following general definition where right-continuity of the filtration is *not* assumed.

Definition 3.2.1 *A stochastic process $X = \{X_t\}$ is called a* **Wiener martingale** *with respect to a filtration $(\mathcal{F}_t)$ if*

(i) X is a Wiener process.

(ii) X_t is $\mathcal{F}_t$-adapted, and $X_t - X_s$ is independent of $\mathcal{F}_s$ for any $0 \le s \le t$.

The definition implies that $(X_t, \mathcal{F}_t)$ is a martingale. The process $(W_t, \mathcal{F}_t^W)$ is a Wiener martingale. There are several filtrations $\mathcal{F}_t \neq \mathcal{F}_t^W$ for which $(W_t, \mathcal{F}_t)$ is a Wiener martingale. We proceed to establish the Markov property of Wiener martingales by using the following simple lemma.

Lemma 3.2.2 *Let X and Y be two independent random variables defined on a probability space $(\Omega, \mathcal{F}, P)$. Let $\mathcal{G}$ be a sub-σ field of $\mathcal{F}$ such that X is independent of $\mathcal{G}$, and Y is measurable with respect to $\mathcal{G}$. Then*

$$P\{X + Y \in B \mid \mathcal{G}\} = P\{X + Y \in B \mid Y\}.$$

for any Borel set B.

Proof Define

$$\mathcal{A} := \{A \in \mathcal{B}(\mathbb{R}^2) : P\{(X,Y) \in A \mid \mathcal{G}\} = P\{(X,Y) \in A \mid Y\}\}.$$

For any two Borel sets B_1 and B_2 of the real line, note that

$$\begin{aligned} P\{(X,Y) \in B_1 \times B_2 \mid \mathcal{G}\} &= E\left(1_{B_1}(X)1_{B_2}(Y) \mid \mathcal{G}\right) \\ &= 1_{B_2}(Y)\, E(1_{B_1}(X)) \\ &= 1_{B_2}(Y)\, E(1_{B_1}(X) \mid Y) \\ &= E\left(1_{B_1}(X)1_{B_2}(Y) \mid Y\right) \\ &= P\{(X,Y) \in B_1 \times B_2 \mid Y\}. \end{aligned}$$

Thus, $\mathcal{A}$ contains all Borel-rectangles.

By the linearity and the monotone convergence theorem for conditional probability, it is easy to see that $\mathcal{A}$ is a Dynkin class. Therefore, $\mathcal{A} = \mathcal{B}(\mathbb{R}^2)$.

For any Borel set B in $\mathbb{R}^1$, the set $A := \{(x,y) : x + y \in B\}$ is a Borel set in $\mathbb{R}^2$, and hence in $\mathcal{A}$, which completes the proof. ∎

Definition 3.2.2 *A process $\{X_t\}$ adapted to $(\mathcal{F}_t)$ has the* **Markov property** *if for any $B \in \mathcal{B}$, and any $0 \le s \le t$,*

$$P\{X_t \in B \mid \mathcal{F}_s\} = P\{X_t \in B \mid X_s\}.$$

The process has the **strong Markov property** *if for any stopping time τ, and Borel set B,*

$$P\{X_{\tau+t} \in B \mid \mathcal{F}_\tau\} = P\{X_{\tau+t} \in B \mid X_\tau\}$$

a.s. on the set $\{\tau < \infty\}$.

Note that, if $(X_t, \mathcal{F}_t)$ is a Wiener martingale, then for any Borel set B in $\mathbb{R}^1$,

$$\begin{aligned} P\{X_t \in B \mid \mathcal{F}_s\} &= P\{X_t - X_s + X_s \in B \mid \mathcal{F}_s\} \\ &= P\{X_t - X_s + X_s \in B \mid X_s\} \\ &= P\{X_t \in B \mid X_s\} \end{aligned}$$

by the above lemma. Thus Wiener martingales possess the Markov property.

Next, we proceed to show that any Wiener martingale has the strong Markov property.

Theorem 3.2.3 *Let $(W_t, \mathcal{F}_t)$ be a Wiener martingale. Let τ be a finite $\mathcal{F}_t$-stopping time. Then,*

(i) the process $Y_t := W_{\tau+t} - W_\tau$ for $t \ge 0$ is a Wiener process, and

(ii) $\sigma(Y_t : t \ge 0)$ is independent of $\mathcal{F}_\tau$.

Proof

Step 1 Let τ have a countable range $\{s_j\}$. For any $B \in \mathcal{F}_\tau$, $0 \le t_1 < t_2 < \cdots < t_k$, and any Borel sets $A_1, \ldots, A_k$,

$$\begin{aligned} &P\{Y_{t_1} \in A_1, \ldots, Y_{t_k} \in A_k, B\} \\ &\quad = \sum_j P\{Y_{t_1} \in A_1, \ldots, Y_{t_k} \in A_k, B, \tau = s_j\} \\ &\quad = \sum_j P\{W_{s_j+t_1} - W_{s_j} \in A_1, \ldots, W_{s_j+t_k} - W_{s_j} \in A_k, B, \tau = s_j\}. \end{aligned}$$

Note that $B \cap \{\tau = s_j\} \in \mathcal{F}_{s_j}$, and $\sigma(W_{s_j+t} - W_{s_j} : t \ge 0)$ is independent of $\mathcal{F}_{s_j}$ by the second requirement in the definition of Wiener martingales. Therefore, the right side above equals

$$\begin{aligned} &\sum_j P\{W_{s_j+t_1} - W_{s_j} \in A_1, \ldots, W_{s_j+t_k} - W_{s_j} \in A_k\} P\{B, \tau = s_j\} \\ &\quad = \sum_j P\{W_{t_1} \in A_1, \ldots, W_{t_k} \in A_k\} P\{B, \tau = s_j\} \\ &\quad = P\{W_{t_1} \in A_1, \ldots, W_{t_k} \in A_k\} P(B). \end{aligned}$$

Thus,

$$P\{Y_{t_1} \in A_1, \dots, Y_{t_k} \in A_k, B\} = P\{W_{t_1} \in A_1, \dots, W_{t_k} \in A_k\}\, P(B). \tag{3.2.1}$$

By taking $B = \Omega$ in Equation (3.2.1), it follows that Y_t is a Wiener process. In Equation (3.2.1), B can be any set in $\mathcal{F}_\tau$ and hence, part (ii) follows.

Step 2 Given τ, construct τ_n with countable range such that $\tau_n \downarrow \tau$. Define $Y_t^n; = W_{\tau_n+t} - W_{\tau_n}$. Equation (3.2.1) holds for the process Y_t^n for any $B \in \mathcal{F}_{\tau_n}$. Note that $\mathcal{F}_\tau \subset \mathcal{F}_{\tau_n}$ since $\tau \leq \tau_n$. Therefore taking $B \in \mathcal{F}_\tau$ in Equation (3.2.1), one obtains

$$P\left\{Y_{t_1}^n \leq a_1, \dots, Y_{t_k}^n \leq a_k, B\right\} = P\{W_{t_1} \leq a_1, \dots, W_{t_k} \leq a_k\}\, P(B). \tag{3.2.2}$$

By the path continuity of W, $Y_t^n \to Y_t$ P-a.s. so that $(Y_{t_1}^n, \dots, Y_{t_k}^n) \to (Y_{t_1}, \dots, Y_{t_k})$ P-a.s. We need to show the equality (3.2.2) for the Y process in the place of Y^n, that is,

$$P\{Y_{t_1} \leq a_1, \dots, Y_{t_k} \leq a_k, B\} = P\{W_{t_1} \leq a_1, \dots, W_{t_k} \leq a_k\}\, P(B). \tag{3.2.3}$$

Step 3 Equation (3.2.3) holds for all $B \in \mathcal{F}_\tau$ satisfying $P(B) = 0$. Fix a $B \in \mathcal{F}_\tau$ with $P(B) > 0$. Define the probability measure Q by $Q(A) = P(A \cap B)/P(B)$. Then Q is absolutely continuous with respect to P and hence, $(Y_{t_1}^n, \dots, Y_{t_k}^n) \to (Y_{t_1}, \dots, Y_{t_k})$ Q-a.s. The corresponding characteristic functions (with respect to Q) therefore converge to that of $(Y_{t_1}, \dots, Y_{t_k})$.

Thus, $Q(Y_{t_1}^n \leq a_1, \dots, Y_{t_k}^n \leq a_k) \to Q(Y_{t_1} \leq a_1, \dots, Y_{t_k} \leq a_k)$ for all continuity points $(a_1, \dots, a_k)$ of the distribution of $(Y_{t_1}, \dots, Y_{t_k})$ with respect to Q. By the definition of Q, Equation (3.2.3) holds for such $(a_1, \dots, a_k)$. Thus, (3.2.3) holds for all vectors $(a_1, \dots, a_k)$ and $B \in \mathcal{F}_\tau$.

By taking $B = \Omega$ in Equation (3.2.3), we get that Y_t is a Wiener process. To prove (ii), observe that the finite-dimensional distributions of the Y process coincide with those of the Wiener process W. Therefore, Equation (3.2.3) can be written as

$$P\{Y_{t_1} \leq a_1, \dots, Y_{t_k} \leq a_k, B\} = P\{Y_{t_1} \leq a_1, \dots, Y_{t_k} \leq a_k\}\, P(B),$$

which completes the proof. ∎

As an application of the strong Markov property of the Wiener process, we again discuss the reflection principle of André.

Fix $T > 0$. Define $M_T = \sup_{0 \leq t \leq T} W_t$. Let $\tau = \inf\{t : M_t \geq a\}$

Theorem 3.2.4 *Let $(W_t, \mathcal{F}_t)$ be a Wiener martingale. For any $a > 0$ and $T \geq 0$,*

$$P(M_T \geq a) = P(\tau \leq T) = 2P(W_T \geq a).$$

Proof By continuity of paths, $W_\tau = a$. Therefore,

$$P\{M_T \geq a, W_T < a\} = P\{\tau < T, W_T - W_\tau < 0\}.$$

By the strong Markov property, $W_{t+\tau} - W_\tau$ is a Wiener process which is independent of $\mathcal{F}_\tau$. Hence,

$$\begin{aligned} P\{\tau < T, W_T - W_\tau < 0\} &= P\left\{\tau < T, W_{(T-\tau)+\tau} - W_\tau < 0\right\} \\ &= P(\tau < T)P(W_{(T-\tau)+\tau} - W_\tau < 0) \\ &= 1/2P(\tau < T) = 1/2P(\tau \leq T) \\ &= 1/2P(M_T \geq a). \end{aligned}$$

Thus,

$$P\{M_T \geq a, W_T < a\} = 1/2P(M_T \geq a).$$

Also,

$$P\{M_T \geq a, W_T \geq a\} = P(W_T \geq a).$$

Adding the two equations,

$$P(M_T \geq a) = 1/2P(M_T \geq a) + P(W_T \geq a),$$

which completes the proof. ■

Remark 3.2.1 Using the above theorem, one can make several exact calculations. For instance, the distribution of τ is given by

$$\begin{aligned} P(\tau \leq t) &= 2P(W_t \geq a) \\ &= \frac{2}{\sqrt{2\pi t}} \int_a^\infty e^{-x^2/2t} dx, \end{aligned}$$

so that by differentiating with respect to t, the density of τ is

$$f_\tau(t) = \frac{a}{\sqrt{2\pi}} t^{-3/2} e^{-a^2/2t}.$$

3.3 Essential Results on Martingales

Some of the most useful and important results in the general theory of martingales are discussed in this section. The credit for much of the impetus to the theory goes to Doob and Meyer. For a full and detailed account of martingales, we refer the interested reader to the books by Doob [14] and Meyer [53].

Definition 3.3.1 *A family of random variables $\mathcal{Z}$ in $L^1(P)$ is said to be* **uniformly integrable** *if*

$$\lim_{c\to\infty} \sup_{Z\in\mathcal{Z}} \int_{\{|Z|>c\}} |Z| dP = 0.$$

A useful equivalent criterion for uniform integrability is given by:

(i) $\sup_{\{Z\in\mathcal{Z}\}} E|Z| < \infty$; that is, $\mathcal{Z}$ is $L^1(P)$-bounded, and

(ii) $\lim\limits_{P(A)\to 0} \sup_{Z\in\mathcal{Z}} \int_A |Z|\, dP = 0$.

If a family of integrable random variables $\{X_t : 0 \le t < T\}$ converges to a limit X a.s. as $t \to T$ where $X \in L^1(P)$, then it doesn't automatically follow that $X_t \to X$ in $L^1(P)$. One needs uniform integrability of $\{X_t : 0 \le t < T\}$ to obtain the $L^1(P)$ convergence. As an application of this concept, we prove the optional sampling theorem in continuous time.

Theorem 3.3.1 *Let $(X_t, \mathcal{F}_t)$ be a martingale with right-continuous paths. Let $(\mathcal{F}_t)$ satisfy the usual conditions. If $S \le T$ are two bounded stopping times, then $E(X_T \mid \mathcal{F}_S) = X_S$ a.s.*

Proof First, note that $X_T \in \mathcal{F}_T$ and $X_S \in \mathcal{F}_S$ since X_t is progressively measurable, and $(\mathcal{F}_t)$ satisfies the usual conditions. We must show that $\int_A X_T dP = \int_A X_S dP$ for all $A \in \mathcal{F}_S$.

There exists a $k > 0$ such that $S \le T \le k - 1$. Define $T_n = \min\{jk/n : jk/n \ge T\}$ for all integers $n \ge 1$. Then, T_n are stopping times, and $T_n \downarrow T$ a.s. as $n \to \infty$. Likewise, one can define stopping times S_n with $S_n \downarrow S$. By Theorem 3.1.8, it follows that

$$\int_A X_{T_n}\, dP = \int_A X_{S_n}\, dP$$

for all $A \in \mathcal{F}_{S_n}$, and hence for all $A \in \mathcal{F}_S \subset \mathcal{F}_{S_n}$; and $\{X_{jk/n} : j = 0, 1, \ldots\}$ is an $\mathcal{F}_{jk/n}$ martingale.

The proof is over if $\int_A X_{T_n}\, dP \to \int_A X_T\, dP$, and $\int_A X_{S_n}\, dP \to \int_A X_S\, dP$ as $n \to \infty$.

Observe that by right-continuity of paths, $X_{T_n} \to X_T$ a.s. as $n \to \infty$. We will show that $\{X_{T_n}\}$ is uniformly integrable. For any fixed n, by Theorem 3.1.8, the pair of random variables $\{X_{T_n}, X_k, \mathcal{F}_{T_n}, \mathcal{F}_k\}$ is a martingale. Therefore, for any $A \in \mathcal{F}_{T_n}$,

$$\int_A |X_{T_n}|\, dP \le \int_A |X_k|\, dP.$$

Thus,

$$\lim_{P(A)\to 0} \sup_n \int_A |X_{T_n}|\, dP = 0. \tag{3.3.1}$$

Also, $E|X_{T_n}| \le E|X_k|$ for all n. This with (3.3.1) is precisely the uniform integrability criterion for $\{X_{T_n}\}$. Likewise, $\{X_{S_n}\}$ is also uniformly integrable. The proof is over. ■

Remark 3.3.1 From the optional sampling theorem, it follows that if $(X_t, \mathcal{F}_t)$ is a martingale with right-continuous paths and $(\mathcal{F}_t)$ satisfies the usual conditions, then for any stopping time τ, $E(X_{\tau\wedge t} \mid \mathcal{F}_{\tau\wedge s}) = X_{\tau\wedge s}$ a.s. In other words, $(X_{\tau\wedge t}, \mathcal{F}_{\tau\wedge t})$ is a martingale.

Definition 3.3.2 *Given a real-valued sequence $\{x_n\}$ and two finite numbers a, b such that $a < b$, the sequence is said to upcross the interval $[a, b]$ at least N times if there exist $0 \le m_1 < n_1 < \cdots < m_N < n_N$ such that $x_{m_i} \le a$ and $x_{n_i} \ge b$ for each $1 \le i \le N$. The sequence $\{x_n\}$ upcrosses precisely N times if it upcrosses $[a, b]$ at least N times but does not upcross $[a, b]$ at least $N + 1$ times.*

Let $U_{[a,b]}$ l denote the number of upcrossings of $[a, b]$ by a given sequence.

Theorem 3.3.2 (Doob's upcrossing inequality) *Let $\{X_n\}$ be a discrete-parameter submartingale adapted to $(\mathcal{F}_n)$. Then the number of upcrossings of $[a, b]$ by $\{X_n\}$ satisfies*

$$E[U_{[a,b]}] \le \sup_n \frac{E(X_n - a)^+}{b - a}. \tag{3.3.2}$$

Proof Set $Y_n = \dfrac{(X_n - a)^+}{b - a}$. Then Y, is a submartingale. Let $\tau_0 = 0$, and define, inductively, the stopping times

$$\sigma_k = \inf\{n \ge \tau_{k-1} : X_n \le a\},$$
$$\tau_k = \inf\{n \ge \sigma_k : X_n \ge b\}$$

for all $k \ge 1$. Fix an $N \in \mathbb{N}$, and let $U^N_{[a,b]}$ be the number of upcrossings of $[a, b]$ by $\{X_{n\wedge N}\}$. Then,

$$\begin{aligned} U^N_{[a,b]} &\le \sum_{k=1}^{N} (Y_{N\wedge\tau_k} - Y_{N\wedge\sigma_k}) \\ &= Y_N - Y_0 - \sum_{k=1}^{N} (Y_{N\wedge\sigma_k} - Y_{N\wedge\tau_{k-1}}) \\ &\le Y_N - \sum_{k=1}^{N} (Y_{N\wedge\sigma_k} - Y_{N\wedge\tau_{k-1}}). \end{aligned}$$

By Doob's optional sampling theorem $E(Y_{N\wedge\sigma_k} - Y_{N\wedge\tau_{k-1}}) \geq 0$ so that

$$E\left(U^N_{[a,b]}\right) \leq EY_N.$$

Allow $N \to \infty$ to obtain

$$E\left(U^N_{[a,b]}\right) \leq \sup_N \frac{E(X_n - a)^+}{b - a}.$$ ■

Next, we prove an important result known as Doob's martingale convergence theorem for a discrete-parameter submartingale.

Theorem 3.3.3 *Let $\{X_n\}$ be an $\mathcal{F}_n$-adapted submartingale. If $\sup_n E[X_n^+] < \infty$, then there exists an $X \in L^1(P)$ such that $X_n \to X$ a.s. as $n \to \infty$.*

Proof From Doob's upcrossing inequality, one can infer that

$$P\{U_{[a,b]} < \infty \;\; \forall \;\; a, b \in \mathbb{Q}\} = 1. \tag{3.3.3}$$

If $\lim_{n\to\infty} X_n(\omega)$ does not exist, then there would exist $a, b \in \mathbb{Q}$ such that

$$\liminf_{n\to\infty} X_n(\omega) < a < b < \limsup_{n\to\infty} X_n(\omega).$$

For such ω's, $U_{[a,b]}(\omega) = \infty$. By (3.3.3), such ω's form a P-null set. Therefore, X_n converges a.s. to a limit. Calling the limit X, note that

$$\begin{aligned} E|X| = E\left(\liminf_{n\to\infty} |X_n|\right) &\leq \liminf_{n\to\infty} E|X_n| \\ &= \liminf_{n\to\infty} E(2X_n^+ - X_n) \\ &\leq 2\sup_n E(X_n^+) - EX_0, \end{aligned}$$

and hence $X \in L^1(P)$. ■

The theorem extends to cadlag submartingales in a straightforward manner. We state it below in a form that we will need later.

Theorem 3.3.4 (**Doob's martingale convergence theorem**) *Let $(X_t, \mathcal{F}_t)$ be a submartingale with right-continuous paths that is $L^1(P)$-bounded, that is, $\sup_{t\geq 0} E|X_t| < \infty$. Then, $X_t \to X_\infty$ P-a.s., as $t \to \infty$, and $E|X_\infty| < \infty$.*

The theorem, in particular, implies that if $(X_t, \mathcal{F}_t)$ is a uniformly integrable right-continuous martingale, then $X_t \to X_\infty$ a.s. as well as in $L^1(P)$. Therefore, for any $s \geq 0$,

$$X_s = \lim_{t\to\infty} E(X_t \mid \mathcal{F}_s) = E(X_\infty \mid \mathcal{F}_s)$$

so that the martingale X_t admits the representation $E(X_\infty \mid \mathcal{F}_t)$. In other words, it is a Doob martingale.

There are thus two kinds of L^1-bounded martingales: (i) those that converge to their limits in L^1 and (ii) those that don't. We illustrate these two kinds by a few examples.

Example 3.3.5 *Let* $S_n = 1 + \sum_{j=1}^{n} \xi_j$ *where* ξ_j *are iid random variables taking the values 1 and −1 with equal probability. Then* $\{S_n\}$ *is a martingale with respect to* $\mathcal{F}_n = \sigma(\xi_1, \ldots, \xi_n)$.

$\{S_n\}$ is not L^1-bounded. For, if $\{S_n\}$ were L^1-bounded, then it has to converge a.s. Since variance $(S_n) = n$, we can conclude that S_n diverges a.s.

Let $T = \inf\{k : S_k = 0\}$. The stopped martingale $\{S_{T\wedge n}\}$ is L^1-bounded since

$$E|S_{T\wedge n}| = ES_{T\wedge n} = ES_0 = 1.$$

Therefore, $\lim_{n\to\infty} S_{T\wedge n}$ exists a.s. Note that

$$\lim_{n\to\infty} S_{T\wedge n} = \begin{cases} S_T & \text{if } T < \infty; \\ \lim_{n\to\infty} S_n & \text{if } T = \infty. \end{cases}$$

Since we already know that $\lim_{n\to\infty} S_n$ doesn't exist on any set of positive measure, the almost sure limit of $S_{T\wedge n}$ is 0. As a by-product, we have also shown that $P(T < \infty) = 1$.

Though $S_{T\wedge n}$ converges to 0 a.s., the convergence is not in the L^1-sense since $E|S_{T\wedge n}| = 1$ for all n.

Example 3.3.6 *In Example 3.1.2, take $f(x) = x^2$. Then, the martingale $X_n = E(f|\mathcal{D}_n)$ is L^1-bounded since it is a positive martingale. It also converges in L^1 since $X_n \leq 1$ for any n, and hence, the dominated convergence theorem can be used.*

Example 3.3.7 *Let $S_n = \sum_{j=1}^{n} \xi_j$ where ξ_j are iid random variables taking the values 1 and −1 with equal probability. Let $\xi_j = 1$ denote the occurrence of heads in the jth toss of a fair coin. Let a gambler start by betting a dollar on heads. If he wins, he adds his winnings to the original bet and bets everything again. If he loses, he loses everything and quits playing.*

To write this mathematically, define V_n in Example 3.2.3 by

$$V_n = \begin{cases} 2^{n-1} & \text{if } S_{n-1} = n-1, \\ 0 & \text{otherwise.} \end{cases}$$

Let $X_n = 1 + (V.X)_n$. Then, X_n is a martingale. Since $X_n \geq 0$, $X_n \to 0$ a.s. as $n \to \infty$. But X_n doesn't converge to 0 in L^1 since $E|X_n| = EX_n = 1$ for all n.

The next result is the Doob martingale inequalities which are of fundamental importance in stochastic analysis. The first inequality is the weak type $(1, 1)$ inequality, while the second is an L^p-inequality. For any process $\{X_t\}$, define

$$X_T^* = \sup_{0\leq t\leq T} |X_t| \quad \text{and} \quad X^* = \sup_t |X_t|.$$

for any $T \geq 0$ with a similar definition for processes set to discrete time. We will first prove inequalities for discrete-parameter submartingales.

Theorem 3.3.8 *Let $X := \{X_n\}$ be a nonnegative submartingale. Then*

(*i*) (**Basic Submartingale Inequality**) *For any $\lambda > 0$,*

$$\lambda P\left\{X_n^* \geq \lambda\right\} \leq E\left(X_n 1_{\left\{X_n^* \geq \lambda\right\}}\right). \tag{3.3.4}$$

(*ii*) (**Doob's L^p Inequality**) *If $E(X_n^p) < \infty$ for a finite $p > 1$, then*

$$E\left[(X_n^*)^p\right] \leq \left(\frac{p}{p-1}\right)^p E(X_n^p). \tag{3.3.5}$$

Proof Define the stopping time $N = \inf\{k : X_k \geq \lambda\}$. Then,

$$\sum_{k=1}^{n} P\{N = k\} = P\left\{X_n^* \geq \lambda\right\}.$$

By the submartingale property of X, we get $\lambda P\{N = k\} \leq E\left(X_n 1_{\{N=k\}}\right)$, so that by summing over k, we obtain

$$\sum_{k=1}^{n} \lambda P\{N = k\} \leq \sum_{k=1}^{n} E\left(X_n 1_{\{N=k\}}\right);$$

that is,

$$\lambda P\left\{X_n^* \geq \lambda\right\} \leq E\left(X_n 1_{\left\{X_n^* \geq \lambda\right\}}\right).$$

To prove part (ii), consider

$$\begin{aligned}
E[(X_n^*)^p] &= \int_0^\infty P\left\{X_n^* > v\right\} p v^{p-1} dv \\
&\leq \int_0^\infty \frac{1}{v} E\left(X_n 1_{\left\{X_n^* \geq v\right\}}\right) p v^{p-1} dv \quad \text{(by part (i))} \\
&= \int_0^\infty \int_\Omega \frac{1}{v} X_n 1_{\left\{X_n^* \geq v\right\}} p v^{p-1} dP dv \\
&= \int_\Omega X_n \int_0^{X_n^*} p v^{p-2} dv\, dP \\
&= \frac{p}{p-1} E\left[X_n (X_n^*)^{p-1}\right] \\
&\leq \frac{p}{p-1} \|X_n\|_{L^p(P)} \|(X_n^*)^{p-1}\|_{L^q(P)}
\end{aligned}$$

by Hölder's inequality. Since $q = \frac{p}{p-1}$, and X_n^* is in $L^p(P)$, the proof is over. ■

Theorem 3.3.9 *Let $(X_t, \mathcal{F}_t)$ be a nonnegative, right-continuous submartingale. Then,*

(i) For any $\lambda > 0$, $\lambda P\{X_T^* > \lambda\} \le EX_T$.

(ii) If for some $1 < p < \infty$, $E(X_T^p) < \infty$, then

$$E(X_T^*)^p \le \left(\frac{p}{p-1}\right)^p E(X_T^p). \tag{3.3.6}$$

Proof Fix any $n \in \mathbb{N}$. Define $Y_k := X_{\frac{kT}{2^n}}$ for $k = 1, \dots, 2^n$. Y_k is a discrete parameter martingale adapted to $(\mathcal{F}_{\frac{kT}{2^n}})$. From the previous theorem,

$$\lambda P\{Y_{2^n}^* > \lambda\} \le EX_T$$

for $\lambda > 0$. By right continuity of paths, $(Y_{2^n}^*) \uparrow (X_T^*)$ as $n \to \infty$. Hence, part (i) follows from the monotone convergence theorem. Likewise, we also have

$$E\left[(Y_{2^n}^*)^p\right] \le \left(\frac{p}{p-1}\right)^p E(X_T^p).$$

Letting $n \to \infty$, part(ii) follows. ■

Remark 3.3.2 If X is a martingale, the basic martingale inequality and the L^p-inequality hold for the submartingale $|X|$.

Corollary 3.3.10 *For $1 < p < \infty$, L^p-bounded martingales converge in L^p.*

Proof If $\{X_t\}$ is an L^p-bounded martingale, it is L^1-bounded, and hence there exists a random variable X_∞ such that a.s. $X_t \to X_\infty$ as $t \to \infty$. By (3.3.6) and L^p-boundedness, $(X^*)^p$ has finite expectation. Since $|X_t - X_\infty|^p \le 2(X^*)^p$, the corollary follows from the dominated convergence theorem. ■

Remark 3.3.3 Doob's type of inequality doesn't hold for $p = 1$. Were there one, then L^1-bounded martingales would converge in the L^1-norm, which we know to be false.

3.4 The Doob-Meyer Decomposition

In this section, we will present one of the fundamental theorems useful in the development of stochastic integration. Throughout this section, let $(\Omega, \mathcal{F}, P)$ be a complete probability space with a filtration $(\mathcal{F}_t)$ that satisfies the usual hypotheses. We start with a result on discrete parameter supermartingales due to Doob.

Theorem 3.4.1 *A $\mathcal{F}_n$-adapted supermartingale $X = \{X_n\}$ has a unique decomposition $X_n = M_n - A_n$ a.s. where M is an $\mathcal{F}_n$-martingale, and A satisfies:*

(i) $A_0 = 0$ a.s.

(ii) $A_{n-1} \le A_n$ a.s.

(iii) A_n is $\mathcal{F}_{n-1}$-measurable for all $n \ge 1$.

Proof Define A_n inductively as follows:

$$A_0 = 0$$
$$A_n = A_{n-1} + \{X_{n-1} - E(X_n|\mathcal{F}_{n-1})\} \quad \text{for } n \geq 1.$$

Since $X_{n-1} \geq E(X_n|\mathcal{F}_{n-1})$, $A_n \geq A_{n-1}$. From the definition, it is clear that A has the required properties. By induction, EA_n is finite for all n. Define $M_n = X_n + A_n$. Then M_n is integrable, and $E[M_n|\mathcal{F}_{n-1}] = M_{n-1}$ so that M is a martingale.

To prove uniqueness, suppose that X admits another such decomposition,

$$X_n = \tilde{M}_n - \tilde{A}_n.$$

Then, from the definition of A_n,

$$\begin{aligned} A_n - A_{n-1} &= X_{n-1} - E(X_n|\mathcal{F}_{n-1}) \\ &= \tilde{A}_n - \tilde{A}_{n-1}. \end{aligned}$$

Moreover, $A_0 = \tilde{A}_0 = 0$, so that $A = \tilde{A}$. ■

Lemma 3.4.2 *If $A = \{A_n\}$ is an adapted sequence satisfying (i) and (ii) of Theorem 3.4.1, and if $A_\infty = \lim A_n$ has finite expectation, then the following are equivalent:*

1. *A_n is $\mathcal{F}_{n-1}$-measurable for each n.*
2. $E\Big[\sum_{j=1}^{\infty} Y_{j-1}(A_j - A_{j-1})\Big] = EY_\infty A_\infty$ *for every bounded $\mathcal{F}_n$-martingale $Y = \{Y_n\}$.*

Proof

Step 1 Assertion (1) implies (2). For, by conditioning on $\mathcal{F}_{j-1}$, one obtains

$$E\left[\sum_{j=1}^{n} A_j(Y_j - Y_{j-1})\right] = 0 = E\left[\sum_{j=1}^{n} A_{j-1}(Y_j - Y_{j-1})\right]$$

for any fixed n. This implies that

$$E\left[\sum_{j=1}^{n} (A_j - A_{j-1})(Y_j - Y_{j-1})\right] = 0,$$

which can be rewritten as

$$E\left[\sum_{j=1}^{n} Y_j(A_j - A_{j-1})\right] = E\left[\sum_{j=1}^{n} Y_{j-1}(A_j - A_{j-1})\right]. \tag{3.4.1}$$

Since $E[Y_j A_{j-1}] = E[A_{j-1} E(Y_j|\mathcal{F}_{j-1})] = E[A_{j-1} Y_{j-1}]$,

$$E\left[\sum_{j=1}^{n} Y_j(A_j - A_{j-1})\right] = E(A_n Y_n).$$

Therefore, Equation (3.4.1) yields

$$E(A_n Y_n) = E\left[\sum_{j=1}^{n} Y_{j-1}(A_j - A_{j-1})\right]$$

for all $n \geq 1$. Since Y is bounded, and $EA_\infty < \infty$, assertion (2) is obtained by taking limits.

Step 2 For the converse, suppose A_n is $\mathcal{F}_n$-adapted. We claim that for any *bounded*, $\mathcal{F}_n$-measurable random variable Z,

$$E(ZA_n) = E[E(Z|\mathcal{F}_{n-1})A_n]. \tag{3.4.2}$$

The proof is by induction. Before we proceed with the induction, it is worthwhile to point out that the symbol ∞ in assertion (2) of the lemma can be replaced by N if a bounded martingale $Y_n : n \leq N$ is considered. To see this, simply modify A by defining $A_j \equiv A_N$ and $Y_j \equiv Y_N$ for all $j > N$.

For $n = 1$, define $Y_0 = E(Z|\mathcal{F}_0)$, and $Y_1 = Z$. In this case, Equation (3.4.2) follows from assertion (2) of the lemma.

Let $n > 1$. Define $Y_j = E(Z|\mathcal{F}_j)$ for all $j = 0, 1, \ldots, n$ so that $Y_n = Z$. From the induction hypothesis,

$$E(Y_j A_j) = E[E(Y_j|\mathcal{F}_{j-1})A_j] = E(Y_{j-1}A_j) \quad \text{for all} \quad 1 \leq j < n.$$

Thus,

$$\begin{aligned} EY_n A_n &= E\left[\sum_{j=1}^{n} Y_{j-1}(A_j - A_{j-1})\right] \\ &= \sum_{j=1}^{n} E(Y_{j-1}A_j) - \sum_{j=0}^{n-1} E(Y_j A_j) \\ &= \sum_{j=1}^{n} E(Y_{j-1}A_j) - \sum_{j=1}^{n-1} E(Y_{j-1}A_j) - E(Y_0 A_0) \\ &= E(Y_{n-1}A_n); \end{aligned}$$

that is, $E(ZA_n) = E[E(Z|\mathcal{F}_{n-1})A_n]$, which completes the induction.

Step 3 Since Z is bounded and A_n is integrable, it is easy to see that

$$E[E(Z|\mathcal{F}_{n-1})A_n] = E[ZE(A_n|\mathcal{F}_{n-1})].$$

Thus, $E(ZA_n) = E[ZE(A_n|\mathcal{F}_{n-1})]$ for all Z, bounded and $\mathcal{F}_n$-measurable. We can conclude that A_n is $\mathcal{F}_{n-1}$-measurable for any $n \geq 1$. ∎

Definition 3.4.1 *An $\mathcal{F}_n$-adapted supermartingale $X = \{X_n\}$ is called a* **potential** *if it satisfies:*

(a) $X_n \geq 0$ a.s. for all n.

(b) $\lim_{n\to\infty} EX_n = 0$.

Using the notation of Theorem 3.4.1, if $X_n = M_n - A_n$ is a nonnegative supermartingale, then $EM_n - EA_n \geq 0$ so that $EA_n \leq EM_0$ for all n. Therefore, $EA_\infty \leq EM_0$. In other words, $A_\infty = \lim_{n\to\infty} A_n$ exists a.s., and has finite expectation. By the dominated convergence theorem, the above limit exists in the $L^1(P)$-norm as well.

If X is a potential, then by using the requirement (b) in the definition, X is uniformly integrable. Since $A_n \leq A_\infty$ for all n, we get the uniform integrability of A. Therefore, $M = X + A$ is uniformly integrable. By the martingale convergence theorem, $M_\infty = \lim_{n\to\infty} M_n$ exists a.s. and in $L^1(P)$. Also, $M_n = E(M_\infty|\mathcal{F}_n)$.

From the above discussion, $X_\infty = \lim_{n\to\infty} X_n$ exists a.s. and in $L^1(P)$. Besides, $EX_\infty = \lim_{n\to\infty} EX_n = 0$, and hence $X_\infty = 0$ a.s. Therefore, $M_\infty = A_\infty$. We summarize the above as

Theorem 3.4.3 *A potential X admits a unique decomposition of the form*

$$X_n = E(A_\infty|\mathcal{F}_n) - A_n \quad \text{for all} \quad n \geq 0$$

where A satisfies (i), (ii) and (iii) of Theorem 3.4.1, and $A_\infty = \lim_{n\to\infty} A_n$ a.s. and in $L^1(P)$.

So far, we have studied processes set to a discrete time parameter. Now, we will study the analogous decomposition for supermartingales defined on a continuous parameter set such as $\mathbb{R}^+$ or $[0, T]$. Surprisingly, the theory carries over only for a special class of supermartingales.

Definition 3.4.2 *A real-valued, $\mathcal{F}_t$-adapted stochastic process $A = \{A_t\}$ is known as an* **increasing process** *with respect to $(\mathcal{F}_t)$ if*

(a) $A_0 = 0$, and A has increasing, right-continuous paths a.s.

(b) $EA_t < \infty$ for all t.

A process A is called an **increasing integrable** process if it is an increasing process, and $\sup_t EA_t < \infty$. If the index set is $[0, T]$ instead of $[0, \infty)$, then an increasing process is automatically integrable as well.

Notation: Let A be an increasing process, and X, a nonnegative stochastic process. Then, for each $\omega \in \Omega$, $\int_0^\infty X_s(\omega)\, dA_s(\omega)$ will denote the Lebesgue-Stieltjes integral on $\mathbb{R}^+$. Likewise, the integral $\int_0^t X_s(\omega)\, dA_s(\omega)$ denotes, for each ω and $t \geq 0$, the Lebesgue-Stieltjes integral $\int_{[0,t]} X_s(\omega)\, dA_s(\omega)$.

Remark 3.4.1 In addition, if X is progressively measurable with respect to a filtration $(\mathcal{F}_t)$, then the process

$$Y_t(\omega) = \int_0^t X_s(\omega)\, dA_s(\omega)$$

is $\mathcal{F}_t$-adapted and right-continuous. Therefore, Y is also progressively measurable. Consequently, for any stopping time τ,

$$Y_\tau = \int_0^\tau X_s\, dA_s$$

is $\mathcal{F}_\tau$-measurable.

Definition 3.4.3 *An integrable increasing process A is called* **natural** *if*

$$\mathrm{E}\int_0^t Y_s\, dA_s = \mathrm{E}\int_0^t Y_{s-}\, dA_s$$

for each $t \in \mathbb{R}^+$, and every nonnegative, bounded, right-continuous martingale Y.

Analogous to Lemma 3.4.2 and Equation (3.4.1), one has the following characterization for natural processes.

Theorem 3.4.4 *Let A be an integrable, increasing process on $\overline{\mathbb{R}^+}$. Then, for every nonnegative, bounded, right-continuous martingale Y,*

(a) $\mathrm{E}\left(\int_0^\infty Y_s\, dA_s\right) = \mathrm{E}\left(A_\infty Y_\infty\right)$, *and*

(b) the process A is natural if and only if

$$\mathrm{E}\left(\int_0^\infty Y_{s-}\, dA_s\right) = E(A_\infty Y_\infty).$$

Proof Fix any $t > 0$. Let $0 = t_0 < t_1 < t_2 \cdots < t_n = t$ be any partition of $[0, t]$. Then

$$\begin{aligned} E(Y_t A_t) &= \sum_{j=1}^n \mathrm{E}\left[Y_t(A_{t_j} - A_{t_{j-1}})\right] \\ &= \sum_{j=1}^n \mathrm{E}\left[Y_{t_j}(A_{t_j} - A_{t_{j-1}})\right], \end{aligned}$$

which converges to $E\int_0^t Y_s\, dA_s$ as the norm of the partition, $\max_j(t_j - t_{j-1}) \to 0$. Here we have used the right-continuity and boundedness of Y as well as the integrability of A in passing to the limit. Allowing $t \to \infty$, the first assertion is proved. The second assertion follows from part (a) and the definition of a natural process. ■

Definition 3.4.4 *Let $X = \{X_t\}$ be a right-continuous, $\mathcal{F}_t$-adapted supermartingale, and let $\mathcal{T}$ denote the collection of all finite stopping times relative to $(\mathcal{F}_t)$. The process X belongs to* **class (D)** *if the collection of random variables $\{X_\tau : \tau \in \mathcal{T}\}$ is uniformly integrable.*

The terminology class (D) was coined by P. A. Meyer, and the letter D in it probably refers to Doob. It is worthwhile to note that any class (D) supermartingale is a uniformly integrable supermartingale, whereas the converse is not true. However, any right-continuous, uniformly integrable martingale is a class (D) martingale.

Definition 3.4.5 *A nonnegative, right-continuous supermartingale is called a* **potential** *if* $\lim_{t\to\infty} EX_t = 0$.

First, we prove the following technical result:

Lemma 3.4.5 *Let $\{U_n\}$ be a uniformly integrable sequence of random variables. Then there exists a subsequence which converges weakly in $L^1(P)$; that is, there exists a random variable U such that*

$$E[U_n Z) \to E(UZ) \ \textit{for any random variable } \ Z \in L^\infty(P).$$

Proof For each fixed k, let us denote $U_n 1_{\{|U_n|\le k\}}$ as $U_{n,k}$. It is obvious that the sequence $\{U_{n,k},\ n \ge 1\}$ is L^2-bounded so that there exists a subsequence which converges weakly in $L^2(P)$. Let us denote the weak limit as L_k. By the Cantor diagonalization procedure, we can extract a single subsequence $\{n_j\}$ such that

$$U_{n_j,k} \to L_k \ \text{ weakly in } \ L^2(P)$$

as $j \to \infty$, for each k. We will write n_j as simply j. Consider

$$\begin{aligned} E|L_k - L_r| &\le \liminf_{j\to\infty} E|U_{j,k} - U_{j,r}| \\ &\le \liminf_{j\to\infty} E\left[|U_j|\, 1_{\{|U_j|>k\wedge r\}}\right] \\ &\le \sup_j \int_{\{|U_j|>k\wedge r\}} |U_j|\, dP, \end{aligned}$$

which tends to zero as $k, r \to \infty$ by the hypothesis. Thus, $\{L_k\}$ is a Cauchy sequence so that L_k converges to L in $L^1(P)$. Take any $Z \in L^\infty(P)$, and consider

$$\begin{aligned} |E(U_j Z - LZ)| &\le E(|U_j - U_{j,k}||Z|) + |E[(U_{j,k} - L_k)Z]| + E(|L_k - L|\,|Z|) \\ &\le \|Z\|_\infty \sup_j \int_{\{|U_j|>k\}} |U_j|\, dP + |E[(U_{j,k} - L_k)Z]| \\ &\quad + \|Z\|_\infty E|L_k - L|. \end{aligned}$$

Allowing $j \to \infty$ and then letting $k \to \infty$, the proof is completed. ■

Theorem 3.4.6 (Doob-Meyer decomposition for potentials) *A potential $X = \{X_t\}$ admits a decomposition of the form*

$$X_t = E[A_\infty|\mathcal{F}_t] - A_t \ \ \textit{for all} \ \ t \in \mathbb{R}^+$$

where A is an integrable increasing process if and only if X is of class (D).

Proof For each $n \in \mathbb{N}$ and $j \geq 0$, consider the process $\{X_{j/2^n}\}$. This is a potential with a discrete-time parameter. By Theorem 3.4.4, we can write

$$X_{j/2^n} = E[A(\infty, n)|\mathcal{F}_{j/2^n}] - A(j/2^n, n)$$

where $A(0, n) = 0$, and $A(j/2^n, n) \leq A((j+1)/2^n, n)$. We also know that $A(j/2^n, n)$ is $\mathcal{F}_{(j-1)/2^n}$-measurable, and $A(\infty, n)$ is integrable where

$$A(\infty, n) = \lim_{j\to\infty} A(j/2^n, n) \text{ a.s.}$$

In the next lemma, we will show that the sequence $\{A(\infty, n)\}$ is uniformly integrable. Assuming it for now, we can conclude that there exists a subsequence $\{A(\infty, n_j)\}$ and a random variable A_∞ such that as $j \to \infty$,

$$E[ZA(\infty, n_j)] \to E[ZA_\infty]$$

for any bounded random variable Z. Therefore, for any fixed $s \in \mathbb{R}^+$,

$$\begin{aligned}\lim_{j\to\infty} E[ZE(A(\infty, n_j) \mid \mathcal{F}_s)] &= \lim_{j\to\infty} E[E(Z \mid \mathcal{F}_s)A(\infty, n_j)] \\ &= E[E(Z \mid \mathcal{F}_s)A_\infty] \\ &= E[ZE(A_\infty \mid \mathcal{F}_s)].\end{aligned}$$

In other words,

$$E(A(\infty, n_j) \mid \mathcal{F}_s) \to E(A_\infty \mid \mathcal{F}_s) \tag{3.4.3}$$

weakly in $L^1(P)$.

Take any two dyadic rationals $s \leq t$. Choose N large enough so that if $n \geq N$, both $A(s, n)$ and $A(t, n)$ are both defined, and $A(s, n) \leq A(t, n)$. Since

$$X_s = E[A(\infty, n) \mid \mathcal{F}_s] - A(s, n) \quad \text{and} \quad X_t = E[A(\infty, n) \mid \mathcal{F}_t] - A(t, n)$$

we obtain

$$E[A(\infty, n_j) \mid \mathcal{F}_s] - X_s \leq E[A(\infty, n_j) \mid \mathcal{F}_t] - X_t$$

for all $n_j \geq N$. By Equation (3.4.3), we have, by letting $j \to \infty$,

$$E[A_\infty \mid \mathcal{F}_s] - X_s \leq E[A_\infty \mid \mathcal{F}_t] - X_t.$$

Let M_t denote a right-continuous version of the uniformly integrable martingale $\{E[A_\infty \mid \mathcal{F}_t]\}$ adapted to $\mathcal{F}_t$. Define $A_t = M_t - X_t$ for all $t \in \mathbb{R}^+$. Then $\{A_t\}$ is right-continuous. Besides, we have shown that it is increasing on dyadic rationals. Therefore, it is right-continuous and increasing on all of $\mathbb{R}^+$.

We know that $\lim t \to \infty A_t$ exists a.s., and $\lim_{t\to\infty} M_t$ exists a.s. and in $L^1(P)$. Therefore $X_\infty = \lim_{t\to\infty} X_t$ exists a.s. It also exists in the $L^1(P)$ sense by the Fatou lemma, and $X_\infty = 0$. Consequently, $A_\infty \in L^1(P)$, and

$$\lim_{t\to\infty} A_t = A_\infty \text{ a.s. and in } L^1(P).$$

It remains to prove that the sequence $\{A(\infty, n)\}$ is uniformly integrable, which is carried out in the following lemma. ■

Lemma 3.4.7 *The sequence* $\{A(\infty, n)\}$ *is uniformly integrable if and only if* $\{X_t\}$ *belongs to class* (D).

Proof

Step 1 Suppose that $\{A(\infty, n)\}$ is uniformly integrable. Under this hypothesis, we know that

$$X_t = E[A_\infty \mid \mathcal{F}_t] - A_t = M_t - A_t.$$

Hence, $0 \le X_\tau \le M_\tau$ for any finite stopping time τ. Since M is a right-continuous, uniformly integrable martingale, it is of class (D). Consequently, X is of class (D).

Step 2 Conversely, suppose that X is of class (D). In order to prove that $\{A(\infty, n)\}$ is uniformly integrable, we first estimate $\int_{\{A(\infty,n)>c\}} A(\infty, n)dP$ for all n and large c. For each $c > 0$ and $n \ge 1$, define

$$T_{n,c} = \inf\left\{\frac{j}{n} : A\left(\frac{j+1}{n}, n\right) > c\right\} \tag{3.4.4}$$

with the convention that infimum of the empty set is ∞. Since $A\left(\frac{j+1}{n}, n\right)$ is $\mathcal{F}_{j/n}$-measurable, $T_{n,c}$ is a stopping time with respect to $(\mathcal{F}_{j/n})$. Note that

(i) $A(\infty, n) > c \iff T_{n,c} < \infty$, and

(ii) $A(T_{n,c}, n) \le c$.

Let T be any stopping time which takes at most a countable number of values, and Z, any integrable random variable. From the definitions of conditional expectation and $\mathcal{F}_T$, we observe that

$$E(Z \mid \mathcal{F}_T) = E(Z \mid \mathcal{F}_t) \quad \text{a.s. on the set} \quad \{T = t\}.$$

From this it follows that

$$X_{T_{n,c}} = E[A(\infty, n) \mid \mathcal{F}_{T_{n,c}}] - A(T_{n,c}, n). \tag{3.4.5}$$

Consider

$$\begin{aligned}\int_{\{A(\infty,n)>c\}} A(\infty,n)\,dP &= \int_{\{T_{n,c}<\infty\}} A(\infty,n)\,dP \text{ by } (i)\\ &= \int_{\{T_{n,c}<\infty\}} [A(T_{n,c},n)dP + X_{T_{n,c}}]\,dP \text{ by } (3.4.5)\\ &= \int_{\{A(\infty,n)>c\}} A(T_{n,c},n)dP + \int_{\{T_{n,c}<\infty\}} X_{T_{n,c}}dP\\ &\le cP\left\{A(\infty,n)>c\right\} + \int_{\{T_{n,c}<\infty\}} X_{T_{n,c}}dP \text{ by } (ii).\end{aligned}$$

Thus, we have

$$\int_{\{A(\infty,n)>c\}} [A(\infty,n)-c]dP \le \int_{\{T_{n,c}<\infty\}} X_{T_{n,c}}dP. \tag{3.4.6}$$

Consider

$$\begin{aligned}cP\left\{A(\infty,n)>2c\right\} &\le \int_{\{A(\infty,n)>2c\}} [A(\infty,n)-c]\,dP\\ &\le \int_{\{A(\infty,n)>c\}} [A(\infty,n)-c]\,dP\\ &\le \int_{\{T_{n,c}<\infty\}} X_{T_{n,c}}dP.\end{aligned}$$

Replacing c by $2c$ in (3.4.6) gives

$$\begin{aligned}\int_{\{A(\infty,n)>2c\}} A(\infty,n)\,dP &\le 2cP\{A(\infty,n)>2c\} + \int_{\{T_{n,2c}<\infty\}} X_{T_{n,2c}}dP\\ &\le \int_{\{T_{n,c}<\infty\}} 2X_{T_{n,c}}dP + \int_{\{T_{n,2c}<\infty\}} X_{T_{n,2c}}dP. \end{aligned} \tag{3.4.7}$$

Step 3 It is easy to see that

$$\begin{aligned}cP\{T_{n,c}<\infty\} &= cP\left\{A(\infty,n)>c\right\}\\ &\le E[A(\infty,n)\\ &= E[E(A(\infty,n)\mid \mathcal{F}_0) - A_0] = EX_0.\end{aligned}$$

Hence, $P\{T_{n,c}<\infty\}$ can be made small (uniformly in n) if c is chosen large enough.

Step 4 Based on (3.4.7), Step 3 and the hypothesis that X belongs to class (D), one may be tempted to conclude that $\{A(\infty,n)\}$ is uniformly integrable. However, the hypothesis doesn't imply the uniform integrability of the family $\{X_{T_{n,c}} : n \ge 1,\ c > 0\}$, since $T_{n,c}$ are not finite a.s.

To overcome this snag, note that the family $\{X_{k\wedge T_{n,c}} : n \geq 1,\ c > 0,\ k \geq 1\}$ is uniformly integrable. Hence, $\{X_{k\wedge T_{n,c}} 1_{\{T_{n,c}<\infty\}} : n \geq 1,\ c > 0,\ k \geq 1\}$ is uniformly integrable as well. Each $X_{T_{n,c}} 1_{\{T_{n,c}<\infty\}}$ is the a.s. limit of the sequence $\{X_{k\wedge T_{n,c}} 1_{\{T_{n,c}<\infty\}} : k \geq 1\}$. Therefore, we can conclude that $\left\{X_{T_{n,c}} 1_{\{T_{n,c}<\infty\}} : n \geq 1,\ c > 0\right\}$ is uniformly integrable.

Given $\epsilon > 0$, there exists δ such that if $P(A) < \delta$, then

$$\int_A X_{T_{n,c}} 1_{\{T_{n,c}<\infty\}} dP < \epsilon.$$

By Step 3, we can find a c_0 large enough so that for all $c > c_0$, then $P\{T_{n,c} < \infty\} < \delta$. By (3.4.7), it follows that

$$\int_{\{A(\infty,n)>2c\}} A(\infty, n) dP < 3\epsilon$$

for all $n \geq 1$ and $c > c_0$. The proof is over. ■

Theorem 3.4.8 *The process A associated with the potential X in the Doob-Meyer decomposition is the unique natural process such that for all $t \geq 0$,*

$$X_t = E[A_\infty \mid \mathcal{F}_t] - A_t.$$

Proof

Step 1 By Theorem 3.4.4, it suffices to show that $E\left(\int_0^\infty Y_{s-} dA_s\right) = E(A_\infty Y_\infty)$ for all nonnegative, bounded, right-continuous $\mathcal{F}_t$-martingales. Consider $\sum_{j=0}^\infty E\left[Y_{j/2^n}\left(A\left(\frac{j+1}{2^n}\right) - A\left(\frac{j}{2^n}\right)\right)\right]$ which equals

$$\sum_{j=0}^\infty \left[Y_{j/2^n} E\left\{A\left(\frac{j+1}{2^n}\right) - A\left(\frac{j}{2^n}\right) | \mathcal{F}_{j/2^n}\right\}\right]$$

$$= \sum_{j=0}^\infty \left[Y_{j/2^n} E\left\{X\left(\frac{j}{2^n}\right) - X\left(\frac{j+1}{2^n}\right) | \mathcal{F}_{j/2^n}\right\}\right]$$

$$= \sum_{j=0}^\infty \left[Y_{j/2^n} E\left\{A\left(\frac{j+1}{2^n}, n\right) - A\left(\frac{j}{2^n}, n\right) | \mathcal{F}_{j/2^n}\right\}\right]$$

$$= \sum_{j=0}^\infty \left[Y_{j/2^n}\left(A\left(\frac{j+1}{2^n}\right) - A\left(\frac{j}{2^n}\right)\right)\right].$$

Since Y is a $\mathcal{F}_t$-martingale and $A\left(\frac{j+1}{2^n}\right)$ is $\mathcal{F}_{j/2^n}$-measurable,

$$E\left[Y_{j/2^n} A\left(\frac{j+1}{2^n}, n\right)\right] = E\left[Y_{j/2^n} EA\left(\frac{j}{2^n}, n\right)\right].$$

Therefore, $\sum_{j=0}^{\infty}\left[Y_{j/2^n}\left(A(\frac{j+1}{2^n}) - A(\frac{j}{2^n})\right)\right]$ is a telescoping sum, and we obtain

$$\sum_{j=0}^{\infty} E\left[Y_{j/2^n}\left(A\left(\frac{j+1}{2^n}\right) - A\left(\frac{j}{2^n}\right)\right)\right] = E[A(\infty, n)Y_\infty]. \tag{3.4.8}$$

Letting $n \to \infty$ along the subsequence $\{n_j\}$ used in the proof of Theorem 3.4.6, we obtain $\lim j \to \infty E[A(\infty, n_j)Y_\infty] = E[A_\infty Y_\infty]$.

Note that by the dominated convergence theorem,

$$E\left(\int_0^\infty Y_{s-}dA_s\right) = \lim_{n\to\infty}\sum_{j=0}^{\infty} E\left[Y_{j/2^n}\left(A\left(\frac{j+1}{2^n}\right) - A\left(\frac{j}{2^n}\right)\right)\right].$$

Thus, by making $n \to \infty$ along the subsequence $\{n_j\}$ in Equation (3.4.8),

$$E\left(\int_0^\infty Y_{s-}\, dA_s\right) = E[A_\infty Y_\infty]$$

so that A is natural.

Step 2 Uniqueness: Let $V = \{V_t\}$ be a natural process associated with X so that

$$X_t = E(V_\infty|\mathcal{F}_t) - V_t.$$

Then, $E\left(\int_0^\infty Y_{s-}dV_s\right) = E[V_\infty Y_\infty]$ for every nonnegative, bounded, right-continuous $\mathcal{F}_t$-martingale Y.

Note that for any $s < t$, both $E[Y_s(V_t - V_s)]$ and $E[Y_s(A_t - A_s)]$ are equal to $E[Y_s(X_s - X_t)]$ so that

$$E[Y_s(V_t - V_s)] = E[Y_s(A_t - A_s)].$$

Hence,

$$\begin{aligned} E(V_\infty Y_{infty}) &= \lim_{n\to\infty}\sum_{j=0}^{\infty} E\left[Y_{j/2^n}\left(V\left(\frac{j+1}{2^n}\right) - V\left(\frac{j}{2^n}\right)\right)\right] \\ &= \lim_{n\to\infty}\sum_{j=0}^{\infty} E\left[Y_{j/2^n}\left(A\left(\frac{j+1}{2^n}\right) - A\left(\frac{j}{2^n}\right)\right)\right] \\ &= E(A_\infty Y_{infty}). \end{aligned}$$

Since Y_∞ can be taken as the indicator function of any set in $\mathcal{F}_\infty$, we conclude that $A_\infty = V_\infty$ a.s. By the Doob-Meyer decomposition, it follows that $A_t = V_t$ for all $t \geq 0$. ∎

The Doob-Meyer decomposition holds for a more general class of potentials. We need the following definition to state this general result.

Definition 3.4.6 *Let $X = \{X_t\}$ be a right-continuous, $\mathcal{F}_t$-adapted supermartingale, and let $\mathcal{T}_c$ denote the collection of all stopping times relative to $(\mathcal{F}_t)$ which are bounded by a positive number c. The process X belongs to* **class (DL)** *if the collection of random variables $\{X_\tau : \tau \in \mathcal{T}_c\}$ is uniformly integrable for all finite $c > 0$.*

Thus class (DL) stands for "locally" in class (D). Every right-continuous martingale belongs to class (DL). The following result is stated without proof. For the proof, see the book by Meyer.

Theorem 3.4.9 *A right-continuous supermartingale X has a Doob-Meyer decomposition:*

$$X_t = M_t - A_t \quad \text{for all} \quad t \in \mathbb{R}^+$$

where M is a right-continuous martingale and A is an increasing process if and only if X is of class (DL). There is only one such decomposition with A as a natural process.

Next, we would like to know if A is continuous when X is a potential of class (DL) with continuous paths. We need the following definition to state the next theorem.

Definition 3.4.7 *Let X be a right-continuous supermartingale of class (DL). Then X is called a* **regular** *process if*

$$\lim_{n\to\infty} EX_{T_n} = EX_T$$

for every increasing sequence $\{T_n\}$ of stopping times which converges to a bounded stopping time T.

It is worthwhile to note that every right-continuous martingale is regular by using the optional sampling theorem.

Theorem 3.4.10 *Let X be a potential of class (DL), and A, its associated natural process. Then A is continuous if and only if X is regular.*

Proof

Step 1 Suppose that A is continuous. Then A is regular by the monotone convergence theorem. Besides, $E[A_\infty \mid \mathcal{F}_t]$, indexed by t, is a right-continuous martingale and hence regular. Therefore X is regular.

Step 2 Conversely, suppose X is regular. Then, A is regular by the Doob-Meyer decomposition. If $T_n \uparrow T$ where T is a bounded stopping time, then $A_{T_n} \leq A_T$ for all n so that $\lim_{n\to\infty} A_{T_n} \leq A_T$. However,

$$E[\lim_{n\to\infty} A_{T_n}] = \lim_{n\to\infty} EA_{T_n} = EA_T$$

so that $A_{T_n} \uparrow A_T$ a.s. as $n \to \infty$. Since the almost sure set may depend on the choice of T_n and T, we cannot conclude that A is left continuous.

From now on, let $T \leq b$ where $b > 0$. For any fixed $c > 0$, let us assume that $A \leq c$ a.s. on $[0, b]$. Otherwise, replace A by $A \wedge c$. For any $n \in \mathbb{N}$, consider the partition of

$[0, b]$ given by the points $t_j = jb/2^n$ for $j = 0, 1, \ldots, 2^n$. Define for $j \geq 1$,

$$Y(t,n) = E(A_{t_j} \mid \mathcal{F}_t) \quad \text{if} \quad t \in I_j = (t_{j-1}, t_j]$$

with $Y(0,n) = 0$. Then $\{Y(t,n)\}$ is a bounded, right-continuous martingale on each I_j. Since A is natural, (and hence right-continuous) we get

$$E\left(\int_0^t Y(s,n)\,dA_s\right) = E\left(\int_0^t Y(s-,n)\,dA_s\right) \tag{3.4.9}$$

for any $t \in [0, b]$.

Step 3 If we can show that along a subsequence $\{n_j\}$,

$$\sup_{t\in[0,b]} |Y(t,n_j) - A_t| \to 0 \quad \text{a.s. as} \quad j \to \infty, \tag{3.4.10}$$

then by Equation (3.4.9), it would follow that

$$E\left(\int_0^a A_s\,dA_s\right) = E\left(\int_0^a A_{s-}\,dA_s\right).$$

Therefore, $E\left[\int_0^a (A_s - A_{s-})\,dA_s\right] = 0$. In other words,

$$E\left[\sum_{s\leq a}(A_s - A_{s-})^2\right] = 0.$$

We can conclude that A has continuous paths a.s. In the next step, we prove (3.4.10).

Step 4 Fix any $\epsilon > 0$. Define the stopping times

$$S_n = b \wedge \inf\{t \in [0,b] : Y(t,n) - A_t > \epsilon\}.$$

Since $Y(b,n) = A_b$ for each n, we obtain

$$\{S_n = b\} = \{Y(t,n) - A_t \leq \epsilon \quad \text{for all} \quad t \in [0,b]\}. \tag{3.4.11}$$

By definition $Y(t,n)$ is decreasing in n which implies that S_n is increasing in n. Let $S = \lim_{n\to\infty} S_n$.

Define a function f_n by $f_n(t) = t_j$ if $t \in I_j$, and $\sigma_n := f_n(S_n)$, where σ_n is a stopping time bounded by b. Clearly, $\lim_{n\to\infty} \sigma_n = S$. Since $\sigma_n \geq S_n$,

$$\begin{aligned}
EA_{\sigma_n} &= \sum_{j=1}^{n} E\left[A_{\sigma_n} 1_{\{S_n \in I_j\}}\right] \\
&= \sum_{j=1}^{n} E\left[Y(S_n, n) 1_{\{S_n \in I_j\}}\right] \\
&= EY(S_n, n)
\end{aligned}$$

by the optional sampling theorem. Using this, (3.4.11), and the simple fact that $S + 1/2^n \geq \sigma_n$, we have

$$\begin{aligned} E[A_{S+1/2^n} - A_{S_n}] &\geq E[A_{\sigma_n} - A_{S_n}] \\ &= E[Y(S_n, n) - A_{S_n}] \\ &\geq \epsilon P\{S_n < b\}. \end{aligned}$$

By the regularity of A and its right-continuity, the left side goes to zero as $n \to \infty$. Hence,

$$\lim_{n\to\infty} P\{S_n < b\} = 0,$$

which implies that

$$\lim_{n\to\infty} P\left\{ \sup_{t\in[0,a]} |Y(t,n) - A_t| > \epsilon \right\} = 0.$$

Therefore, there exists a subsequence $\{n_j\}$ such that (3.4.10) holds. ■

3.5 The Meyer Process for L^2-martingales

Let us recall that an $\mathcal{F}_t$-adapted process M is called an L^2-martingale (or a square integrable martingale) if M is an L^2-bounded martingale; that is, $\sup_t E(M_t^2) < \infty$. If the time parameter set is $[0, T]$, then we require $\sup_{t\in[0,T]} E(M_t^2) < \infty$. The latter reduces to $EM_T^2 < \infty$.

In what follows, $[0, \infty)$ is taken as the time parameter set, though the results carry over when the parameter set is $[0, T]$ *mutatis mutandis*. Any martingale M that we consider will be replaced by a right-continuous version of it. Since two different right-continuous versions coincide for all t, except on a P-null set, the ensuing results are independent of the version. The filtration $(\mathcal{F}_t)$ is assumed to satisfy the usual conditions. All martingales considered in this section are adapted to $(\mathcal{F}_t)$. We now associate with a given L^2-martingale M a certain natural process called the Meyer process or the conditional variation process.

Consider an L^2-martingale $M = \{M_t\}$. It is uniformly integrable, which implies that $M_\infty = \lim_{t\to\infty} M_t$ exists a.s. and in the $L^1(P)$-sense.

Also $M_t = E(M_\infty \mid \mathcal{F}_t)$ a.s. By the Fatou lemma, $E(M_\infty^2) < \infty$. Hence, $\{M_t^2,\ t \in [0,\infty]\}$ is a submartingale which yields $\lim_{t\to\infty} E(M_t^2) = E(M_\infty^2)$. By the Doob inequality,

$$E[\sup_{t\in[0,T]} M_t^2] \leq 4E(M_T^2)$$

for any $T > 0$. Allowing $T \to \infty$, we obtain

$$E[\sup_{t\geq 0} M_t^2] \leq 4E(M_\infty^2). \tag{3.5.1}$$

Therefore, $\{-M_t^2,\ t \in [0,\infty]\}$ is a uniformly integrable supermartingale.

Define $X_t = E(M_\infty^2 \mid \mathcal{F}_t) - M_t^2$, where we take the right-continuous version of the martingale $E(M_\infty^2 \mid \mathcal{F}_t)$. It is clear that X is a nonnegative right-continuous supermartingale, and

$$X_\infty = \lim_{t\to\infty} X_t$$

exists a.s. Since $E(M_\infty^2 \mid \mathcal{F}_t) \to M_\infty^2$ a.s. and $M_t^2 \to M_\infty^2$ a.s. as $t \to \infty$, we have

$$X_\infty = 0 \text{ a.s.}$$

By uniform integrability, $\lim_{t\to\infty} EX_t = 0$ so that X is a potential. In fact, X belongs to class (D) because it is dominated by a martingale of class (D):

$$0 \leq X_t \leq E(M_\infty^2 \mid \mathcal{F}_t).$$

The Doob-Meyer decomposition applies to the potential X so that there exists a unique natural process $A = \{A_t\}$ satisfying

$$X_t = E(A_\infty \mid \mathcal{F}_t) - A_t.$$

From the discussion above, we have

Definition 3.5.1 *The* **Meyer process** *or the* **conditional variation process** *of an L^2-martingale M, denoted by $\langle M\rangle$ or $\langle M, M\rangle$, is defined by*

$$\langle M\rangle_t - \langle M\rangle_0 = A_t$$

where $\langle M\rangle_0 = M_0^2$.

We should have called the above process the Doob-Meyer process. It is simply for our own convenience that we have shortened it to Meyer process. It is worthwhile to note that $\langle M\rangle_t$ is the unique (up to a version) natural process such that $M_t^2 - \langle M\rangle_t$ is a martingale. Also, if $0 \leq s \leq t$,

$$\begin{aligned} E[(M_t - M_s)^2 \mid \mathcal{F}_s] &= E[(M_t^2 - M_s^2) \mid \mathcal{F}_s] \\ &= E[(\langle M\rangle_t - \langle M\rangle_s) \mid \mathcal{F}_s]. \end{aligned}$$

Suppose that the L^2-martingale M has continuous paths. Then $M^2 = \{M_t^2\}$ is a nonnegative continuous submartingale and is therefore a regular process. For, let T be a stopping time bounded by b. For any sequence of stopping times $\{T_n\}$ such that $T_n \uparrow T$ a.s.,

$$M_{T_n}^2 \leq \sup_{t\in[0,b]} M_t^2 \quad \forall\ n.$$

By the Doob inequality

$$E\left(\sup_{t\in[0,b]} M_t^2\right) \leq 4E(M_b^2)$$

so that $\{M_{T_n}^2\}$ and M_T form a uniformly integrable family. By path continuity, it follows that $E(M_{T_n}^2) \to E(M_T^2)$ as $n \to \infty$.

The martingale $E(M_\infty^2 \mid \mathcal{F}_t)$ is also a regular process. Hence, X is regular. By Theorem 3.4.10, the Meyer process $\langle M\rangle_t$ is continuous. Thus we have proved

Proposition 3.5.1 *If M is an L^2-martingale with continuous paths, then the Meyer process $\langle M\rangle_t$ has continuous paths.*

Remark 3.5.1 The converse of Proposition 3.5.1 is not true. The well-known counterexample is provided by a right-continuous version of a Poisson process $N = \{N_t\}$ with parameter $\lambda > 0$. Since N is a process with independent increments and for all $0 \leq s \leq t$,

$$E(N_t - N_s) = \lambda(t - s),$$

it follows that $N_t - \lambda t$ is an L^2-martingale. One can easily check that

$$E[(N_t - N_s)^2 \mid \mathcal{F}_s^N] = \lambda(t - s).$$

Therefore, $A_t = \lambda t$ will be the Meyer process for N.

Theorem 3.5.2 *Let $A = \{A_t\}$ be any continuous, integrable, increasing process. Let Π be a partition of $[0, t]$. Define*

$$A_t^\Pi = \sum_{j=0}^{n-1} E[(A_{t_{j+1}} - A_{t_j}) \mid \mathcal{F}_{t_j}].$$

Then for all t, $E|A_t^\Pi - A_t| \to 0$ as $\|\Pi\| \to 0$ where $\|\Pi\| = \max_j(t_{j+1} - t_j)$.

Proof

Step 1 Suppose that $E(A_t^2) < \infty$. Then,

$$E\left[(A_t - A_t^\Pi)^2\right] = E\left[\sum_j (A_{t_{j+1}} - A_{t_j}) - A_t^\Pi\right]^2$$

$$= E\left\{\sum_j [(A_{t_{j+1}} - A_{t_j}) - E[(A_{t_{j+1}} - A_{t_j}) \mid \mathcal{F}_{t_j}]]^2\right\}.$$

Let us denote $A_{t_{j+1}} - A_{t_j}$ as Δ_j. Expanding the square on the right side, and observing that

$$E\left\{\Delta_i - (\Delta_i \mid \mathcal{F}_{t_i})\right\}\left\{\Delta_j - (\Delta_j \mid \mathcal{F}_{t_j})\right\} = 0$$

if $i < j$, we obtain

$$\begin{aligned} E\left[\left(A_t - A_t^{\Pi}\right)^2\right] &= E\left\{\sum_j \left[\Delta_j - E\left(\Delta_j \mid \mathcal{F}_{t_j}\right)\right]^2\right\} \\ &= E\left\{\sum_j \Delta_j^2\right\} - E\left\{\sum_j \left[tE\left(\Delta_j \mid \mathcal{F}_{t_j}\right)\right]^2\right\} \\ &\le E\left\{\sum_j \left(tA_{t_{j+1}} - A_{t_j}\right)^2\right\} \\ &\le E\left[\sup_{\Pi}\left(A_{t_{j+1}} - A_{t_j}\right) A_t\right]. \end{aligned}$$

Noting that A_s is uniformly continuous on $[0, t]$, and $A_t \in L^2(P)$, the dominated convergence theorem allows us to conclude that the right side goes to 0 as $|\Pi| \to 0$.

Step 2 Let $E|A_t| < \infty$. For $s \in [0, t]$, define $B_s = A_s \wedge n$, and $C_s = A_s - B_s$. Just as we defined A_t^{Π}, let us define B_t^{Π} and C_t^{Π}. Then

$$\begin{aligned} |A_t - A_t^{\Pi}| &\le |B_t - B_t^{\Pi}| + |C_t - C_t^{\Pi}| \\ &\le |B_t - B_t^{\Pi}| + C_t + C_t^{\Pi}. \end{aligned}$$

Note that $EC_t^{\Pi} = EC_t$. Given any $\epsilon > 0$, we can make $EC_t < \epsilon$ if n is large enough. Besides, $E|B_t - B_t^{\Pi}|$ can be made small by making the partition fine enough. Thus, $E|A_t^{\Pi} - A_t| \to 0$. ■

If M is an L^2-martingale, we can apply the above result to $A_t = \langle M\rangle_t - \langle M\rangle_0$. We thus have the following result.

Theorem 3.5.3 *Let M be an L^2-martingale with $\langle M\rangle$ being continuous. Suppose that $\Pi_n = \{t_j^n\}$ is a finite partition of $[0, t]$. Then*

$$\sum_{\Pi_n} E[(M_{t_{j+1}^n} - M_{t_j^n})^2 \mid \mathcal{F}_{t_j^n}] \to \langle M\rangle_t - \langle M\rangle_0$$

in $L^1(P)$ as $|\Pi_n| \to 0$.

The above result provides us the reason for denoting the Meyer process as the conditional variation process.

Definition 3.5.2 *Let M and N be two L^2-martingales. Then the* **conditional covariation** *(or simply the covariation) process between M and N is defined by*

$$\langle M \rangle = \frac{1}{4}[\langle M+N \rangle_t - \langle M-N \rangle].$$

From the definition and the previous theorem, it follows that if $\langle M \rangle$ and $\langle N \rangle$ are continuous, then

$$\sum_{\Pi_n} E[(M_{t^n_{j+1}} - M_{t^n_j})(N_{t^n_{j+1}} - N_{t^n_j}) \mid \mathcal{F}_{t^n_j}] \to \langle M, N \rangle_t - M_0 N_0$$

in $L^1(P)$ as $|\Pi_n| \to 0$. From this, one can conclude that for any two continuous L^2-martingales,

$$\sum_{\Pi_n} E[(M_{t^n_{j+1}} - M_{t^n_j})(N_{t^n_{j+1}} - N_{t^n_j})] \to \langle M, N \rangle_t - M_0 N_0 \tag{3.5.2}$$

in $L^1(P)$ as $|\Pi_n| \to 0$.

Let M be as above, and A, a continuous, integrable, increasing process. Using (3.5.2), we can define $\langle M, V \rangle$ as follows:

$$\langle M, A \rangle_t = L^1 - \lim_{n\to\infty} \sum_{\Pi_n} E[(M_{t^n_{j+1}} - M_{t^n_j})(A_{t^n_{j+1}} - A_{t^n_j})].$$

It is clear that the right side is 0 so that $\langle M, A \rangle_t = 0$. Formally one writes this as $dMdA = 0$. Likewise, $dMdM = d\langle M \rangle$. We summarize these facts in the following multiplication table:

	dM	dA
dM	$d\langle M \rangle$	0
dA	0	0

One can also infer that $M_t N_t - \langle M, N \rangle_t$ is a continuous martingale when M and N are continuous L^2-martingales. As shown in Chapter 10, the last statement holds even when M and N are L^2-cadlag-martingales.

3.6 Local Martingales

Definition 3.6.1 *A real-valued $\mathcal{F}_t$-adapted process $M = \{M_t\}$ is called a* **local martingale** *with respect to $(\mathcal{F}_t)$ if there exists an increasing sequence $\{\tau_n\}$ of stopping times such that $\tau_n \uparrow \infty$ a.s. (if the time interval is $[0, T]$ instead of $[0, \infty)$, then $\tau_n \uparrow T$ a.s.), and for each n, $M_{t\wedge\tau_n}$ is an $\mathcal{F}_t$-martingale.*

We will call the sequence of stopping times $\{\tau_n\}$ a **localizing sequence**. In general, a local property is defined as follows:

Definition 3.6.2 *Let X be a stochastic process. A property π is called* local *or is said to* hold locally *if there exists a localizing sequence $\{\tau_n\}$ such that $X_{t\wedge\tau_n}$ has the property π for each n.*

The phrases *locally square integrable* and *locally bounded* are to be understood in the above sense. A locally square integrable martingale will also be called a local L^2-martingale.

We will first show that the statement "$M_{t\wedge\tau_n}$ is an $\mathcal{F}_t$-martingale" is equivalent to the statement "$M_{t\wedge\tau_n}$ is an $\mathcal{F}_{t\wedge\tau_n}$-martingale."

Lemma 3.6.1 *Suppose that a process M is $\mathcal{F}_t$-adapted and τ is a stopping time with respect to $\mathcal{F}_t$. Then*

1. *$\{M_{t\wedge\tau}\}$ is $\mathcal{F}_t$-adapted if and only if $\{M_{t\wedge\tau}\}$ is $\mathcal{F}_{t\wedge\tau}$-adapted.*
2. *$\{M_{t\wedge\tau}\}$ is an $\mathcal{F}_t$-martingale if and only if $\{M_{t\wedge\tau}\}$ is an $\mathcal{F}_{t\wedge\tau}$-martingale.*

Proof The "if" part of the first statement is straightforward since $\mathcal{F}_{t\wedge\tau} \subset \mathcal{F}_t$. To prove the converse, consider the set $A := \{M_{t\wedge\tau} \in B\}$ for any Borel set B in $\mathbb{R}$. We are given that $A \in \mathcal{F}_t$, and we need to show that $A \in \mathcal{F}_{t\wedge\tau}$.

For any $r \geq 0$, let $F := A \cap \{t \wedge \tau \leq r\}$.

If $t \leq r$, then $t \wedge \tau \leq r$ so that $F = A \in \mathcal{F}_t \subset \mathcal{F}_r$.

On the other hand, if $t > r$, then $t \wedge \tau \leq r$ is equivalent to $\tau \leq r$. Therefore, $F = A \cap \{\tau \leq r\}$. If $\tau(\omega) \leq r, t \wedge \tau(\omega) = \tau(\omega) = r \wedge \tau(\omega)$. Hence, $F = \{M_{r\wedge\tau} \in B\} \cap \{\tau \leq r\}$ is in $\mathcal{F}_r$, since by hypothesis, $\{M_{r\wedge\tau} \in B\} \in \mathcal{F}_r$.

Thus in both cases, $F \in \mathcal{F}_r$, and the proof of the first statement is completed.

The "only if" part of the second assertion of the lemma follows easily. For, given that $E(M_{t\wedge\tau}|\mathcal{F}_s) = M_{s\wedge\tau}$ for any $0 \leq s \leq t$, it is clear that

$$\begin{aligned} E(M_{t\wedge\tau}|\mathcal{F}_{s\wedge\tau}) &= E[E(M_{t\wedge\tau}|\mathcal{F}_s)|\mathcal{F}_{s\wedge\tau}] \\ &= E(M_{s\wedge\tau}|\mathcal{F}_{s\wedge\tau}) \\ &= M_{s\wedge\tau}. \end{aligned}$$

Conversely, given that $E(M_{t\wedge\tau}|\mathcal{F}_{s\wedge\tau}) = M_{s\wedge\tau}$, one can write, for any $A \in \mathcal{F}_s$,

$$\int_A E(M_{t\wedge\tau}|\mathcal{F}_s)\, dP = \int_{A\cap\{\tau\leq s\}} M_{t\wedge\tau}\, dP + \int_{A\cap\{\tau>s\}} M_{t\wedge\tau}\, dP. \tag{3.6.1}$$

For the first integral on the right side, note that $t \wedge \tau = s \wedge \tau$ since $\tau \leq s$. Hence,

$$\int_{A\cap\{\tau\leq s\}} M_{t\wedge\tau}\, dP = \int_{A\cap\{\tau\leq s\}} M_{s\wedge\tau}\, dP.$$

For the second integral on the right side of Equation (3.6.1), if we can show that $A \cap \{\tau > s\} \in \mathcal{F}_{s\wedge\tau}$, then by the hypothesis,

$$\int_{A\cap\{\tau>s\}} M_{t\wedge\tau}\, dP = \int_{A\cap\{\tau>s\}} M_{s\wedge\tau}\, dP.$$

The above two equalities and Equation (3.6.1) would then complete the proof.

To show that $A \cap \{\tau > s\} \in \mathcal{F}_{s\wedge\tau}$, consider for any $r \geq 0$, the event $A \cap \{\tau > s\} \cap \{s \wedge \tau \leq r\}$. If $r < s$, the set is empty. If $r \geq s$, the event is equal to $A \cap \{\tau > s\}$, and $A \cap \{\tau > s\} \in \mathcal{F}_r$. Thus, in either case, $A \cap \{\tau > s\} \cap \{s \wedge \tau \leq r\} \in \mathcal{F}_r$. ■

The next lemma gives a sufficient condition for a local martingale to be a martingale. A sharper condition is given in Section 6.7 of Chapter 6.

Lemma 3.6.2 *Let $M := \{M_t\}$ be a uniformly integrable local martingale. Then M is a martingale.*

Proof Let $\{T_n\}$ be a localizing sequence for M so that $E[M_{t\wedge T_n}|\mathcal{F}_s] = M_{s\wedge T_n}$ for any fixed n and all $s \leq t$. Clearly, $\lim_{n\to\infty} M_{t\wedge T_n} = M_t$ a.s., and $\lim_{n\to\infty} M_{s\wedge T_n} = M_s$ a.s. These are also limits in $L^1(P)$ by the assumption of uniform integrability. Hence, $E[M_t|\mathcal{F}_s] = M_s$. ■

The above result implies in particular that a bounded local martingale is a martingale. Though the next lemma holds in general ([60]), we will prove it for continuous processes.

Lemma 3.6.3 *Let M be a continuous, local local martingale. Then M is a continuous local martingale.*

Proof Since M is a local local martingale, there exists a localizing sequence $\{T_n\}$ such that for each n, $\{M_{t\wedge T_n}\}$ is a local martingale. Define for each $n \in \mathbb{N}$

$$\sigma_n = \begin{cases} \inf\{t : |M_t| \geq n\}, \\ \infty \ \ (\text{or} \ \ T) \text{ if the set is empty.} \end{cases}$$

Clearly $\sigma_n \uparrow \infty$ (or T) a.s., and hence $\tau_n := T_n \wedge \sigma_n$ increases to ∞ (or T). For each fixed n, $\{M_{t\wedge\tau_n}\}$ is a bounded local martingale and is therefore a martingale by Lemma 3.6.2. ■

Exercises

1. Suppose that a player starts by betting a dollar on heads based on a fair coin. If he wins, he adds the winnings to the previous bet and bets again. If he loses a bet, he loses everything and quits playing. Let X_n denote the gambler's wealth at the end of

n bets. Show that $\{X_n\}$ is a martingale. Is it a uniformly integrable martingale? What is $\lim_{n\to\infty} X_n$ a.s., if it exists?

2. Let $\{X_t\}_{0\le t\le\infty}$ be an adapted cadlag process. If, for any stopping time T, $X_T \in L^1(P)$ and $E(X_T) = 0$, prove that $\{X_t\}_{0\le t\le\infty}$ is a uniformly integrable martingale.
3. Suppose that $\{M_t\}$ is a continuous local martingale with $M_0 = 0$ and $\langle M\rangle_\infty \le 1$. Show that for any $r \ge 0$,

$$P\{\sup_t M_t \ge r\} \le e^{-r^2/2}.$$

4. For any process Y and a stopping time T, let $Y_t^T = Y_{t\wedge T}$. Let M, N be two L^2-bounded, square integrable, continuous martingales. Prove that

$$\langle M^T, N\rangle = \langle M, N^T\rangle = \langle M^T, N^T\rangle = \langle M, N\rangle^T.$$

5. Two L^2-bounded, cadlag martingales M and N are orthogonal if $E(M_T N_T) = 0$ for every stopping time T, and if $M_0 N_0 = 0$. If $T = \infty$, then $M_T = M_\infty$. Show that M, N are orthogonal if and only if MN is a martingale started at 0.
6. Let X be a submartingale. Show that $\sup_t E(|X_t|) < \infty$ is equivalent to $\sup_t E(X_t^+) < \infty$.
7. Suppose that M is a local martingale and $\sup_{0\le s\le t} |M_s| \in L^1(P)$ for each $t > 0$. Show that M is a martingale.
8. If M is a positive local martingale, show that M is a supermartingale.
9. Let M be a continuous martingale. Show that M converges a.s. on the set $\{\sup_t M_t < \infty\}$.
10. Let X be a nonnegative supermartingale. Define $T = \inf\{t \ge 0 : X_t = 0\}$. If

$$\Lambda = \{\omega : X_t(\omega) = 0 \text{ for all } t \ge T(\omega)\},$$

show that $P(\Lambda) = 1$.

11. Let W be a Wiener process. Show that $P(A) = 0$ or 1 for any $A \in \mathcal{F}_{0+}^W$. This result is known as Blumenthal's 0-1 law.

4 Analytical Tools for Brownian Motion

In this chapter, we introduce the semigroup associated with Brownian motion. The infinitesimal generator for the semigroup is identified along with its domain. The heat equation is shown to be the partial differential equation associated with Brownian motion. Motivated by this example, a concise treatment of semigroups of operators is given. The latter will especially be useful when we discuss the martingale problem and the Kolmogorov forward and backward equations.

4.1 Introduction

We begin by recalling the basic properties of an important class of unbounded linear operators known as closed operators defined in a Banach space B. An unbounded linear operator T in a Hilbert space B is a pair $(\mathcal{D}(T), T)$ consisting of a linear subspace $\mathcal{D}(T) \subset B$ and a linear operator $T : \mathcal{D}(T) \to B$.

It is worthwhile to recall that the **graph** of a linear operator T in B is the subspace

$$\Gamma(T) = \{(x, Tx) : x \in \mathcal{D}(T)\} \subset B \times B.$$

Definition 4.1.1 *An operator T is said to be a* **closed** *operator if its graph $\Gamma(T)$ is a closed subset of $B \times B$; that is, if $x_n \in \mathcal{D}(T)$ such that $x_n \to x$ and $Tx_n \to y$, then $x \in \mathcal{D}(T)$ and $Tx = y$.*

An operator T is called **closable** if the closure of its graph, namely $\overline{\Gamma(T)}$, is a graph, that is, $(0, y) \in \overline{\Gamma(T)}$ implies that $y = 0$. If T is closable, construct $\overline{T}$ as follows: If $x_n \in \mathcal{D}(T)$ with $x_n \to x$ and $Tx_n \to y$, then include x in $\mathcal{D}(\overline{T})$ and define $\overline{T}x := y$. Then $\overline{T}$ is the closure of T and $\Gamma(\overline{T}) = \overline{\Gamma(T)}$.

Example 4.1.1 *Any bounded operator T defined on a subspace $\mathcal{D}(T) \subset B$ is closable.*

Proof If $\mathcal{D}(T)$ is a closed set, then T is a closed operator by continuity of T.

If $\mathcal{D}(T)$ is not a closed set, we can extend T to the closure of $\mathcal{D}(T)$ as follows: Let $x_n \in \mathcal{D}(T)$ with $x_n \to x$. Observe that by boundedness of T, $\{Tx_n\}$ is a Cauchy sequence in B, and hence, convergent. Let $Tx_n \to y$. Extend T by defining $Tx := y$. This extension is a bounded operator with the same norm on a closed domain. Thus, T is closed. ■

Remark 4.1.1 We can extend a bounded operator T on a closed domain $\mathcal{D}(T)$ to the whole of B without changing its norm.

A useful result on closed operators is the *closed graph theorem*, which states that a closed operator T is bounded if $\mathcal{D}(T)$ is closed in B.

4.2 The Brownian Semigroup

Let $\hat{\mathbf{C}}$ denote the space of all continuous functions defined on $\mathbb{R}^d$ that vanish at infinity. Let the space be equipped with the norm $\|f\| = \sup_{x\in\mathbb{R}^d} |f(x)|$. Then $\hat{\mathbf{C}}$ is a Banach space, and any function in $\hat{\mathbf{C}}$ is bounded and uniformly continuous. These two properties of the space are essential for the discussion that follows.

Define the linear transformation T_t on $\hat{\mathbf{C}}$ by $T_tf(x) = E^xf(B_t)$, where E^x denotes expectation when the Brownian motion, B_t, is started at x, that is, $B_0 = x$. In other words,

$$T_tf(x) = \int_{\mathbb{R}^d} f(y) \frac{1}{(2\pi t)^{d/2}} \exp\left\{-\frac{|x-y|^2}{2t}\right\} dy.$$

By the bounded convergence theorem, it is easy to show that T_tf is in $\hat{\mathbf{C}}$ upon noting that $T_tf(x)$ can also be written as $Ef(x+W_t)$, where W_t is a standard Wiener process. Thus T_t is a linear operator from $\hat{\mathbf{C}}$ into $\hat{\mathbf{C}}$.

We will use the notation $p(t,x,y)$ for $\frac{1}{(2\pi t)^{d/2}} \exp\left\{-\frac{|x-y|^2}{2t}\right\}$. It is worthwhile to note that $T_tf(x)$ can be expressed in a variety of ways. We collect them below:

$$T_tf(x) = E^xf(B_t) = Ef(x+W_t) = Ef(x+\sqrt{t}Z) = \int_{\mathbb{R}^d} f(y)p(t,x,y)dy,$$

where Z is a standard normal vector with mean vector 0 and variance-covariance matrix as the $d \times d$ identity matrix.

We prove below three basic properties of the family of operators $\{T_t\}$.

1. It is clear that $T_0 = I$, where I is the identity operator. Observe that for any x,

$$|T_tf(x)| \le E|f(x+W_t)| \le \|f\|$$

so that $\|T_tf\| \le \|f\|$. In other words, T_t is a *contraction* operator. A simpler observation is that T_t is a *positive* semigroup, since $T_tf \ge 0$ if $f \ge 0$.

2. For any positive t, s, it is well known that

$$\int_{\mathbb{R}^d} p(t,x,y)p(s,y,z)dy = p(t+s,x,z).$$

This is a particular case of what are, in general, known as Chapman-Kolmogorov equations in the theory of Markov processes. It can be verified by evaluating the integral directly. Using this, we get

$$\begin{aligned} T_t T_s f(x) &= \int_{\mathbb{R}^d} T_s f(y) p(t,x,y)dy \\ &= \int_{\mathbb{R}^d}\int_{\mathbb{R}^d} f(z)p(s,y,z)dz p(t,x,y)dy \\ &= \int_{\mathbb{R}^d} f(z)\left(\int_{\mathbb{R}^d} p(t,x,y)p(s,y,z)dy\right)dz \\ &= \int_{\mathbb{R}^d} f(z)p(t+s,x,z)dz \\ &= T_{t+s}f(x). \end{aligned}$$

Thus, $T_t T_s f(x) = T_{t+s} f(x)$. This is known as the *semigroup* property of the family of operators $\{T_t\}$.

3. Let $f \in \hat{\mathbf{C}}$. We claim that $T_t f \to f$ in $\hat{\mathbf{C}}$ as $t \to 0$.

Proof Indeed, by uniform continuity, given $\epsilon > 0$, there exists $\delta > 0$ such that

$$|f(x-y) - f(x)| < \epsilon/2 \quad \text{if} \quad |y| < \epsilon. \tag{4.2.1}$$

Using this δ, for $y = (y_1, \ldots, y_d)$, define

$$p_\delta(t) := \int_{|y|\geq\delta} p(t,0,y)dy.$$

Then,

$$\begin{aligned} p_\delta(t) &\leq \prod_{j=1}^{d}\int_{|y_j|\geq\delta} p(t,0,y_j)\,dy_j \\ &= \left(2\int_{\delta/\sqrt{t}}^{\infty}\frac{1}{\sqrt{2\pi}}\exp\left\{-\frac{z^2}{2}\right\}dz\right)^d, \end{aligned}$$

which tends to 0 as $t \to 0^+$. Hence, there exists a ρ such that

$$p_\delta(t) < \frac{\epsilon}{4\|f\|} \quad \forall\ 0 < t < \rho. \tag{4.2.2}$$

Using (4.2.1) and (4.2.2),

$$\begin{aligned}|T_tf(x) - f(x)| &\leq \int_{\mathbb{R}^d} |f(y) - f(x)|\, p(t,x,y)dy \\ &= \int_{\mathbb{R}^d} |f(x-y) - f(x)|\, p(t,0,y)dy \\ &\leq 2\|f\|p_\delta(t) + \int_{\{y:|y|<\delta\}} |f(x-y) - f(x)|\, p(t,0,y)dy \\ &\leq 2\|f\|p_\delta(t) + \epsilon/2\end{aligned}$$

so that $\|T_tf - f\| \leq \epsilon$ if $t < \rho$. ■

From this third property, one can readily obtain, for any $0 \leq s \leq t$, $\lim_{t\downarrow s} T_tf = T_sf$ since T_tf can be written $T_{t-s}T_sf$. This is known as the *strong continuity* of the semigroup $\{T_t\}$. In general,

Definition 4.2.1 *A family $\{S_t : t \geq 0\}$ of bounded linear operators on a Banach space B is called a* **semigroup** *if $S_{t+s} = S_tS_s$ for all $s, t \in B$, and $S_0 = I$.*

The semigroup is said to be **strongly continuous** *if $\lim_{t\to 0} S_tf = f$ for every $f \in B$.*

The semigroup is a **contraction** *if $\|S_t\| \leq 1$ for all $t \geq 0$.*

We will call T_t as the Brownian semigroup. It is a **Feller semigroup**, that is, $\{T_t\}$ is a strongly continuous, positive, contraction semigroup defined on $\hat{\mathbf{C}}$ such that $T_t\hat{\mathbf{C}} \subset \hat{\mathbf{C}}$ for all $t \geq 0$. Next, we define the infinitesimal generator of a semigroup. Typically, a generator is an unbounded operator, and hence is not defined everywhere on B.

Definition 4.2.2 *The* **infinitesimal generator** *of a semigroup $\{S_t\}$ on B is the linear operator A given by*

$$Af = \lim_{t\to 0} \frac{S_tf - f}{t}.$$

The domain of A, denoted D_A, is the subspace of all $f \in B$ for which the above limit exists.

It is usually difficult to find D_A, and hence one has to be contented with a subspace of D_A. For the Brownian semigroup T_t, we will find the generator on a subspace $\hat{\mathbf{C}}^2$ consisting of all twice continuously differentiable functions f on $\mathbb{R}^d$ such that f and its first and second order derivatives are in $\hat{\mathbf{C}}$.

Theorem 4.2.1 *The infinitesimal generator of the Brownian semigroup is given by $\frac{1}{2}\Delta$ on the space $\hat{\mathbf{C}}^2$.*

Proof Let $f \in \hat{C}^2$. $T_t f(x) = Ef(x + \sqrt{t}Z)$ where Z is a d-dimensional $N(0, I)$ vector. By the Taylor formula,

$$f(x + \sqrt{t}Z) = f(x) + \sum_{i=1}^{d} f_i'(x)\sqrt{t}Z_i + \frac{t}{2}\sum_{i,j=1}^{d} f_{ij}''(x)Z_iZ_j$$
$$+\frac{t}{2}\sum_{i,j=1}^{d}\left\{f_{ij}''(\xi) - f_{ij}''(x)\right\} Z_iZ_j$$

where ξ is between x and $x + \sqrt{t}Z$. Taking expectation,

$$T_t f(x) = f(x) + \frac{t}{2}\Delta f(x) + \frac{t}{2}R_t \tag{4.2.3}$$

where $R_t = \sum_{i,j=1}^{d} E\left\{f_{ij}''(\xi) - f_{ij}''(x)\right\} Z_iZ_j$. We will show that $R \to 0$ as $t \to 0^+$. Given $\epsilon > 0$, there exists $\delta > 0$ such that

$$\max_{i,j}|f_{ij}''(y) - f_{ij}''(x)| < \frac{\epsilon}{2d^2}$$

if $|x - y| < \delta$. For any $C > 0$,

$$R \leq 2\max_{i,j}\|f_{ij}''\| \sum_{i,j=1}^{d} E\left(|Z_i|\,|Z_j|1_{\{|Z|>C\}}\right) + \sum_{i,j=1}^{d} E\left(|f_{ij}''(\xi) - f_{ij}''(x)|\,|Z_i|\,|Z_j|1_{\{|Z|\leq C\}}\right).$$

Choose C large enough so that the first term on the right is less than $\epsilon/2$. If $t < (\frac{\delta}{C})^2$, then the second term on the right is less than $\epsilon/2$. Thus $R_t \to 0$. Using this in Equation (4.2.3), the proof is completed. ■

Thus, the generator of the Brownian semigroup is an extension of the differential operator $\frac{1}{2}\Delta$ defined on the subspace $\hat{C}^2$.

Different processes may share the same infinitesimal generator A. Such situations arise, for instance, by imposing various boundary conditions on a process. These cases have to be distinguished by the differing domains of the generator A. The basic semigroup theory is given in the next section to facilitate the discussion of such problems.

4.3 Resolvents and Generators

Let B be a Banach space. We will first define the integral of a B-valued function over an interval. Let $\phi : (a, b) \to B$ where (a, b) is an interval in $\mathbb{R}$. Let $P = \{a = t_0 < t_1 < t_2 \cdots < t_n = b\}$ denote a partition of (a, b). The norm of P is defined by $\|P\| = \max_j(t_j - t_{j-1})$. A Riemann sum for ϕ corresponding to the partition P is given by $R(\mathbf{s}, P) = \sum_{j=1}^{n} \phi(s_j)(t_j - t_{j-1})$ where $s_j \in (t_{j-1}, t_j)$ and $\mathbf{s} = (s_1, \ldots, s_n)$.

Definition 4.3.1 *The function ϕ is said to be integrable in the sense of Bochner if the Riemann sums converge in the norm to a limit as $\|P\|$ goes to zero, that is, given $\epsilon > 0$, there exists $\delta > 0$ such that $|R(s, P)| < \epsilon$ for all P satisfying $\|P\| < \delta$ and for all choice of evaluation points s.*

Let $\{S_t : t \geq 0\}$ be a strongly continuous positive contraction semigroup of operators defined on a Banach space B. Let A denote its generator. The semigroup property of S_t and the definition of its generator lead us to guess that $S_t = e^{At}$. But A is seldom bounded, and therefore e^{At} cannot be defined. However, for any $\lambda > 0$, we can define an operator

$$R_\lambda f = \int_0^\infty e^{-\lambda t} S_t f dt \tag{4.3.1}$$

and expect that R_λ would equal $(\lambda - A)^{-1}$ on the range of $\lambda - A$. The above integral converges since $\|S_t f\| \leq \|f\|$. The family of operators R_λ with $\lambda > 0$ defined by (4.3.1) is known as the **resolvent** of the semigroup S_t.

Example 4.3.1 *Consider the one-dimensional standard Brownian motion. The resolvent for the semigroup T_t can be explicitly obtained in this case. We will use moment generating functions of normal and exponential random variables.*

Indeed, for any $\lambda > 0$ and $|\beta| < \sqrt{2\lambda}$, consider

$$\begin{aligned}
\lambda \int_0^\infty e^{-\lambda t} \int_{-\infty}^\infty e^{\beta y} p(t, x, y)\, dy\, dt &= \lambda \int_0^\infty e^{-\lambda t} e^{\beta x + \frac{\beta^2}{2} t} \\
&= e^{\beta x} \lambda \int_0^\infty e^{-(\lambda - \frac{\beta^2}{2})t} dt \\
&= e^{\beta x} \frac{2\lambda}{2\lambda - \beta^2} \\
&= e^{\beta x} \frac{\sqrt{2\lambda}}{\sqrt{2\lambda} + \beta} \frac{\sqrt{2\lambda}}{\sqrt{2\lambda} - \beta},
\end{aligned}$$

which is readily seen as the moment generating function $E(e^{\beta Y})$ of the random variable $Y = x + X_1 - X_2$ where X_1 and X_2 are independent exponential random variables with parameter $\sqrt{2\lambda}$. In fact, $X_1 - X_2$ is a double-exponential random variable with parameter $\sqrt{2\lambda}$. The density function of Y is given by

$$p_Y(y) = \frac{\sqrt{2\lambda}}{2} e^{-\sqrt{2\lambda}|y-x|} \quad \text{for} \quad -\infty < y < \infty.$$

We have thus shown that

$$\lambda \int_{-\infty}^\infty e^{\beta y} \int_0^\infty e^{-\lambda t} p(t, x, y)\, dt\, dy = \int_{-\infty}^\infty e^{\beta y} \frac{\sqrt{2\lambda}}{2} e^{-\sqrt{2\lambda}|y-x|}\, dy.$$

Therefore, we get for all y,

$$\lambda \int_0^\infty e^{-\lambda t} p(t, x, y)\, dt = \frac{\sqrt{2\lambda}}{2} e^{-\sqrt{2\lambda}|y-x|}. \tag{4.3.2}$$

Recalling that for any $f \in \hat{C}$, *and* $\lambda > 0$,

$$\begin{aligned} R_\lambda f(x) &= \int_0^\infty e^{-\lambda t} \int_{-\infty}^\infty f(y) p(t, x, y)\, dy \\ &= \int_{-\infty}^\infty f(y) \int_0^\infty e^{-\lambda t} p(t, x, y)\, dt\, dy \\ &= \int_{-\infty}^\infty f(y) \frac{1}{\sqrt{2\lambda}} e^{-\sqrt{2\lambda}|y-x|}\, dy. \end{aligned}$$

Typically, one cannot obtain the resolvent explicitly. For $d \geq 3$, the expression for the resolvent of the Brownian semigroup is complicated and involves modified Bessel functions. Luckily, one can bypass the need for such explicit expressions and simply use the standard identities for a resolvent.

Lemma 4.3.2 *If* $f \in D(A)$, *then*

(*i*) $S_t f \in D(A)$ *for each* $t \geq 0$, *and*

$$\frac{dS_t f}{dt} = AS_t f = S_t Af. \tag{4.3.3}$$

(*ii*) $S_t f - f = \int_0^t S_s Af ds = \int_0^t AS_s f ds.$

Proof Consider

$$\begin{aligned} \lim_{h\to 0^+} \frac{1}{h}(S_h S_t f - S_t f) &= \lim_{h\to 0^+} \frac{S_t(S_h f - f)}{h} \\ &= S_t Af \end{aligned}$$

since S_t is a bounded operator, and $f \in D(A)$. We have thus shown that $S_t f \in D(A)$ and $AS_t f = S_t Af$. Since $S_h S_t = S_{h+t}$, we have also shown that the right-hand derivative of $S_t f$ exists and equals $S_t Af$.

If $h < 0$, then write $h = -\delta$ and denote $\frac{(S_\delta f - f)}{\delta}$ as $A_\delta f$ so that

$$\begin{aligned} \frac{1}{h}(S_{h+t} f - S_t f) &= S_{t-\delta} A_\delta f \\ &= S_{t-\delta}(A_\delta f - Af) + S_{t-\delta}(Af - S_\delta Af) + S_t Af. \end{aligned}$$

Since S is a contraction semigroup, the first two terms on the right side go to zero as $\delta \to 0$. Hence, the left-hand derivative of $S_t f$ also exists at any $t > 0$ and equals $S_t Af$. Part (i) is thus proved.

To prove part (ii), observe that the function $F(t) := \int_0^t S_s Afds$ is a differentiable B-valued function of t. Clearly, $F'(t) = S_t Af$. By part (i), $\frac{d(S_t f - f)}{dt} = S_t Af$. Both F and $S_t f - f$ are zero at $t = 0$, and their derivatives are equal. Hence, $S_t f - f = \int_0^t S_s Afds$. ■

Remark 4.3.1 If $f \in B$, then one can easily show that for any $t > 0$, $\int_0^t S_s f\, ds \in D(A)$. In this situation, it follows that

$$S_t f - f = A\left(\int_0^t S_s f ds\right).$$

Theorem 4.3.3 *Let R_λ be the resolvent of a strongly continuous contraction semigroup S_t on a Banach space B. Then, for any $f \in B$ and $\lambda > 0$,*

$$R_\lambda = (\lambda - A)^{-1} f$$

where A is the generator of the semigroup.

Proof

Step 1 Let $(\lambda - A)R_\lambda f = f$ for all $f \in B$. Indeed, with $A_h = \frac{S_h - I}{h}$, we have

$$\begin{aligned} A_h R_\lambda f &= \frac{1}{h}\int_0^\infty e^{-\lambda t}(S_{t+h} f - S_t f)\, dt \\ &= \frac{e^{\lambda h}}{h} R_\lambda f - \frac{e^{\lambda h}}{h}\int_0^h e^{-\lambda t} S_t f dt - \frac{1}{h} R_\lambda f \\ &= \frac{(e^{\lambda h} - 1)}{h} R_\lambda f - \frac{e^{\lambda h}}{h}\int_0^h e^{-\lambda t} S_t f dt. \end{aligned}$$

As $h \to 0^+$, the right side converges to $\lambda R_\lambda f - f$. Therefore, $R_\lambda f \in D(A)$, and

$$AR_\lambda f = \lambda R_\lambda f - f;$$

that is,

$$(\lambda - A)R_\lambda f = f, \tag{4.3.4}$$

which also shows that the range of $(\lambda - A)$ is B.

Step 2 Let $R_\lambda(\lambda - A)f = f$ for all $f \in D(A)$.

For $f \in D(A)$, and $h > 0$, we have

$$R_\lambda A_h f = A_h \int_0^\infty e^{-\lambda t} S_t f dt$$

so that, as $h \to 0^+$, we get $R_\lambda Af = AR_\lambda f$, since $R_\lambda f \in D(A)$ by Step 1. Hence,

$$\begin{aligned} R_\lambda(\lambda - A)f &= \lambda R_\lambda f - AR_\lambda f \\ &= (\lambda - A)R_\lambda f \end{aligned}$$

so that

$$R_\lambda(\lambda - A)f = f \tag{4.3.5}$$

Thus, $(\lambda - A) : D(A) \to B$ is one-to-one. It is an onto map by Step 1. Equations (4.3.4) and (4.3.5) finish the proof. ∎

Remark 4.3.2 For any $\lambda > 0$, the operator $\lambda - A : D(A) \to B$ is bijective, so that the range of R_λ is equal to $D(A)$.

Proposition 4.3.4 *Let A be the generator of a strongly continuous contraction semigroup S_t on a Banach space B. Then A is a closed operator and $D(A)$ is dense in B.*

Proof Since R_λ is a bounded operator, and $R_\lambda^{-1} = \lambda - A$, we get the closedness of $\lambda - A$. Hence, A is closed.

For any $f \in B$, we have

$$\begin{aligned} \|\lambda R_\lambda f - f\| &\le \lambda \int_0^\infty e^{-\lambda t} \|S_t f - f\| dt \\ &= \int_0^\infty e^{-u} \|S_{u/\lambda} f - f\| du, \end{aligned}$$

which, by the bounded convergence theorem, tends to zero as $\lambda \to \infty$. Thus,

$$\lim_{\lambda \to \infty} \lambda R_\lambda f = f.$$

By the above remark, $\lambda R_\lambda f \in D(A)$ for all $\lambda > 0$ which finishes the proof. ∎

So far, we have assumed that A is the generator of a strongly continuous contraction semigroup. We state below the Hille-Yosida theorem, which gives necessary and sufficient conditions for an operator A to generate a strongly continuous contraction semigroup.

Theorem 4.3.5 *A linear operator A defined in a Banach space B is the generator of a strongly continuous contraction semigroup on B if and only if:*

(i) $D(A)$ is dense in B.

(ii) A is **dissipative,** *that is, $\|\lambda f - Af\| \ge \lambda \|f\|$ for every $f \in D(A)$ and $\lambda > 0$.*

(iii) For some $\lambda > 0$, the range of $(\lambda - A)$ is B.

Very rarely does one get A and $D(A)$ concretely. However, for the Brownian semigroup, one can determine A and $D(A)$.

Example 4.3.6 *Consider the semigroup T_t for the one-dimensional Brownian motion. We already know that the generator A is given by $Af = f''$ for all $f \in \hat{C}^2$. We will show that $D(A) = f \in \hat{C}^2$.*

Proof We already know that $\hat{C}^2 \subset D(A)$. To prove the reverse containment, take any $g \in D(A)$. Then $g = R_\lambda f$ for some $f \in \hat{C}$. Since

$$R_\lambda f(x) = \int_{-\infty}^\infty f(y) \frac{1}{\sqrt{2\lambda}} e^{-\sqrt{2\lambda}|y-x|} dy,$$

we can differentiate inside the integral with respect to x by using the dominated convergence theorem.

$$\frac{dR_\lambda f(x)}{dx} = \int_{-\infty}^{\infty} f(y)\operatorname{sgn}\{y-x\}\, e^{-\sqrt{2\lambda}|y-x|}\, dy \tag{4.3.6}$$

$$= \int_x^{\infty} f(y)e^{-\sqrt{2\lambda}(y-x)}\, dy - \int_{-\infty}^{x} f(y)e^{-\sqrt{2\lambda}(x-y)}\, dy. \tag{4.3.7}$$

From (4.3.6), it is clear that $|\frac{dR_\lambda f(x)}{dx}| \leq \sqrt{2\lambda}R_\lambda|f|(x)$ and hence $\frac{dR_\lambda f}{dx} \in \hat{\mathbf{C}}$. From (4.3.7),

$$\frac{d^2R_\lambda f(x)}{dx^2} = -2f(x) + \sqrt{2\lambda}\int_{-\infty}^{\infty} f(y)e^{-\sqrt{2\lambda}|y-x|}\, dy$$
$$= -2f(x) + 2\lambda R_\lambda f(x)$$

so that $R_\lambda f \in \hat{\mathbf{C}}^2$. ■

Example 4.3.7 *If we define $S_t f(x) := Ef(x+|W_t|)$ for $f \in \hat{\mathbf{C}}$, where W_t is a standard one-dimensional Wiener process, then one can show that the infinitesimal generator is again given by $Af = \frac{\Delta}{2}f$. However,*

$$D(A) = \left\{f \in \hat{\mathbf{C}}^2 : f'(0+) = 0\right\}.$$

Thus it is possible for two processes to have the same generator. However, $D(A)$ varies to distinguish the processes.

In the next proposition, A and $D(A)$ are identified for the multidimensional Brownian motion. To facilitate the discussion, we recall the notion of weak derivatives.

Definition 4.3.2 *Let u be any locally integrable function in $\mathbb{R}^d$, and α be any d-dimensional multi-index. Then a locally integrable function v is called the αth* **weak or distributional derivative** *of u if it satisfies*

$$\int_{\mathbb{R}^d} v\phi\, dx = (-1)^{|\alpha|}\int_{\mathbb{R}^d} u\, D^\alpha\phi\, dx$$

for all $\phi \in C_0^\infty(\mathbb{R}^d)$. This is also written as

$$\langle v,\phi\rangle = (-1)^{|\alpha|}\langle u, D^\alpha\phi\rangle.$$

Lemma 4.3.8 *Let T_t be the semigroup for the Brownian motion in $\mathbb{R}^d$ with the resolvent denoted by R_λ. Then*

(i) The semigroup T_t maps $\hat{\mathbf{C}}$ to $\hat{\mathbf{C}}^\infty \cap L^1(\mathbb{R}^d)$.

(ii) The resolvent R_λ maps $\hat{\mathbf{C}}_0^\infty$ to $\hat{\mathbf{C}}^\infty$. (Besides, for any $f \in \hat{\mathbf{C}}_0^\infty$, $R_\lambda f$ and its derivatives of any order are in $L^1\mathbb{R}^d$).

Proof The first statement follows by noting that $p(t,x,y)$, as a function of x, has bounded derivatives of all orders for any fixed t. Besides, any such derivative is integrable. Hence the dominated convergence theorem allows us to finish the proof.

For (ii), write $R_\lambda f(x) = \int_0^\infty e^{\lambda t} \int_{\mathbb{R}^d} f(x-y)p(t,0,y)\,dy\,dt$. One can make use of the boundedness of the derivatives of f and the bounded convergence theorem to show the differentiability of $R_\lambda f$.

If K denotes the support of f, then

$$\begin{aligned}\int_{\mathbb{R}^d} |R_\lambda f(x)|dx &= \int_0^\infty e^{\lambda t} \int_{\mathbb{R}^d} Ef(x+W_t)\,dx\,dt \\ &\leq \|f\|_\infty m(K) \int_0^\infty e^{\lambda t}\,dt \\ &< \infty,\end{aligned}$$

where m denotes the d-dimensional Lebesgue measure. The integrability of the derivatives of $R_\lambda f$ can likewise be established. ■

Remark 4.3.3 It should be pointed out that if $f \in \hat{C}$, then $R_\lambda f$ is not infinitely differentiable. In fact, we have already proved that $R_\lambda f$ is only twice differentiable for the Brownian semigroup in dimension $d = 1$. This restriction is due to the divergence of the time integral, in the vicinity of $t = 0$.

Proposition 4.3.9 *The generator A for the Brownian semigroup T_t in dimensions $d \geq 2$ is given by $\Delta/2$, where the Laplacian is understood in the weak sense. The domain*

$$D(A) = \left\{ f \in \hat{C} : \frac{\Delta}{2} f \in \hat{C} \text{ where } \Delta \text{ is in the weak sense} \right\}.$$

Proof

Step 1 Let G stand for $\Delta/2$ taken in the weak sense, while A denotes the generator of the semigroup T_t. Let $f \in \hat{C}$. Then, for any $\lambda > 0$,

$$\begin{aligned}\langle GR_\lambda f, \phi\rangle &= \langle R_\lambda f, \frac{\Delta}{2}\phi\rangle \\ &= \langle f, R_\lambda A\phi\rangle \\ &= \langle f, AR_\lambda \phi\rangle \\ &= \langle f, \lambda R_\lambda \phi - \phi\rangle \\ &= \langle \lambda R_\lambda f - f, \phi\rangle.\end{aligned}$$

Thus,

$$GR_\lambda f = \lambda R_\lambda f - f = AR_\lambda f; \tag{4.3.8}$$

that is, G coincides with A on the range of R_λ. In other words, $G = A$ on $D(A)$.

Step 2 From Equation (4.3.8), one can infer that

$$D(A) \subset D_1 = \{f \in \hat{C} : Gf \in \hat{C}\}.$$

We need to show equality.

Take any $f \in D_1$. Then, for any $\lambda > 0$, $\lambda f - Gf \in \hat{C}$. Let us denote $\lambda f - Gf$ as g. Then, $R_\lambda g \in D(A)$. Since A coincides with G on $D(A)$, we get

$$(\lambda - G)R_\lambda g = g = (\lambda - G)f;$$

that is,

$$(\lambda - G)(R_\lambda g - f) = 0. \tag{4.3.9}$$

Let h denote $R_\lambda g - f$. We will show that $h = 0$.

For any given $\phi \in C_0^\infty$, consider its resolvent. By Lemma 4.3.8, $R_\lambda \phi$ and its derivatives are in L^1. Hence, given any $\epsilon > 0$, there exists an $N > 0$ such that support of $\phi \subset B_N$, and

$$\int_{B_N^c} (|R_\lambda \phi(x)| + |\nabla R_\lambda \phi(x)|)\, dx < \frac{\epsilon}{\|h\|_\infty}$$

where $B_N = \{x : |x| \le N\}$. Let η_N be a nonnegative C_0^∞ function defined on $\mathbb{R}^d$ such that $\eta_N = 1$ in the ball B_N, and $|\nabla \eta_N| + |\frac{\Delta}{2}\eta_N| \le 1$. For simplicity of notation, we will call it η. Clearly,

$$\langle h, \phi \rangle = \langle h\eta, \phi \rangle = \langle h, \eta(\lambda - \frac{\Delta}{2})R_\lambda \phi \rangle.$$

Using

$$\eta \Delta R_\lambda \phi = \Delta \eta R_\lambda \phi - R_\lambda \phi \Delta \eta - 2\nabla \eta \cdot \nabla R_\lambda \phi,$$

we get the bound

$$\begin{aligned} |\langle h, \phi \rangle| &\le |\langle h, (\lambda - \frac{\Delta}{2})\eta R_\lambda \phi \rangle| + \epsilon \\ &= |\langle h, (\lambda - G)\eta R_\lambda \phi \rangle| + \epsilon \\ &\le |\langle (\lambda - G)h, \eta R_\lambda \phi \rangle| + \epsilon \\ &= \epsilon \end{aligned}$$

by using (4.3.9). Since ϵ is arbitrary, we get

$$\langle h, \phi \rangle = 0$$

for all $\phi \in C_0^\infty$. Therefore, $h = 0$; that is, $f = R_\lambda g$ so that $f \in D(A)$. ■

In general, the infinitesimal generator A and its domain $D(A)$ are difficult to ascertain. However, we do not need the full generator in stochastic analysis.

4.4 Pregenerators and Martingales

Let $\{S_t\}$ be a strongly continuous contraction semigroup on a Banach space B with generator A. A linear operator L defined on a subspace D of $D(A)$ is known as a **pregenerator** of S_t if the closure of L is A; that is, $L = A|_D$, and $\bar{L} = A$. We call D a **core** for A.

In several problems, pregenerators have simpler descriptions. Often, they turn out to be differential operators. One can construct the semigroup from L. We will first show that $\hat{C}_0^{\infty}$ is a core for the Brownian semigroup.

Note that dissipativity of A implies that of L. Let us denote the range of an operator by $\mathcal{R}$.

Lemma 4.4.1 *Let L be a pregenerator of A. Then $\overline{\mathcal{R}(\lambda - L)} = \mathcal{R}(\lambda - \bar{L})$ for all $\lambda > 0$.*

Proof If the pair (f, g) lies in the graph of $\bar{L}$, then there exists a sequence $\{f_n\}$ in D such that $f_n \to f$ and $Lf_n \to g$ as $n \to \infty$. For any $\lambda > 0$,

$$(\lambda - \bar{L})f = \lambda f - g = \lim_{n\to\infty} (\lambda f_n - Lf_n) \in \overline{\mathcal{R}(\lambda - L)}.$$

Thus,

$$\mathcal{R}(\lambda - \bar{L}) \subset \overline{\mathcal{R}(\lambda - L)}.$$

To prove the equality of the two sets, note that $\mathcal{R}(\lambda - L) \subset \mathcal{R}(\lambda - \bar{L})$, and hence, it suffices to prove that $\mathcal{R}(\lambda - \bar{L})$ is a closed set.

If $\{g_n\}$ is a sequence in $\mathcal{R}(\lambda - \bar{L})$, each g_n can be written as $(\lambda - \bar{L})f_n$. Let $g_n \to g$. Since $\bar{L}$ is dissipative, $\{f_n\}$ is Cauchy. There exists an f such that $f_n \to f$. Therefore, $\bar{L}f_n = \lambda f_n - g_n \to \lambda f - g$. Thus $\bar{L}f = \lambda f - g$ so that $g \in \mathcal{R}(\lambda - \bar{L})$. The proof is thus completed. ■

Given an operator L, we would like to know if L is the pregenerator of a strongly continuous contraction semigroup S_t. For this, first and foremost, we need the closability of L. The next lemma gives sufficient conditions for it.

Lemma 4.4.2 *Let L be a dissipative linear operator on B with $D(L)$ dense in B. Then L is closable.*

Proof By dissipativity of L, we get that $(\lambda - L)^{-1}$ exists for any $\lambda > 0$ and $\|(\lambda - L)^{-1}\| \leq \frac{1}{\lambda}$. Let $f_n \in D(L)$ such that $f_n \to 0$ and $Lf_n \to g$. We need to show that $g = 0$. Since $D(L)$ is dense in B, there exists a sequence $\{g_n\}$ in $D(L)$ such that $g_n \to g$. Then,

$$\begin{aligned}\|g_n\| &\leq \frac{1}{\lambda}\|(\lambda - L)g_n\| \\ &\leq \|f_n\| + \|(g_n - f_n)\| + \frac{1}{\lambda}\|Lg_n\|.\end{aligned}$$

By letting $\lambda \to \infty$ and then $n \to \infty$, we get that $g_n \to 0$. ■

Using the lemmas given above, we can restate the Hille-Yosida Theorem as follows:

Theorem 4.4.3 *A linear operator L on a Banach space B is the pregenerator of a strongly continuous contraction semigroup on B if and only if*

(i) $D(L)$ is dense in B.

(ii) L is dissipative.

(iii) $\mathcal{R}(\lambda - L)$ is dense in B for some $\lambda > 0$.

Thus, a subspace D is a core for a generator A of a strongly continuous contraction semigroup provided that (i) D is dense in B and (ii) $\mathcal{R}(\lambda - A|_D)$ is dense in B for some $\lambda > 0$.

Proposition 4.4.4 *The space $\hat{C}_0^\infty(\mathbb{R}^d)$ is a core for the generator of the Brownian semigroup.*

Proof

Step 1 Let B denote $\hat{\mathbf{C}}$, and $D := \hat{\mathbf{C}}^\infty$. From Lemma 4.3.8, $T_t : B \to D$. Therefore, $T_t : D \to D$. The subspace D is dense in B. Also $D \subset D(A)$. Given any $f \in D$, define

$$f_n := \frac{1}{n}\sum_{j=1}^{n^2} e^{-\lambda j/n} T_{j/n} f.$$

Then, $f_n \in D$. Therefore,

$$\begin{aligned}\lim_{n\to\infty} (\lambda - A) f_n &= \lim_{n\to\infty} \frac{1}{n}\sum_{j=1}^{n^2} e^{-\lambda j/n} T_{j/n}(\lambda - A) f \\ &= \int_0^\infty e^{-\lambda t} T_t(\lambda - A) f dt \\ &= f.\end{aligned}$$

Thus, $\mathcal{R}(\lambda - A|_D)$ is dense in D, and hence dense in B. Thus, D is a core for A.

Step 2 Given $f \in D$, consider the function $f_N = f\eta_N$ where η_N is as in step 2 of the proof of Proposition 4.3.9. Then $f_N \in \hat{C}_0^\infty(\mathbb{R}^d)$ and $f_N \to f$ and $Af_N \to Af$ in B. The proof is thus completed. ∎

Next, we will show the connection of infinitesimal generators to martingales in the simple situation of a d-dimensional Brownian motion.

Proposition 4.4.5 *Let $\{W_t\}$ be a standard d-dimensional Brownian motion defined on $(\Omega, \mathcal{F}, \mathcal{F}_t, P)$, where $\mathcal{F}_t$ denotes the natural filtration. If A denotes the generator of the Brownian semigroup, then for any $f \in D(A)$,*

$$M_t = f(W_t) - f(0) - \int_0^t Af(W_s)\, ds$$

is an $\mathcal{F}_t$-martingale.

Proof Since f and Af are bounded, $M_t \in L^1(P)$ for all $t \geq 0$. By the Markov property, $E(g(W_t) \mid \mathcal{F}_s) = E_{X_s}(g(W_t)) = T_{t-s}g(X_s)$ for any continuous, bounded function g, and for all $s \leq t$. Therefore,

$$\begin{aligned}
E(M_t \mid \mathcal{F}_s) &= M_s + E\left\{f(W_t) - f(W_s) - \int_s^t Af(W_u)\, du \mid \mathcal{F}_s\right\} \\
&= M_s + E\left\{f(W_t) - f(W_s) - \int_0^{t-s} Af(W_{u+s})\, du \mid \mathcal{F}_s\right\} \\
&= M_s + T_{t-s}f(W_s) - f(W_s) - \int_0^{t-s} T_uAf(W_s)\, du \\
&= M_s
\end{aligned}$$

by using part (ii) of Lemma 4.3.2. ■

Proposition 4.4.5 holds if A and $D(A)$ are replaced by a pregenerator L and its domain, $D(L)$. By using a pregenerator, we can work with a differential operator L which has a simpler domain. This is the idea behind the formulation of the martingale problem by Stroock and Varadhan. The martingale problem will be discussed in detail in a separate chapter.

Exercises

1. Let S_t be a strongly continuous semigroup on on a Banach space B, with infinitesimal generator A. Show that for all $t > 0$ and $f \in B$, $\int_0^t S_sfds \in D(A)$, and that

$$S_tf - f = A\left(\int_0^t S_sfds\right).$$

2. Let S_t on $\hat{C}$ be defined by $S_tf(x) = f(x + t)$. Show that S_t is a strongly continuous contraction semigroup. Find its generator.
3. Let A be a dissipative operator on a Banach space B. For all $\lambda > 0$, prove that A is a closed operator if and only if range of $\lambda - A$ is closed.

5 Stochastic Integration

Stochastic calculus begins with the notion of stochastic integrals invented by Kiyosi Itô in 1944. Stochastic integrals play a central and important role in the construction of diffusion processes. We start this chapter by constructing stochastic integrals with respect to Brownian motion. Even though Brownian paths are of unbounded variation on any non-empty open interval, the Itô integral can be defined using an L^2 isometry. The essential features of stochastic integrals are discussed in detail. The theorem on change of variables is known as the Itô formula (or the Itô lemma), and its proof is given in full. Examples are given to illustrate the importance and usefulness of the Itô formula. Important applications of the Itô formula such as Lévy's characterization of Brownian motion, the Burkhölder-Davis-Gundy inequality, and the martingale representation theorem are given in Section 5.6. The final section of this chapter is devoted to the Girsanov theorem on change of measures.

5.1 The Itô Integral

In this section we will construct the stochastic integral step by step, starting with integrals of simple, bounded, adapted integrands. Our approach removes the abstraction inherent in extension of isometries, and helps us to explicitly identify the class of integrands for which the Itô integral can be defined.

Step 1 Let $\{W_t\}$ be an $\mathcal{F}_t$-adapted Wiener martingale. Consider the process $H = \sum_{j\geq 0} H_j \, 1_{[t_j, t_{j+1})}$ where $0 = t_0 < t_1 < \cdots$, and for each j, H_j is measurable with respect to $\mathcal{F}_{t_j}$ and $E(H_j^2) < \infty$. We will call any process that admits such a representation an L^2–step process. It is natural to define for any $t > 0$,

$$\int_0^t H_s \, dW_s = \sum_j H_j (W_{t_{j+1}\wedge t} - W_{t_j \wedge t}). \tag{5.1.1}$$

Equivalently,

$$\int_0^t H_s \, dW_s = \sum_{j=0}^{n-1} H_j \left(W_{t_{j+1}} - W_{t_j}\right) + H_n \left(W_t - W_{t_n}\right) \quad \text{if} \quad t_n \leq t < t_{n+1}. \tag{5.1.2}$$

We will denote $\int_0^t H_s \, dW_s$ by I_t. The following three properties hold for the above integral. From Equation (5.1.1), the first property follow easily.

(i) *Linearity of the integral*: for any two L^2-step processes H, G, and real numbers a, b,

$$\int_0^t (aH_s + bG_s) \, dW_s = a \int_0^t H_s \, dW_s + b \int_0^t G_s \, dW_s.$$

(ii) *Martingale Property*: The process $I_t = \int_0^t H_s \, dW_s$ is an $\mathcal{F}_t$-martingale.

(iii) **The Itô Isometry**: $E\{I_t^2\} = \int_0^t E(H_s^2) \, ds$.

Proof We will use Equation (5.1.2) in proving the martingale property. Let H be an L^2-step function as above. Take any $s < t$. Let s, t be such that $t_m \leq s < t_{m+1}$ and $t_n \leq t < t_{n+1}$. Then,

$$E\left(I_t \mid \mathcal{F}_s\right) = I_s + J_s$$

where

$$J_s = E\left[H_m \left(W_{t_{m+1}} - W_s\right) + \sum_{m+1}^{n-1} H_j \left(W_{t_{j+1}} - W_{t_j}\right) + H_n \left(W_t - W_{t_n}\right) \mid \mathcal{F}_s\right].$$

Clearly, the first term on the right-side,

$E[H_m(W_{t_{m+1}} - W_s) \mid \mathcal{F}_s]$, is equal to $H_m E[(W_{t_{m+1}} - W_s) \mid \mathcal{F}_s]$, which is equal to zero 0.

Next, let us consider the conditional expectation of a typical summand in the second term:

$$\begin{aligned} E[H_j(W_{t_{j+1}} - W_{t_j}) \mid \mathcal{F}_s] &= E[E\left\{H_j(W_{t_{j+1}} - W_{t_j}) \mid \mathcal{F}_{t_j}\right\} \mid \mathcal{F}_s] \\ &= E[H_j E\left\{(W_{t_{j+1}} - W_{t_j}) \mid \mathcal{F}_{t_j}\right\} \mid \mathcal{F}_s] \\ &= 0. \end{aligned}$$

Likewise, $E[H_n(W_t - W_{t_n}) \mid \mathcal{F}_s] = 0$. Thus, $J_s = 0$, and the martingale property of the integral follows.

Next, we will prove the isometry. Note that

$$E\left\{\left(\int_0^t H_s\,dW_s\right)^2\right\} = E\left(\sum_j H_j\left(W_{t_{j+1}\wedge t} - W_{t_j\wedge t}\right)\right)^2$$
$$= S_1 + S_2$$

where $S_1 = \sum_j E\left\{H_j^2(W_{t_{j+1}\wedge t} - W_{t_j\wedge t})^2\right\}$, and

$$S_2 = 2\sum_{i<j} E\left\{H_iH_j(W_{t_{i+1}\wedge t} - W_{t_i\wedge t})(W_{t_{j+1}\wedge t} - W_{t_j\wedge t})\right\}.$$

The terms of the sum S_1 vanish if $t \leq t_j$. If $t > t_j$,

$$E\left\{H_j^2(W_{t_{j+1}\wedge t} - W_{t_j\wedge t})^2\right\} = E\left(E\left\{H_j^2\left(W_{t_{j+1}\wedge t} - W_{t_j}\right)^2 |\mathcal{F}_{t_j}\right\}\right)$$
$$= E\left(H_j^2 E\left\{\left(W_{t_{j+1}\wedge t} - W_{t_j}\right)^2 |\mathcal{F}_{t_j}\right\}\right)$$
$$= E\left(H_j^2\right)\left(t_{j+1}\wedge t - t_j \wedge t\right).$$

Hence, $S_1 = \sum_j E(H_j^2)(t_{j+1}\wedge t - t_j \wedge t) = \int_0^t E(H_s^2)\,ds$.

The terms of the sum S_2 vanish if $t \leq t_j$. For any $t > t_j$,

$$E\left\{H_iH_j\left(W_{t_{i+1}\wedge t} - W_{t_i\wedge t}\right)\left(W_{t_{j+1}\wedge t} - W_{t_j\wedge t}\right)\right\}$$
$$= E\left\{H_iH_j\left(W_{t_{i+1}\wedge t} - W_{t_i\wedge t}\right)E\left(W_{t_{j+1}\wedge t} - W_{t_j\wedge t}|\mathcal{F}_{t_j}\right)\right\}$$
$$= 0.$$

■

Step 2 Let H be an adapted, bounded process with continuous paths. Define $H_n(t) = H_{[nt]/n}$ for all $t \geq 0$. Since H_n is an L^2-step process, $\int_0^t H_n(s)\,dW_s$ is defined by Step 1.

$$E\left\{\left(\int_0^t H_n(s)\,dW_s - \int_0^t H_m(s)\,dW_s\right)^2\right\} = E\left\{\left(\int_0^t (H_n(s) - H_m(s))\,dW_s\right)^2\right\}$$
$$= E\int_0^t (H_n(s) - H_m(s))^2\,ds$$

which tends to zero as $n, m \to \infty$ by the bounded convergence theorem and the continuity of paths of H. Thus, for any $t \geq 0$, the sequence $\left\{\int_0^t H_n(s)\,dW_s : n \geq 1\right\}$ is Cauchy in $L^2(P)$. Its limit is denoted $\int_0^t H_s\,dW_s$ since $\lim_{n\to\infty} E\int_0^t (H_n(s) - H_s)^2\,ds = 0$.

Let $\{G_n\}$ be another sequence of L^2-step processes such that $\lim_{n\to\infty} E\int_0^t (G_n(s) - H_s)^2\,ds = 0$. It follows that

$\lim_{n\to\infty} E\int_0^t (G_n(s) - H_n(s))^2\,ds = 0$. Therefore,

$$\lim_{n\to\infty} E\left\{\left(\int_0^t H_n(s)\,dW_s - \int_0^t G_n(s)\,dW_s\right)^2\right\} = 0. \tag{5.1.3}$$

Since the $L^2(P)$-limit of $\int_0^t H_n(s)\,dW_s$ is $\int_0^t H_s\,dW_s$, it follows that the $L^2(P)$ – $\lim_{n\to\infty}\int_0^t G_n(s)\,dW_s = \int_0^t H_s\,dW_s$. Thus, $\int_0^t H_s\,dW_s$ is well defined.

The three properties of integrals given in Step 1 continue to hold. In fact, linearity of the integral follows easily. To prove the martingale property of $\left\{\int_0^t H_s\,dW_s : t \geq 0\right\}$, note that for each n, and $0 \leq s \leq t$,

$$E\left(\int_0^t H_n(u)\,dW_u|\mathcal{F}_s\right) = \int_0^s H_n(u)\,dW_u. \tag{5.1.4}$$

Clearly, by the Jensen inequality,

$$L^2(P) - \lim_{n\to\infty} E\left(\int_0^t H_n(u)\,dW_u - \int_0^t H_u\,dW_u|\mathcal{F}_s\right) = 0,$$

and $\int_0^s H_n(u)\,dW_u \to \int_0^s H_u\,dW_u$ in $L^2(P)$. Therefore, upon taking the $L^2(P)$-limit in equation (5.1.4), the proof is completed.

The isometry is quite simple to prove. In fact,

$$\begin{aligned} E\left[\left(\int_0^t H_s\,dW_s\right)^2\right] &= \lim_{n\to\infty} E\left[\left(\int_0^t H_n(s)\,dW_s\right)^2\right] \\ &= \lim_{n\to\infty} E\left(\int_0^t H_n^2(s)\,ds\right) \\ &= E\left(\int_0^t H_s^2\,ds\right) \end{aligned}$$

by the bounded convergence theorem.

Step 3 Let H be an adapted, bounded, jointly measurable process. Define $H_n(t) = n\int_{t-1/n}^t H_s\,ds$ where, for $s < 0$, H_s is taken to be zero. Then, H_n is an adapted, bounded, process with continuous paths.

Fix any $\omega \in \Omega$. We will suppress ω in what follows. Define $A(t) = \int_0^t H_s\,ds$ so that $A(t)$ is absolutely continuous as a function of t. Therefore, $A'(t)$ exists for almost everywhere (a.e.) t. For such t,

$$A'(t) = \lim_{n\to\infty} H_n(t).$$

Besides, $A(t) = \int_0^t A'(s)\,ds$ for all t so that $A'(t) = H_t$ for a.e. t. Thus, $H_n(t,\omega) \to H(t,\omega)$ for a.e. t.

Let $t > 0$ be any fixed time. Define

$$B_\omega = \left\{ s \in [0,t] : \lim_{n\to\infty} H_n(s,\omega) = H_s(\omega) \right\}.$$

The set $B_\omega \in \mathcal{B}(\mathbb{R}^+)$. If m denotes the Lebesgue measure on $\mathbb{R}$, then $m(B_\omega) = t$, and so $\int_\Omega m(B_\omega)\, dP = t$.

Thus, $\{(s,\omega) : s \in [0,t] \text{ and } \lim_{n\to\infty} H_n(s,\omega) = H_s(\omega)\}$ is an almost sure set with respect to the product measure $m \times P$. By the bounded convergence theorem, $\lim_{n\to\infty} E\int_0^t (H_n(s) - H_s)^2\, ds = 0$. Therefore, $\{H_n\}$ is a Cauchy sequence in $L^2([0,t] \times \Omega, m \times P)$. Equivalently, $\{\int_0^t H_n(s)\, dW_s\}$ is Cauchy in $L^2(P)$, and hence, converges to a random variable that we denote as $\int_0^t H_s\, dW_s$.

As in the previous step, the integral is well defined. It has linearity, the martingale property and satisfies the isometry

$$E\left\{ \left(\int_0^t H_s\, dW_s \right)^2 \right\} = E\int_0^t H_s^2\, ds.$$

Step 4 Let H be an adapted, jointly measurable process satisfying the condition $E\int_0^t H_s^2\, ds < \infty$ for all $t > 0$. Define

$$G_n(s) = H_s\, 1_{\{|H_s| \le n\}}.$$

Clearly, $|G_n(s) - H_s| \le |H_s| \in L^2(m \times P)$, and the set $\left\{(s,\omega) : \lim_{n\to\infty} G_n(s) = H_s\right\}$ is an $m \times P$ almost sure set. By using the dominated convergence theorem, $\lim_{n\to\infty} E\int_0^t (G_n(s) - H(s))^2\, ds = 0$.

As in Step 3, $\int_0^t H_s\, dW_s$ can be defined as the $L^2(P)$-limit of $\int_0^t G_n(s)\, dW_s$. If $\{F_n\}$ is a sequence of adapted, bounded, jointly measurable processes such that $\lim_{n\to\infty} E\int_0^t (F_n(s) - H(s))^2\, ds = 0$, then the $L^2(P)$-limit of $\int_0^t F_n(s)\, dW_s$ exists and coincides with $\int_0^t H_s dW_s$. Thus the integral is well-defined. The integral has linearity, martingale property, and isometry.

Before we proceed to Step 5, the following property of stochastic integrals is proved.

Lemma 5.1.1 *Let H be an adapted, measurable process satisfying the condition $E\int_0^t H_s^2\, ds < \infty$ where $t > 0$ is fixed. Then for every $R > 0$, and $\epsilon > 0$,*

$$P\left\{ \left| \int_0^t H_s\, dW_s \right| > \epsilon \right\} \le \frac{R}{\epsilon^2} + P\left\{ \int_0^t H_s^2\, ds > R \right\} \tag{5.1.5}$$

Proof For all $0 \le s \le t$, define

$$H_R(s) = \begin{cases} H_s & \text{if } \int_0^s H_u^2\, du \le R, \\ 0 & \text{otherwise.} \end{cases}$$

Let $S := \sup\left\{s \le t : \int_0^s H_u^2\,du \le R\right\}$. Then,

$$\int_0^t H_R^2(s)\,ds = \int_0^t H_s^2\, 1_{\left\{\int_0^s H_u^2\,du \le R\right\}} ds$$
$$= \int_0^S H_s^2\,ds \le R$$

so that $E\int_0^t H_R^2(s)\,ds \le R$ and

$$\left\{\omega : \sup_{s\le t}\left|H_s - H_R(s)\right| > 0\right\} = \left\{\omega : \int_0^t H_s^2\,ds > R\right\}.$$

Therefore,

$$P\left\{\left|\int_0^t H_s\,dW_s\right| > \epsilon\right\} \le P\left\{\left|\int_0^t H_R(s)\,dW_s\right| > \epsilon\right\} + P\left\{\sup_{s\le t}\left|H_s - H_R(s)\right| > 0\right\}$$
$$\le \frac{E\left(\int_0^t H_R(s)\,dW_s\right)^2}{\epsilon^2} + P\left\{\int_0^t H_s^2\,ds > R\right\}$$
$$\le \frac{R}{\epsilon^2} + P\left\{\int_0^t H_s^2\,ds > R\right\}.$$

■

Step 5 Let H be an adapted process with $\int_0^t H_s^2\,ds < \infty$ a.s. for all $t > 0$. Define

$$H_n(s) = \begin{cases} H_s & \text{if } \int_0^s H_u^2\,du \le n, \\ 0 & \text{otherwise.} \end{cases}$$

Then H_n is an adapted process with $E\int_0^t H_n(s)^2\,ds < \infty$ so that $\int_0^t H_n(s)\,dW_s$ is defined by Step 4. Note that

$$\int_0^t (H_n(s) - H_s)^2\,ds = \int_0^t H_s^2 1_{\left\{\int_0^s H_u^2\,du > n\right\}} ds.$$

Hence, for any given $\epsilon > 0$, $P\{\int_0^t (H_n(s) - H_s)^2\,ds > \epsilon\} \le P\{\int_0^t H_s^2\,ds > n\}$, which tends to zero as $n \to \infty$. In other words, $\int_0^t (H_n(s) - H_s)^2\,ds \to 0$ in probability as $n \to \infty$.

For any $m \le n$, $\sup_{0\le s\le t}\left|H_n(s) - H_m(s)\right| = 0$ if $\int_0^t H_s^2\,ds \le m$. Therefore,

$$P\left\{\left|\int_0^t H_n(s)\,dW_s - \int_0^t H_m(s)\,dW_s\right| > \epsilon\right\} \le P\left\{\int_0^t H_s^2\,ds > m\right\},$$

which goes to zero as $m \to \infty$; that is, $\{\int_0^t H_n(s)\,dW_s\}$ is Cauchy in probability, and hence has a limit in probability. This limit is denoted $\int_0^t H_s\,dW_s$.

We will now show that this limit is well defined. Let $\{G_n\}$ be a sequence of adapted processes such that for each n, $E\int_0^t G_n(s)\,ds < \infty$, and $\lim_{n\to\infty}\int_0^t (G_n(s) - H_s)^2\,ds = 0$ in probability. Then

$$\lim_{n\to\infty}\int_0^t (G_n(s) - H_n(s))^2\,ds = 0 \text{ in probability.}$$

Using Lemma 5.1.1,

$$P\left\{\left|\int_0^t (G_n(s) - H_n(s))\,dW_s\right| > \epsilon\right\} \le \frac{R}{\epsilon^2} + P\left\{\left|\int_0^t (G_n(s) - H_n(s))^2\,ds > R\right.\right\}.$$

For any $\delta > 0$, taking $R = \epsilon^2\delta$,

$$P\left\{\left|\int_0^t (G_n(s) - H_n(s))\,dW_s\right| > \epsilon\right\} \le \delta + P\left\{\left|\int_0^t (G_n(s) - H_n(s))^2\,ds > \epsilon^2\delta\right.\right\}.$$

Taking the limit as $n \to \infty$,

$$\limsup_{n\to\infty} P\left\{\left|\int_0^t (G_n(s) - H_n(s))\,dW_s\right| > \epsilon\right\} \le \delta.$$

By the arbitrariness of δ, $\int_0^t G_n(s)\,dW_s - \int_0^t H_n(s)\,dW_s \to 0$ in probability. Since $\int_0^t H_n(s)\,dW_s$ converges in probability to $\int_0^t H_s\,dW_s$, we conclude that $\int_0^t G_n(s)\,dW_s$ converges to $\int_0^t H_s\,dW_s$ in probability.

The integral constructed in this step possesses linearity. However, *the martingale property and isometry of the integral hold only locally*. In other words, the properties hold up to a suitable stopping time. To see this, for any $n > 0$, define the stopping time

$$\tau_n = \inf\left\{t : \int_0^t H_u^2\,du > n\right\}.$$

Let $M_t^n = \int_0^{t\wedge\tau_n} H_u\,dW_u$. Note that $M_t^n = \int_0^t H_u 1_{\{u<\tau_n\}}\,dW_u$. The integrand $H_u 1_{\{u<\tau_n\}}$ is $\mathcal{F}_u$ adapted, and $E[\int_0^t H_u^2 1_{\{u<\tau_n\}} du] < \infty$. Hence, by Step 4, for any $0 \le s \le t$, we get

$$E(M_t^n|\mathcal{F}_s) = E\left(\int_0^t H_u 1_{\{u<\tau_n\}}\,dW_u \,|\, \mathcal{F}_s\right) = \int_0^s H_u 1_{\{u<\tau_n\}}\,dW_u = M_s^n. \qquad (5.1.6)$$

Since $\tau_n \uparrow \infty$ a.s. as $n \to \infty$, the equality (5.1.6) yields the local martingale property of the stochastic integral. The isometry holds for the integral stopped at τ_n; that is, $E(M_t^n)^2 = E\int_0^{t\wedge\tau_n} H_s^2\,ds$. The construction of stochastic integrals is thus completed.

We summarize the results proved thus far in the following two theorems. Before stating them, we introduce the following important notation:

Let $\mathcal{M}_W^2$ denote the space of all processes $\{H_t\}$ that are $\mathcal{B}(\mathbb{R}^+) \times \mathcal{F}$-measurable, and $(\mathcal{F}_t)$-adapted with $E\int_0^t H_s^2\,ds < \infty$ for any $t > 0$. The space $\mathcal{M}_W^2[0,T]$ will denote the set of all measurable, $(\mathcal{F}_t)$-adapted processes that satisfy $E\int_0^T H_s^2\,ds < \infty$.

Let $\mathcal{L}_W^2$ denote the class of all measurable processes $\{H_t\}$ which are $(\mathcal{F}_t)$-adapted, with $\int_0^t H_s^2\,ds < \infty$ a.s. for any $t > 0$. The space $\mathcal{L}_W^2[0,T]$ will denote the set of all measurable, $(\mathcal{F}_t)$-adapted processes that satisfy $\int_0^T H_s^2\,ds < \infty$ a.s.

Theorem 5.1.2 *Let $\{W_t, \mathcal{F}_t\}$ be a standard one-dimensional Wiener martingale. Let $\{H_t\} \in \mathcal{M}_W^2$. Then the process $\{\int_0^t H_s\,dW_s\}$ is a square integrable $\mathcal{F}_t$-martingale, and $E(\int_0^t H_s\,dW_s)^2 = E\int_0^t H_s^2\,ds$ for all $t > 0$.*

The continuity in t of the martingale $\{\int_0^t H_s\,dW_s\}$ will be proved in the next section.

Theorem 5.1.3 *Let $\{W_t, \mathcal{F}_t\}$ be a standard one-dimensional Wiener martingale. Let $\{H_t\} \in \mathcal{L}_W^2$. Then, the process $\{\int_0^t H_s\,dW_s\}$ is a locally square integrable $\mathcal{F}_t$- local martingale.*

Evaluation of a stochastic integral can be hard, and the integrals do not satisfy the usual rules of calculus such as the power rule for integration.

Example 5.1.4 *We will evaluate the Itô integral $\int_0^1 W_s\,dW_s$. The integrand $\{W_s\}$ is adapted, and $E\int_0^1 W_s^2\,ds = \int_0^1 s\,ds < \infty$. Let $s_j = j/n$ for $j = 0, 1, \ldots, n$. Define*

$$H_n(s) = \begin{cases} 0 & \text{if } 0 \le s < s_1; \\ W_{s_j} & \text{if } s_j \le s < s_{j+1} \text{ for } j = 1, \ldots, n. \end{cases}$$

$$\begin{aligned} E\int_0^1 (H_n(s) - W_s)^2\,ds &= E\sum_{j=0}^{n-1}\int_{s_j}^{s_{j+1}} (W_{s_j} - W_s)^2\,ds \\ &- \sum_{j=0}^{n-1}\int_{s_j}^{s_{j+1}} (s - s_j)\,ds \\ &= \sum_{j=0}^{n-1} (s_{j+1} - s_j)^2/2 = \frac{1}{2n} \to 0 \end{aligned}$$

as $n \to \infty$. *Let* $(\Delta W)_j$ *denote* $W_{s_{j+1}} - W_{s_j}$.

$$\begin{aligned}\int_0^1 H_n(s)\,dW_s &= \sum_{j=0}^{n-1} W_{s_j}(\Delta W)_j = \sum_{j=0}^{n-1} W_{s_j}W_{s_{j+1}} - \sum_{j=0}^{n-1} W_{s_j}^2\\ &= \sum_{j=0}^{n-1}(W_{s_j} - W_{s_{j+1}})W_{s_{j+1}} + \sum_{j=0}^{n-1} W_{s_{j+1}}^2 - \sum_{j=0}^{n-1} W_{s_j}^2\\ &= \sum_{j=0}^{n-1}(W_{s_j} - W_{s_{j+1}})(\Delta W)_j + \sum_{j=0}^{n-1} W_{s_j}(W_{s_j} - W_{s_{j+1}}) + W_1^2\\ &= -\sum_{j=0}^{n-1}[(\Delta W)_j]^2 - \int_0^1 H_n(s)\,dW_s + W_1^2.\end{aligned}$$

Thus $2\int_0^1 H_n(s)\,dW_s = -\sum_{j=0}^{n-1}[(\Delta W)_j]^2 + W_1^2$. *In Chapter 2, we showed that the quadratic variation* $\sum_{j=0}^{n-1}[(\Delta W)_j]^2$ *converges in probability to one as* $n \to \infty$. *Besides,*

$$E\left(\sum_{j=0}^{n-1}\left[(\Delta W)_j\right]^2\right)^2 = nE\left[(\Delta W)_1\right]^2 + \binom{n}{2}\left(E\left[(\Delta W)_1\right]^2\right)^2$$

since $\{(\Delta W)_j : j = 1, 2, \ldots, n\}$ *are independent normal random variables. Hence,* $E(\sum_{j=0}^{n-1}[(\Delta W)_j]^2)^2 \le 4$. *Therefore,* $\sum_{j=0}^{n-1}[(\Delta W)_j]^2$ *converges in* $L^2(P)$ *to* 1, *which allows us to obtain* $2\int_0^1 H_n(s)\,dW_s \xrightarrow{L^2(P)} -1 + W_1^2$. *Thus,*

$$\int_0^1 W_s\,dW_s = \frac{W_1^2}{2} - \frac{1}{2}.$$

Remark 5.1.1 One can easily infer from the above example that the Riemann sum $\sum_{j=0}^{n-1} W_{s_{j+1}}(\Delta W)_j \xrightarrow{L^2(P)} \frac{W_1^2}{2} + \frac{1}{2}$. Thus, in the present context, the limit of Riemann sums vary with the choice of evaluation points for the integrand.

5.2 Properties of the Integral

In this section, the essential properties of the stochastic integral are listed along with complete proofs. Some of the proofs hinge on the martingale inequalities presented in Chapter 3. Let $\{W_t, \mathcal{F}_t\}$ be a Wiener martingale where $\{\mathcal{F}_t\}$ satisfies the usual conditions. Let $H \in \mathcal{M}_W^2[0, T]$.

1. $E\int_0^t H_s\,dW_s = 0$. for all $t \ge 0$. This property readily follows from the martingale property of the integral.
2. The process $\{\int_0^t H_s\,dW_s : t \ge 0\}$ is an $\mathcal{F}_t$-martingale with continuous paths.

Proof It suffices to show continuity of the sample paths on $[0, T]$ for any $T > 0$. If H is an L^2-step process, by Equation (5.1.1), the path continuity of the integral is quite obvious. Hence, from the Doob martingale inequalities,

$$P\left\{\sup_{0\leq t\leq T}\left|\int_0^t H_s\,dW_s\right| > a\right\} \leq \frac{1}{a^2}\int_0^T EH^2(s)\,ds \tag{5.2.1}$$

$$E\left\{\sup_{0\leq t\leq T}\left|\int_0^t H_s\,dW_s\right|^2\right\} \leq 4\int_0^T EH^2(s)\,ds \tag{5.2.2}$$

For a general H, let H_n be a sequence of L^2-step functions such that $\lim_{n\to\infty} E\int_0^T |H_n(s) - H_s|^2\,ds = 0$. Choose a subsequence $\{n_k\}$ such that $E\int_0^T |H_{n_k}(s) - H_s|^2\,ds \leq 1/2^k$. Then, $E\int_0^T |H_{n_{k+1}}(s) - H_{n_k}(s)|^2\,ds \leq 3/2^k$. By Equation (5.2.1) applied to the L^2-step process $H_{n_{k+1}} - H_{n_k}$,

$$P\left\{\sup_{0\leq t\leq T}\left|\int_0^t H_{n_{k+1}}(s)\,dW_s - \int_0^t H_{n_k}(s)\,dW_s\right| > \frac{1}{k^2}\right\} \leq \frac{3k^4}{2^k}.$$

Using the Borel-Cantelli lemma, the series

$$\sum_{k=1}^{\infty}\left(\int_0^t H_{n_{k+1}}(s)\,dW_s - \int_0^t H_{n_k}(s)\,dW_s\right)$$

converges almost surely uniformly in t. For any given ω in an almost sure set, there exists an N such that for all $n \geq m \geq N$, we have

$$\sup_{0\leq t\leq T}\left|\sum_{k=m}^{n}\left(\int_0^t H_{n_{k+1}}(s)\,dW_s - \int_0^t H_{n_k}(s)\,dW_s\right)\right| \leq \sum_{k=m}^{n}\frac{1}{k^2},$$

which tends to 0 as $m \to \infty$. In other words, the series

$$\int_0^t H_{n_1}(s)\,dW_s + \sum_{k=1}^{\infty}\left(\int_0^t H_{n_{k+1}}(s)\,dW_s - \int_0^t H_{n_k}(s)\,dW_s\right)$$

converges uniformly with respect to t in $[0, T]$ a.s.. Therefore, the sum of this series has continuous paths. By the choice of H_n, it is a version of the stochastic integral $\int_0^t H_s\,dW_s$. ■

3. For any $a > 0, p > 1$, and $T > 0$,

$$P\left\{\sup_{0\le t\le T}\left|\int_0^t H_s\,dW_s\right| > a\right\} \le \frac{1}{a^2}\int_0^T EH^2(s)\,ds$$

$$E\left\{\sup_{0\le t\le T}\left|\int_0^t H_s\,dW_s\right|^2\right\} \le 4\int_0^T EH^2(s)\,ds.$$

Since the process $\{\int_0^t H_s\,dW_s : 0 \le t \le T\}$ is a continuous martingale, $\sup_{0\le t\le T}|\int_0^t H_s\,dW_s|$ is a random variable. Hence, the inequalities follow from the martingale property of stochastic integrals.

4. For any $s \le t$, $E((\int_0^t H_u\,dW_u - \int_0^s H_u\,dW_u)^2 \mid \mathcal{F}_s) = E(\int_s^t H_u^2\,du \mid \mathcal{F}_s)$. Equivalently, the process $(\int_0^t H_u\,dW_u)^2 - \int_0^t H_u^2\,du$ is a $\mathcal{F}_t$-martingale.

Proof Let H be an L^2-step function as in Step 1 of Section 5.1. Let

$$0 = t_0 < t_1 \cdots < t_{m-1} \le s < t_m \cdots < t_n = t.$$

Then: $E((\int_0^t H_u\,dW_u - \int_0^s H_u\,dW_u)^2 \mid \mathcal{F}_s)$

$$= E\left(\left\{H_{t_{m-1}}(W_{t_m} - W_s) + \sum_{j=m}^{n-1} H_{t_j}(W_{t_{j+1}} - W_{t_j})\right\}^2 \middle| \mathcal{F}_s\right)$$

$$= E\left(\left\{H_{t_{m-1}}^2(W_{t_m} - W_s)^2 + \sum_{j=m}^{n-1} H_{t_j}^2(W_{t_{j+1}} - W_{t_j})^2\right\} \middle| \mathcal{F}_s\right)$$

$$= E\left(\int_s^t H_u^2 du \mid \mathcal{F}_s\right)$$

where we have used the independence of the increments of the Wiener process.

For a general H, let $\{H_n\}$ be a sequence of L^2-step processes such that $\lim_{n\to\infty} E\int_0^T (H_n(s) - H_s)^2\,ds = 0$. Then, by L^2-convergence of the corresponding stochastic integrals, we have for any $A \in \mathcal{F}_s$ that

$$E\left[1_A\left(\int_0^t H_u\,dW_u - \int_0^s H_u\,dW_u\right)^2\right] = \lim_{n\to\infty}\left[1_A\left(\int_0^t H_n(u)\,dW_u - \int_0^s H_n(u)\,dW_u\right)^2\right]$$

$$= \lim_{n\to\infty} E\left(1_A\int_s^t H_n^2(u)\,du\right)$$

since $1_A\int_s^t H_n(u)\,dW_u = \int_s^t 1_A H_n(u)\,dW_u$; continuing,

$$= E\left(1_A\int_s^t H_u^2\,du\right),$$

which finishes the proof. ■

5. Let σ and τ be two stopping times with $\sigma \leq \tau$. Then for all $t \geq 0$,

$$E\left(\int_0^{t\wedge\tau} H_s\,dW_s \mid \mathcal{F}_{t\wedge\sigma}\right) = \int_0^{t\wedge\sigma} H_s\,dW_s \text{ a.s.}$$

A simple observation from the above equality is that $\int_0^{t\wedge\sigma} H_s\,dW_s$ is $\mathcal{F}_{t\wedge\sigma}$-measurable. The property is a direct consequence of the Doob optional sampling theorem.

6. Let σ and τ be as above. Then for any $t \geq 0$,

$$E\left(\left(\int_0^{t\wedge\tau} H_s\,dW_s - \int_0^{t\wedge\sigma} H_s\,dW_s\right)^2 \mid \mathcal{F}_{t\wedge\sigma}\right) = E\left(\int_{t\wedge\sigma}^{t\wedge\tau} H_s^2\,ds \mid \mathcal{F}_{t\wedge\sigma}\right) \text{ a.s.}$$

Proof From Property 4, it follows that $(\int_0^t H_u\,dW_u)^2 - \int_0^t H_u^2\,du : t \leq T\}$ is a martingale. We will call this martingale $\{M_t\}$. By the Doob optional sampling theorem,

$$E\left(\left(\int_0^{t\wedge\tau} H_s\,dW_s\right)^2 \mid \mathcal{F}_{t\wedge\sigma}\right) = M_{t\wedge\sigma} + E\left(\int_0^{t\wedge\tau} H_s^2\,ds \mid \mathcal{F}_{t\wedge\sigma}\right). \quad (5.2.3)$$

Expanding $E((\int_0^{t\wedge\tau} H_s\,dW_s - \int_0^{t\wedge\sigma} H_s\,dW_s)^2 \mid \mathcal{F}_{t\wedge\sigma})$, we get

$$E\left(\left(\int_0^{t\wedge\tau} H_s\,dW_s\right)^2 + \left(\int_0^{t\wedge\sigma} H_s\,dW_s\right)^2 - 2\left(\int_0^{t\wedge\sigma} H_s\,dW_s\right)\left(\int_0^{t\wedge\tau} H_s\,dW_s\right) \middle| \mathcal{F}_{t\wedge\sigma}\right).$$

The first term in the above display equals $M_{t\wedge\sigma} + E(\int_0^{t\wedge\tau} H_s^2\,ds \mid \mathcal{F}_{t\wedge\sigma})$ by Equation (5.2.3). The second term is equal to $(\int_0^{t\wedge\sigma} H_s\,dW_s)^2$, and

$$E\left[\left(\int_0^{t\wedge\sigma} H_s\,dW_s\right)\left(\int_0^{t\wedge\tau} H_s\,dW_s\right) \middle| \mathcal{F}_{t\wedge\sigma}\right] = \left(\int_0^{t\wedge\sigma} H_s\,dW_s\right)^2.$$

Using these, one arrives at

$$E\left(\left(\int_0^{t\wedge\tau} H_s\,dW_s - \int_0^{t\wedge\sigma} H_s\,dW_s\right)^2 \mid \mathcal{F}_{t\wedge\sigma}\right) = E\left(\int_{t\wedge\sigma}^{t\wedge\tau} H_s^2\,ds \mid \mathcal{F}_{t\wedge\sigma}\right).$$

■

7. Let σ and τ be as above. Let $G \in M_W^2[0,T]$. Then for any $t \geq 0$,

$$E\left(\left(\int_0^{t\wedge\tau} H_u\,dW_u - \int_0^{t\wedge\sigma} H_u\,dW_u\right)\left(\int_0^{t\wedge\tau} G_u\,dW_u - \int_0^{t\wedge\sigma} G_u\,dW_u\right) \mid \mathcal{F}_{t\wedge\sigma}\right)$$
$$= E\left(\int_{t\wedge\sigma}^{t\wedge\tau} H_u G_u\,du \mid \mathcal{F}_{t\wedge\sigma}\right) \text{ a.s.}$$

In particular, by taking $\tau = t, \sigma = s$ and $s \leq t$, the process

$$\left(\int_0^t H_u\, dW_u\right)\left(\int_0^t G_u\, dW_u\right) - \int_0^t H_u G_u\, du$$

is a $\mathcal{F}_t$-martingale.

Proof $E((\int_{t\wedge\sigma}^{t\wedge\tau}(H_s - G_s)\, dW_s)^2 \mid \mathcal{F}_{t\wedge\sigma}) = E(\int_{t\wedge\sigma}^{t\wedge\tau}(H_s - G_s)^2\, ds|\mathcal{F}_{t\wedge\sigma})$ by Property 6. On the other hand, expanding the left side above, and using Property 6 for the square terms, we get the above result. ■

8. For any stopping time τ, and $t \geq 0$, $\int_0^{t\wedge\tau} H_s\, dW_s = \int_0^t 1_{\{s\leq\tau\}} H_s\, dW_s$ a.s.

   Proof Consider $E(\int_0^{t\wedge\tau} H_s\, dW_s - \int_0^t 1_{\{s\leq\tau\}} H_s\, dW_s)^2$, which upon expansion

$$\begin{aligned} &= E\left[\int_0^{t\wedge\tau} H_s^2\, ds\right] + E\left[\int_0^{t\wedge\tau} 1_{\{s\leq\tau\}} H_s^2\, ds\right] \\ &\quad - 2E\left[\left(\int_0^{t\wedge\tau} H_s\, dW_s\right)\left(\int_0^t 1_{\{s\leq\tau\}} H_s\, dW_s\right)\right] \\ &= 2E\left[\int_0^{t\wedge\tau} H_s^2\, ds\right] - 2E\left[\left(\int_0^{t\wedge\tau} H_s\, dW_s\right)\left(\int_0^t 1_{\{s\leq\tau\}} H_s\, dW_s\right)\right]. \end{aligned}$$

   We can write the term $E[(\int_0^{t\wedge\tau} H_s\, dW_s)(\int_0^t 1_{\{s\leq\tau\}} H_s\, dW_s)]$ as

$$\begin{aligned} &= E\left[\left(\int_0^{t\wedge\tau} H_s\, dW_s\right)\left(\int_0^{t\wedge\tau} 1_{\{s\leq\tau\}} H_s\, dW_s + \int_{t\wedge\tau}^t 1_{\{s\leq\tau\}} H_s\, dW_s\right)\right] \\ &= E\left(\int_0^{t\wedge\tau} H_s^2 1_{\{s\leq\tau\}}\, ds\right) + E\left[E\left(\int_0^{t\wedge\tau} H_s\, dW_s\right)\left(\int_{t\wedge\tau}^t 1_{\{s\leq\tau\}} H_s\, dW_s\right)\Big|\mathcal{F}_{t\wedge\tau}\right]. \end{aligned}$$

   by Property 7; continuing, the above term

$$\begin{aligned} &= E\int_0^{t\wedge\tau} H_s^2\, ds + E\left[\left(\int_0^{t\wedge\tau} H_s\, dW_s\right) E\left(\int_{t\wedge\tau}^t 1_{\{s\leq\tau\}} H_s\, dW_s|\mathcal{F}_{t\wedge\tau}\right)\right] \\ &= E\int_0^{t\wedge\tau} H_s^2\, ds \end{aligned}$$

   where we have used Property 5. Thus, $E(\int_0^{t\wedge\tau} H_s\, dW_s - \int_0^t 1_{\{s\leq\tau\}} H_s\, dW_s)^2 = 0$, which completes the proof. ■

9. Let H and G be any two processes in $\mathcal{M}_W^2[0, T]$. Define the set

$$A = \left\{\omega : H_t(\omega) = G_t(\omega)\ \forall\ 0 \leq t \leq T\right\}.$$

   Then $\int_0^T H_t\, dW_t = \int_0^T G_t\, dW_t$ a.s. on A.

Proof Define

$$\tau = \inf\{t : H_t \neq G_t\} \wedge T.$$

Let X_t denote $\int_0^t (H_s - G_s)\, dW_s$. By Property 8,

$$X_\tau = \int_0^T 1_{[0,\tau]}(s)(H_s - G_s)\, dW_s \text{ a.s.}$$

For each ω, by the definition of τ,

$$(H_s(\omega) - G_s(\omega))1_{[0,\tau(\omega)]}(s) = (H_{\tau(\omega)}(\omega) - G_{\tau(\omega)}(\omega))1_{\{\tau(\omega)\}}(s).$$

Hence, $\int_0^T (H_s - G_s)^2 1_{[0,\tau]}(s)\, ds = 0$. Hence, $EX_\tau^2 = 0$, which implies that $X_\tau = 0$ a.s. For $\omega \in A$, we have $\tau(\omega) = T$ so that $X_{\tau(\omega)}(\omega) = X_T(\omega)$ on A. Therefore $X_T = 0$ a.s. on A. ■

10. If $H \in \mathcal{L}_W^2$, then the process $\{\int_0^t H_s\, dW_s : 0 \leq t \geq 0\}$ is an $\mathcal{F}_t$-local martingale with continuous paths.

Proof The adapted process H_n defined in Step 5 of Section 5.1 satisfies $E\int_0^t H_n^2(s)\, ds < \infty$, so that by Property 2 proved above, $\{\int_0^t H_n(s)\, dW_s\}$ has continuous paths. We will denote it as $I(H_n)(t)$. Define the set

$$A_n = \left\{\omega : \int_0^T H_u^2\, du \leq n\right\}.$$

Then $A_n \uparrow A$ where $A = \cup_1^\infty A_n$. If $\omega \in A_m$, then $H_n(t,\omega) - H_m(t,\omega)$ for all $n \geq m$. Therefore, for almost all (a.a.) $\omega \in A_m$,

$$I(H_n)(t,\omega) = I(H_m)(t,\omega) \ \forall \ t \text{ and } \mathrm{n} \geq \mathrm{m}.$$

Hence, for all $\omega \in A$, that is, for a.a. ω, $\lim_{n\to\infty} I(H_n)(t,\omega)$ exists. Define for any $t \geq 0$

$$I(H)(t,\omega) = \begin{cases} \lim_{n\to\infty} I(H_n)(t,\omega) & \text{if } \omega \in A, \\ 0 & \text{otherwise.} \end{cases}$$

Then, $I(H)(t,\omega)$ has continuous paths for $\omega \in A$. Moreover, as in Step 5 in Section 5.1,

$$I(H_n)(t) \to \int_0^t H_s\, dW_s \text{ as } n \to \infty,$$

where limit is in the sense of convergence in probability. Therefore, for each $t \geq 0$,

$$\int_0^t H_s\, dW_s = I(H)(t) \text{ a.s.}$$

The existence of a continuous version of $\int_0^t H_s\, dW_s$ has thus been proved. ■

11. **(Dominated convergence theorem)** Let $\{H_n\}$ be a sequence of processes in $\mathcal{L}_W^2$. Let G be another process in $\mathcal{L}_W^2$ such that $H_n(t) \leq G(t)$ for all $n \in \mathbb{N}$ and $t \geq 0$. Suppose that $H_n(t) \to H(t)$ a.s. for each t as $n \to \infty$. Then $\int_0^t H_n(s)\, dW_s \to \int_0^t H(s)\, dW_s$ uniformly on compact time intervals in probability.

Proof Pick any finite time interval $[0, T]$. Then $\int_0^T (H_n(s) - H(s))^2\, ds \to 0$ a.s. by the Lebesgue dominated convergence theorem. Fix any $\epsilon > 0$. Define, for each n,

$$\tau_n := T \wedge \inf\{t \geq 0 : \int_0^t (H_n(s) - H(s))^2\, ds > \epsilon^4\}.$$

Let $M_n(t)$ denote $\int_0^{t\wedge\tau_n} (H_n(s) - H(s))\, dW_s$. By the Fatou lemma and Doob's inequality,

$$E[\sup_{0\leq t\leq \tau_n} M_n^2(t)] \leq 4\epsilon^4.$$

Define the set $A := \{\sup_{0\leq t\leq T} |M_n(t)| > \epsilon\}$. Then,

$$\begin{aligned} P(A) &= P\{A \cap (\tau_n < T)\} + P\{A \cap (\tau_n = T)\} \\ &\leq P\{\tau_n < T\} + P\left\{\sup_{0\leq t\leq \tau_n} |M_n(t)| > \epsilon\right\} \\ &\leq P\left\{\int_0^T (H_n(s) - H(s))^2\, ds > \epsilon^4\right\} + \frac{4\epsilon^4}{\epsilon^2} \end{aligned}$$

by the Chebyshev inequality. Letting $n \to \infty$ and then $\epsilon \to 0$, the proof is completed. ■

An extension of stochastic integrals. Let $H \in \mathcal{M}_W^2$. Let us denote the stochastic integral $\int_0^t H_s\, dW_s$ by M_t. Suppose that $\{G_s\}$ is another adapted process satisfying $GH \in \mathcal{M}_W^2$. Then the stochastic integral $\int_0^t G_s H_s\, dW_s$ is well defined, and we denote it as $\int_0^t G_s\, dM_s$. It is a square integrable martingale.

If the above conditions on H and G are replaced by $H \in \mathcal{L}_W^2$ and $GH \in \mathcal{L}_W^2$, then the stochastic integral $\int_0^t G_s\, dM_s$ is defined by $\int_0^t G_s H_s\, dW_s$. The process is a local martingale.

5.3 Vector-valued Processes

We turn our attention to an extension of stochastic integrals to vector-valued processes.

Definition 5.3.1 *Let $W = (W_1, \ldots, W_n)$ be an n-dimensional $\mathcal{F}_t$-Wiener martingale, that is, each W_j is an independent $\mathcal{F}_t$-Wiener martingale. Suppose $H = (H_1, \ldots, H_n)$ is an adapted n-dimensional process satisfying $E\int_0^t |H_s|^2\, ds < \infty$ for all $t \geq 0$. Then, $\int_0^t H_s\, dW_s$ is defined by $\sum_{j=1}^n \int_0^t H_j(s)\, dW_j(s)$.*

Let us denote the stochastic integral defined above by M_t. Then $\{M_t\}$ is a continuous L^2-martingale. We can denote $\int_0^t |H_s|^2\, ds$ by $\langle M\rangle_t$, so that by Section 5 of Chapter 3, $E[M_t^2] = E\langle M\rangle_t$. By Property 5, $M_t^2 - \langle M\rangle_t$ is a martingale.

If $\{N_t\}$ is another such L^2-martingale given by $N_t = \int_0^t G_s\, dW_s$, then we denote as $\langle M, N\rangle_t$ the process $\int_0^t (H_s, G_s)\, ds$ where (H_s, G_s) stands for the inner product in dimension n. Clearly, $\{M_t N_t - \langle M, N\rangle_t\}$ is a martingale.

If the process $H = (H_1, \ldots, H_n)$ is adapted and satisfies the weaker condition that $\int_0^t |H_s|^2\, ds < \infty$ a.s. for all $t \geq 0$, then the integral $\int_0^t H_s\, dW_s$ is a continuous $\mathcal{F}_t$-local martingale.

A process $M_t = (M_1(t), \ldots, M_d(t))$ taking values in $\mathbb{R}^d$ is called a d-dimensional martingale with respect to a filtration $\mathcal{F}_t$ if $(M_j(t), \mathcal{F}_t)$ is a martingale for each $j = 1, \ldots, d$. We assume that the filtration $(\mathcal{F}_t)$ satisfies the usual conditions.

Let $\mathcal{A}$ denote the class of d-dimensional, adapted processes $\{A_t\}$ where $A_t = (A_1(t), \ldots, A_d(t))$ satisfies the following conditions:

(i) $A_j(0) = 0$ a.s.

(ii) The function $t \to A_j(t, \omega)$ is, for a.a. ω, continuous and of bounded variation in every finite interval.

(iii) $E|A_j|_t < \infty$ for every t, where $|A_j|_t$ is the total variation of $A_j(s)$ in the interval $[0, t]$.

By $\mathcal{A}_{\text{loc}}$, we shall mean the class of all d-dimensional processes A satisfying the conditions (i) and (ii) above.

Definition 5.3.2 *An $\mathcal{F}_t$-adapted, d-dimensional process $\{X_t\}$ is called a continuous* **semimartingale** *if X has continuous paths and has the form*

$$X_t = X_0 + M_t + A_t \tag{5.3.1}$$

for all t a.s., where (i) $E|X_0| < \infty$, (ii) M is a continuous $\mathcal{F}_t$-martingale with $M_0 = 0$ a.s., and (iii) $A \in \mathcal{A}$.

If in the decomposition (5.3.1), M is a continuous local martingale and A belongs to $\mathcal{A}_{\text{loc}}$, then X is called a continuous **local semimartingale**.

5.4 The Itô Formula

We proceed to the derivation of the Itô formula, which is the change of variable formula in stochastic analysis. To understand this result, let us first consider a deterministic function u defined on $\mathbb{R}^+$ which is differentiable. Let f be any function in $C_b^1(\mathbb{R})$. Then, by the fundamental theorem of calculus,

$$\begin{aligned} f(u(t)) - f(u(0)) &= \int_0^t (f \circ u)'(s)\, ds \\ &= \int_0^t f'(u(s))u'(s)\, ds \\ &= \int_0^t f'(u(s))\, du(s). \end{aligned}$$

Thus, $f(u(t)) = f(u(0)) + \int_0^t f'(u(s))\, du(s)$. The equality continues to hold if u is simply a continuous function with bounded variation on each finite time interval. Since Brownian paths are of unbounded variation on each finite time interval, the function u cannot be taken as a Brownian motion. The change of variable formula for the Brownian motion reads as follows and is a particular application of the Itô formula

$$f(W_t) = f(W_0) + \int_0^t f'(W_s)\, dW_s + \frac{1}{2}\int_0^t f''(W_s)\, ds$$

where f is assumed to be in $C_b^2(\mathbb{R})$. The formula is stated in terms of a stochastic integral, and the extra term on the right side arises from the finiteness of the quadratic variation of W. The boundedness of f and its derivatives is not needed and is assumed in this discussion merely for simplicity.

We will first prove the Itô formula for a particular class of continuous semimartingales described below.

In the semimartingale decomposition of X, the martingale part of each coordinate is given by a stochastic integral as in Definition 5.3.1. In otherwords, we will assume that there is an underlying n-dimensional Wiener martingale $(W_t, \mathcal{F}_t : t \geq 0)$. Then, $M = (M_1, \ldots, M_d)$ is such that each M_i is a square integrable $\mathcal{F}_t$-martingale of the form $M_i = \sum_{j=1}^n \int_0^t H_{i,j}(s)\, dW_j(s)$ where $H_i = (H_{i,1}, \ldots, H_{i,n})$ is an n-dimensional adapted process satisfying $E\int_0^t |H_i|^2(s)\, ds < \infty$ for each $t > 0$.

Theorem 5.4.1 *Let $X_t = X_0 + M_t + A_t$ be a d-dimensional, continuous semimartingale as described above. Let f be a real-valued function such that $f \in C_b^2(\mathbb{R}^d)$. Then,*

$$\begin{aligned} f(X_t) = f(X_0) &+ \sum_{j=1}^d \int_0^t \frac{\partial f}{\partial x_j}(X_s)\, dM_j(s) + \sum_{j=1}^d \int_0^t \frac{\partial f}{\partial x_j}(X_s)\, dA_j(s) \\ &+ \frac{1}{2}\sum_{i,j=1}^d \int_0^t \frac{\partial^2 f}{\partial x_i \partial x_j}(X_s)\, d\langle M_i, M_j\rangle_s. \end{aligned} \tag{5.4.1}$$

Proof We will take $d = 1$, since the idea of the proof remains the same in the multidimensional case. We need to show that

$$f(X_t) = f(X_0) + \int_0^t f'(X_s)\, dM(s) + \int_0^t f'(X_s)\, dA(s) + \frac{1}{2}\int_0^t f''(X_s)\, d\langle M\rangle_s. \qquad (5.4.2)$$

Step 1 First, note that the first integral on the right side of Equation (5.4.2) is well defined since f' is bounded. Hence, it is a continuous martingale. Define for each integer $n \geq 1$ the stopping time

$$\tau_n = \begin{cases} \inf\{t : |M_t| > n \text{ or } |A|_t > n \text{ or } \langle M\rangle_t > n\} & \text{if } |X_0| \leq n; \\ 0 & \text{if } |X_0| > n. \end{cases}$$

Since $E\langle M\rangle_t < \infty$, $E[\sup_{0\leq s\leq t}|M_s|] < \infty$ for any $t > 0$. This, along with the integrability of X_0, and $|A|_t$ yield that $\tau_n \uparrow \infty$ a.s. If we prove the equality (5.4.2) till time τ_n a.s. on the set $\{\omega : \tau_n(\omega) > 0\}$, then by letting $n \to \infty$, we get (5.4.2) in full. Therefore, without loss of generality, we may assume that $X_0, M_t, |A|_t$, and $\langle M\rangle_t$ are all bounded by a constant, say, K for all $\omega \in \Omega$ and $t \geq 0$.

Step 2 For any $t \geq 0$, let $\pi = \{t_0 = 0, t_1, \ldots, t_n = t\}$ be a partition of the interval $[0, t]$. Then,

$$\begin{aligned} f(X_t) - f(X_0) &= \sum_{j=1}^n [f(X_{t_j}) - f(X_{t_{j-1}})] \\ &= \sum_{j=1}^n f'(X_{t_{j-1}})(X_{t_j} - X_{t_{j-1}}) + \frac{1}{2}\sum_{j=1}^n f''(\eta_j)(X_{t_j} - X_{t_{j-1}})^2, \end{aligned} \qquad (5.4.3)$$

where η_j is between $X_{t_{j-1}}$ and X_{t_j}. The first term of (5.4.3) equals

$$\begin{aligned} &= \sum_{j=1}^n f'(X_{t_{j-1}})(A_{t_j} - A_{t_{j-1}}) + \sum_{j=1}^n f'(X_{t_{j-1}})(M_{t_j} - M_{t_{j-1}}) \\ &= I_1^\pi + I_2^\pi. \end{aligned}$$

Define $|\pi| = \max_j(t_j - t_{j-1})$. As $|\pi| \to 0$, $I_1^\pi \to \int_0^t f'(X_s)\, dA_s$ a.s. The process $f'(X)$ is a bounded, adapted, continuous process. Therefore, by using Step 2 in the construction of the stochastic integral in Section 5.1, we can conclude that $I_2^\pi \to \int_0^t f'(X_s)\, dM_s$ in $L^2(P)$ as $|\pi| \to 0$. Thus, the first term on the right side of (5.4.3) converges to $\int_0^t f'(X_s)\, dX_s$.

Step 3 To find the limit of the second term on the right side of (5.4.3), consider $\sum_{j=1}^{n} f''(\eta_j)(X_{t_j} - X_{t_{j-1}})^2$, which

$$
\begin{aligned}
&= \sum_{j=1}^{n} f''(\eta_j)(A_{t_j} - A_{t_{j-1}})^2 + 2\sum_{j=1}^{n} f''(\eta_j)(A_{t_j} - A_{t_{j-1}})(M_{t_j} - M_{t_{j-1}}) \\
&\quad + \sum_{j=1}^{n} f''(\eta_j)(M_{t_j} - M_{t_{j-1}})^2 \\
&= J_1^\pi + J_2^\pi + J_3^\pi .
\end{aligned}
$$

Since $|A|_t \leq K$, we get

$$
|J_1^\pi| + |J_2^\pi| \leq 2K \, \|f''\|_\infty (\max_j |A_{t_j} - A_{t_{j-1}}| + \max_j |M_{t_j} - M_{t_{j-1}}|),
$$

which converges to 0 a.s. as $|\pi| \to 0$ by the path continuity of A and M. Thus, $J_1^\pi + J_2^\pi$ tends to 0 as $|\pi| \to 0$.

It remains to prove that $J_3^\pi \to \int_0^t f''(X_s)\, d\langle M\rangle_s$ in some sense as $|\pi| \to 0$. We will establish the $L^1(P)$-convergence of J_3^π. For this, we need a bound which is proved in the next step.

Step 4 We will show that

$$
E\left\{\left[\sum_{j=1}^{n} (M_{t_j} - M_{t_{j-1}})^2\right]^2\right\} \leq 6K^3. \tag{5.4.4}
$$

Indeed, upon expansion, the left side of equation (5.4.4)

$$
\begin{aligned}
&= E\left(\sum_{j=1}^{n} (M_{t_j} - M_{t_{j-1}})^4\right) + 2E\sum_{i=1}^{n-1}\sum_{j=i+1}^{n} (M_{t_i} - M_{t_{i-1}})^2 (M_{t_j} - M_{t_{j-1}})^2 \\
&= T_1 + T_2.
\end{aligned}
$$

Note that by using the boundedness of M and $\langle M\rangle$,

$$
\begin{aligned}
T_1 &\leq 4K^2 E \sum_{j=1}^{n} (M_{t_j} - M_{t_{j-1}})^2 \\
&\leq 4K^2 E \sum_{j=1}^{n} (\langle M\rangle_{t_j} - \langle M\rangle_{t_{j-1}}) \\
&\leq 4K^3.
\end{aligned}
$$

Using the simple observation that $E(M_{t_{j-1}} M_{t_j} \mid \mathcal{F}_{t_{j-1}}) = M^2_{t_{j-1}}$, one obtains

$$T_2 = 2E\left(\sum_{i=1}^{n-1}(M_{t_i} - M_{t_{i-1}})^2 E\left\{\sum_{j=i+1}^{n}(M_{t_j} - M_{t_{j-1}})^2 \,|\mathcal{F}_{t_{i-1}}\right\}\right)$$

$$= 2E\left(\sum_{i=1}^{n-1}(M_{t_i} - M_{t_{i-1}})^2 E\left\{\sum_{j=i+1}^{n}(M^2_{t_j} - M^2_{t_{j-1}}) \,|\mathcal{F}_{t_{i-1}}\right\}\right)$$

since $E[(M_{t_j} - M_{t_{j-1}})^2 \,|\mathcal{F}_{t_{i-1}}] = E[E\{(M_{t_j} - M_{t_{j-1}})^2 \,|\mathcal{F}_{t_{j-1}}\} \mid \mathcal{F}_{t_{i-1}}]$; continuing,

$$T_2 \leq 2E\sum_{i=1}^{n-1}(M_{t_i} - M_{t_{i-1}})^2 E(M_t^2 \mid \mathcal{F}_{t_{j-1}})$$

$$\leq 2K^2 E\sum_{i=1}^{n-1}(M_{t_i} - M_{t_{i-1}})^2)$$

$$\leq 2K^3.$$

Step 5 We introduce $J_4^\pi = \frac{1}{2}\sum_{j=1}^{n} f''(X_{t_{j-1}})(M_{t_j} - M_{t_{j-1}})^2$ as an approximation to J_3^π. In fact, using the estimate obtained in the previous step and the Cauchy-Schwarz inequality,

$$E\left|J_3^\pi - J_4^\pi\right| \leq \frac{1}{2}\sqrt{6K^3}\left(E\max_j \left|f''(\eta_j) - f''\left(X_{t_{j-1}}\right)\right|^2\right)^{1/2}.$$

This bound tends to zero as $|\pi| \to 0$ by the bounded convergence theorem, which applies since f'' is continuous and bounded. Thus, $J_3^\pi - J_4^\pi \to 0$ in $L^1(P)$ as $|\pi| \to 0$.

Step 6 Define $J_5^\pi = \frac{1}{2}\sum_{j=1}^{n} f''(X_{t_{j-1}})(\langle M\rangle_{t_j} - \langle M\rangle_{t_{j-1}})$. We claim that $E\left[|J_4^\pi - J_5^\pi|^2\right] \to 0$ as $|\pi| \to 0$. Indeed, consider $E[(J_4^\pi - J_5^\pi)^2]$, which

$$= \frac{1}{4}E\left(\left[\sum_{j=1}^{n} f''\left(X_{t_{j-1}}\right)\left\{\left(M_{t_j} - M_{t_{j-1}}\right)^2 - \left(\langle M\rangle_{t_j} - \langle M\rangle_{t_{j-1}}\right)\right\}\right]^2\right)$$

$$= \frac{1}{4}E\left[\sum_{j=1}^{n}\left(f''\left(X_{t_{j-1}}\right)\right)^2\left\{\left(M_{t_j} - M_{t_{j-1}}\right)^2 - \left(\langle M\rangle_{t_j} - \langle M\rangle_{t_{j-1}}\right)\right\}^2\right]$$

since the cross-product terms vanish by the following observation: Let N_k denote $(M_{t_k} - M_{t_{k-1}})^2 - (\langle M\rangle_{t_k} - \langle M\rangle_{t_{k-1}})$ for any k. For any $i < j$,

$$E(N_i N_j) = E\left(N_i E\left\{N_j \mid \mathcal{F}_{t_{j-1}}\right\}\right) = 0,$$

since the above conditional expectation is zero.

Thus we obtain the bound

$$
\begin{aligned}
E\left[(J_4^\pi - J_5^\pi)^2\right] &\le \frac{1}{2}\|f''\|_\infty^2 E\left\{\sum_{j=1}^{n}\left[\left(M_{t_j}-M_{t_{j-1}}\right)^4 + \left(\langle M\rangle_{t_j} - \langle M\rangle_{t_{j-1}}\right)^2\right]\right\} \\
&\le \frac{1}{2}\|f''\|_\infty^2 \left\{E\left(\max_j \left(M_{t_j}-M_{t_{j-1}}\right)^2 \sum_{j=1}^{n}\left[\left(M_{t_j}-M_{t_{j-1}}\right)^2\right]\right)\right. \\
&\quad \left. + E\left[\langle M\rangle_t \max_j \left(\langle M\rangle_{t_j} - \langle M\rangle_{t_{j-1}}\right)\right]\right\}. \qquad (5.4.5)
\end{aligned}
$$

By the Cauchy-Schwarz inequality and the bound obtained in Step 4, the first term on the right side can be bounded above by

$$
\begin{aligned}
&\frac{1}{2}\|f''\|_\infty^2 \left[E\left(\max_j (M_{t_j}-M_{t_{j-1}})^4\right)\right]^{1/2}\left[E\left(\sum_{j=1}^{n}\left(M_{t_j}-M_{t_{j-1}}\right)^2\right)^2\right]^{1/2} \\
&\le \frac{1}{2}\|f''\|_\infty^2 6K^3 \left[E(\max_j \left(M_{t_j}-M_{t_{j-1}}\right)^4)\right]^{1/2},
\end{aligned}
$$

which converges to 0 by the bounded convergence theorem. The last term on the right side of (5.4.5) converges to zero by the bounded convergence theorem.

Step 7 By the definition of J_5^π, it follows that $J_5^\pi \to \frac{1}{2}\int_0^t f''(X_s)d\langle M\rangle_s$ a.s. Therefore, by Steps 5 and 6, if $\{\pi_n\}$ is a sequence of partitions of $[0,t]$, with $|\pi_n| \to 0$, then there exists a subsequence $\{\pi_{n_j}\}$ such that

$$
J_3^{\pi_{n_j}} \to \frac{1}{2}\int_0^t f''(X_s)\, d\langle M\rangle_s,
$$

$I_1^{\pi_{n_k}} \to \int_0^t f'(X_s)\, dA_s$ a.s., and $I_2^{\pi_{n_k}} \to \int_0^t f'(X_s)\, dM_s$ a.s.

Thus, Equation (5.4.2) holds for a fixed time t. Since both sides are continuous in t a.s., the theorem is proved for all $t \ge 0$ a.s. ■

The following theorem is a particular case of the Itô formula given in Theorem (5.4.1), in which we replace the d-dimensional continuous semimartingale X_t by a $(d+1)$-dimensional continuous semimartingale with the following form:

(i) $X_t = (V_t, M_t)$ where $\{V_t\}$ is a real-valued, $\mathcal{F}_t$-adapted process with $E|V_0| < \infty$, and $\{V_t - V_0\} \in \mathcal{A}$.

(ii) M_t is a d-dimensional square integrable $\mathcal{F}_t$-martingale as in Theorem (5.4.1).

Theorem 5.4.2 *Let X be a $(d+1)$-dimensional, continuous semimartingale as described above. Let f be a real-valued function such that $f \in C_b^2(\mathbb{R}^{d+1})$. Then,*

$$f(V_t, M_t) = f(V_0, M_0) + \int_0^t \frac{\partial f}{\partial v}(V_s, M_s)\,dV_s + \sum_{j=1}^d \int_0^t \frac{\partial f}{\partial x_j}(V_s, M_s)\,dM_j(s)$$
$$+ \frac{1}{2}\sum_{i,j=1}^d \int_0^t \frac{\partial^2 f}{\partial x_i \partial x_j}(V_s, M_s)\,d\langle M_i, M_j\rangle_s. \tag{5.4.6}$$

5.5 An Extension of the Itô Formula

A simple, useful extension of the Itô formula is given below for a function of a continuous *local* semimartingale X. The function and its first two derivatives need not be bounded. The time interval is taken to be $[0, T]$. In the semimartingale decomposition of X, we will assume that the local martingale part of each coordinate is given by a stochastic integral as given below: Suppose there is an underlying n-dimensional Wiener martingale $(W_t, \mathcal{F}_t : t \geq 0)$. Then, $M = (M_1, \ldots, M_d)$ is such that each M_i is a square integrable, $\mathcal{F}_t$-adapted local martingale of the form $M_i = \sum_{j=1}^n \int_0^t H_{i,j}(s)\,dW_j(s)$, where $H_i := (H_{i,1}, \ldots, H_{i,n})$ is an n-dimensional adapted process satisfying $\int_0^T |H_i|^2(s)\,ds < \infty$ a.s.

Theorem 5.5.1 *Let $X_t = X_0 + M_t + A_t$ be a d-dimensional, continuous local semimartingale as described above. Let f be a real-valued function such that $f \in C^2(\mathbb{R}^d)$. Then, for any $0 \leq t \leq T$,*

$$f(X_t) = f(X_0) + \sum_{j=1}^d \int_0^t \frac{\partial f}{\partial x_j}(X_s)\,dM_j(s) + \sum_{j=1}^d \int_0^t \frac{\partial f}{\partial x_j}(X_s)\,dA_j(s)$$
$$+ \frac{1}{2}\sum_{i,j=1}^d \int_0^t \frac{\partial^2 f}{\partial x_i \partial x_j}(X_s)\,d\langle M_i, M_j\rangle_s. \tag{5.5.1}$$

Proof The proof is given for $d = 1$, since the result in general can be proved along similar lines. Since $f(X_s), f'(X_s)$, and $f''(X_s)$ are continuous functions in s, each of the integrals $\int_0^T |f'(X_s)|\,d|A|_s$, $\int_0^T |f''(X_s)|\,d\langle M\rangle_s$, and $\int_0^T f'(X_s)^2\,d\langle M\rangle_s$, is finite a.s. Therefore, the right side of equation (5.5.1) has a meaning. For instance, the stochastic integral $\int_0^t f'(X_s)\,dM_s$ is a continuous local martingale.

Define for each integer $n \geq 1$ the stopping times

$$\sigma_n = T \wedge \inf\left\{t \leq T : |M_t| \geq n, |A|_t \geq n \text{ or } \int_0^t f'(X_s)^2\,d\langle M\rangle_t \geq n\right\}$$

and

$$S_n = \begin{cases} T & \text{if } |X_0| \leq n, \\ 0 & \text{otherwise.} \end{cases}$$

Let $\tau_n = \sigma_n \wedge S_n$. Define $X_0^n := X_0 I_{\{|X_0| \le n\}}$, $M_t^n := M_{t\wedge\tau_n}$, and $A_t^n := A_{t\wedge\tau_n}$. Let $X_t^n := X_0^n + M_t^n + A_t^n$.

Clearly, X_t^n is a continuous semimartingale with $|X_t^n| \le 3n$. Let U and V be bounded open intervals such that $[-3n, 3n] \subset U \subset \bar{U} \subset V$. Let φ be a $C_b^2(\mathbb{R})$ function such that

$$\varphi(x) = \begin{cases} 1 & \text{if } x \in \bar{U}; \\ 0 & \text{if } x \in V^c. \end{cases}$$

Define $g(x) = f(x)\varphi(x)$. It is clear that $g \in C_b^2(\mathbb{R})$. Moreover, for $x \in [-3n, 3n]$, $g(x), g'(x)$, and $g''(x)$ coincide with $f(x), f'(x)$, and $f''(x)$, respectively. The Itô formula given in Theorem 5.4.1 applied to the function g and the semimartingale X_t^n yields

$$f(X_t^n) = f(X_0^n) + \int_0^t f'(X_s^n)\, dM_s^n + \int_0^t f'(X_s^n)\, dA_s^n + \frac{1}{2}\int_0^t f''(X_s^n)\, d\langle M^n\rangle_s. \tag{5.5.2}$$

Since $X_t^n \to X_t$, and $X_0^n \to X_0$ a.s., we have $f(X_t^n) \to f(X_t)$, and $f(X_0^n) \to f(X_0)$. Since $\int_0^t f'(X_s^n)\, dM_s^n = \int_0^{t\wedge\tau_n} f'(X_s)\, dM_s = \int_0^t f'(X_s) 1_{\{s<\tau_n\}}\, dM_s$, and $\tau_n \uparrow T$ a.s. , we get $\int_0^t f'(X_s^n)\, dM_s^n \to \int_0^t f'(X_s)\, dM_s$ in probability as $n \to \infty$. The stochastic integral $\int_0^t f'(X_s)\, dM_s$ is a local martingale.

For almost all ω, there is an $N(\omega)$ such that for all $n \ge N(\omega)$,

$$\int_0^t f'(X_s^n)\, dA_s^n = \int_0^t f'(X_s)\, dA_s.$$

Hence, $\int_0^t f'(X_s^n)\, dA_s^n \to \int_0^t f'(X_s)\, dA_s$ a.s. as $n \to \infty$. Likewise, one can conclude that a.s.

$$\int_0^t f''(X_s^n)\, d\langle M^n\rangle_s \to \int_0^t f''(X_s)\, d\langle M\rangle_s.$$

Allowing $n \to \infty$ in (5.5.2), we thus obtain the Itô formula (5.5.1). ■

One can obtain a statement analogous to Theorem 5.4.2 when f is not necessarily a bounded function, $X_t = (V_t, M_t)$ with $E|V_0| < \infty$, $V_t - V_0 \in \mathcal{A}_{\text{loc}}$, M_t is a $\mathcal{F}_t$-adapted, d-dimensional local martingale as in the above theorem.

We illustrate the usefulness of the Itô formula with an example.

Example 5.5.2 *Let $\{W_t\}$ be a standard one-dimensional Wiener process and $(\mathcal{F}_t)$ be its natural filtration. A simple application of the Itô formula to the function $f(x) = x^2$ yields*

$$W_t^2 = 2\int_0^t W_s\, dW_s + t.$$

Hence, $\int_0^t W_s\,dW_s = \frac{W_t^2}{2} - \frac{t}{2}$. *The ease with which this equality is obtained should be contrasted with the method used in Example 5.1.4.*

Example 5.5.3 *In the context of the above example, we will apply the Itô formula for the function* $f(t,x) = \exp\{x - t/2\}$. *Denoting partial derivatives by subscripts, one has* $f(t,x) = f_x(t,x) = f_{xx}(t,x)$ *and* $f_t(t,x) = -1/2f(t,x)$. *Then,*

$$\begin{aligned} f(t,W_t) &= 1 + \int_0^t f_s(s,W_s)\,ds + \int_0^t f_x(s,W_s)\,dW_s + \frac{1}{2}\int_0^t f_{xx}(s,W_s)\,ds \\ &= 1 - \frac{1}{2}\int_0^t f(s,W_s)\,ds + \int_0^t f(s,W_s)\,dW_s + \frac{1}{2}\int_0^t f(s,W_s)\,ds \\ &= 1 + \int_0^t f(s,W_s)\,dW_s. \end{aligned}$$

We will call $\exp W_t - t/2$ *the* **stochastic exponential** *of* W *and denote it by* $\mathcal{E}(W)_t$. *Just as the exponential function* $u(t) = e^t$ *satisfies the equation* $u(t) = 1 + \int_0^t u(s)\,ds$, *the stochastic exponential* $Y_t = \mathcal{E}(W)_t$ *satisfies the equation* $Y_t = 1 + \int_0^t Y_s\,dW_s$.

The process $\mathcal{E}(W)_t$ *is an* L^2*-martingale since*

$$E\int_0^t f^2(s,W_s)\,ds = \int_0^t e^{-s}Ee^{2W_s}\,ds = \int_0^t e^s\,ds < \infty.$$

Likewise, for any fixed constant k, *the process* $\mathcal{E}(kW)_t$ *satisfies the equation* $Y_t = 1 + k\int_0^t Y_s\,dW_s$.

5.6 Applications of the Itô Formula

First, we present Lévy's characterization of a Brownian motion. The proof given below is due to Kunita and Watanabe and uses the Itô formula. In this result, we assume that the filtration $\mathcal{F}_t$ satisfies the usual hypotheses.

Theorem 5.6.1 (Lévy's characterization) *Let* M_t *be a n-dimensional continuous local martingale adapted to* $\mathcal{F}_t$. *Let* $M_0 = 0$, *and for each* t, $\langle M\rangle_t = t\,I$, *where* I *is the n-dimensional identity matrix. Then*

(i) M_t *is a Wiener process.*

(ii) For all $s < t$, $\sigma(M_v - M_u : s \le u \le v \le t)$ *is independent of* $\mathcal{F}_s$.

Proof Let $[0,T]$ be a given time interval. Let $\{\tau_n\}$ be a localizing sequence of stopping times for M_t such that $\tau_n \uparrow T$ a.s. We know that

$$E\left\{M_{t\wedge\tau_n} \mid \mathcal{F}_s\right\} = M_{s\wedge\tau_n} \quad \text{for each } n, \text{ and } s < t. \tag{5.6.1}$$

Since $\langle M^j\rangle_t = t$ for all t and $j = 1,2,\ldots,n$, we have $E(M_t^j)^2 < \infty$. Therefore, $\{M_t^j : 0 \le t \le T\}$ is a uniformly integrable family of random variables. This allows us to let $n \to \infty$ in (5.6.1) so that each M^j is a square integrable martingale on $[0, T]$ for any fixed T.

We are given that M_t has continuous paths and $M_0 = 0$. To prove (i), it therefore suffices to show that for all $u \in \mathbb{R}^d$ and $s < t$

$$E\left\{e^{i(u, M_t - M_s)} \mid \mathcal{F}_s\right\} = e^{-\frac{|u|^2}{2}(t-s)},$$

where $(u, M_t - M_s)$ stands for $\sum_{j=1}^n u_j (M_t^j - M_s^j)$ and $|u|^2 = \sum_{j=1}^n u_j^2$. Applying the Itô formula to the function $f(x) = e^{i(u,x)}$, we obtain

$$e^{i(u, M_t - M_s)} = 1 + i \sum_{j=1}^n u_j \int_s^t e^{i(u, M_r - M_s)}\, dM_r^j - \frac{|u|^2}{2} \int_s^t e^{i(u, M_r - M_s)}\, dr. \tag{5.6.2}$$

Since $\int_s^t e^{i(u, M_r - M_s)}\, dM_r^j$ as a process indexed by t is a martingale,

$$E\left\{\int_s^t e^{i(u, M_r - M_s)}\, dM_r^j \mid \mathcal{F}_s\right\} = 0 \text{ a.s.}$$

Therefore, integrating Equation (5.6.2) over any set $A \in \mathcal{F}_s$ with respect to P,

$$\int_A e^{i(u, M_t - M_s)}\, dP = P(A) - \frac{|u|^2}{2} \int_s^t \int_A e^{i(u, M_r - M_s)}\, dP\, dr.$$

Denoting $\int_A e^{i(u, M_t - M_s)}\, dP$ as $g(t)$, we can rewrite the above equation as

$$g(t) = P(A) - \frac{|u|^2}{2} \int_s^t g(r)\, dr.$$

Thus, $g(t) = P(A) e^{-\frac{|u|^2}{2}(t-s)}$; that is,

$$\int_A e^{i(u, M_t - M_s)}\, dP = \int_A e^{-\frac{|u|^2}{2}(t-s)}\, dP$$

for all $A \in \mathcal{F}_s$ so that

$$E\left\{e^{i(u, M_t - M_s)} \mid \mathcal{F}_s\right\} = e^{-\frac{|u|^2}{2}(t-s)}. \tag{5.6.3}$$

Thus, (i) is proved.

To prove part (ii), take any partition $\{s = a_0 < a_1 < \cdots < a_k = t\}$ of $[s, t]$ and k number of vectors $\{u^j\}$ in $\mathbb{R}^n$. Then,

$$\begin{aligned} E\left(\Pi_{j=1}^{k} e^{i(u^j, M_{a_j} - M_{a_{j-1}})} \mid \mathcal{F}_s\right) &= E\left(\Pi_{j=1}^{k-1} e^{i(u^j, M_{a_j} - M_{a_{j-1}})} E\left[e^{i(u^k, M_{a_k} - M_{a_{k-1}})} \mid \mathcal{F}_{a_{k-1}}\right] \mid \mathcal{F}_s\right) \\ &= e^{-\frac{|u^k|^2}{2}(a_k - a_{k-1})} E\left(\Pi_{j=1}^{k-1} e^{i(u^j, M_{a_j} - M_{a_{j-1}})} \mid \mathcal{F}_s\right) \\ &= \Pi_{j=1}^{k} e^{-\frac{|u^j|^2}{2}(a_j - a_{j-1})} \end{aligned}$$

by induction and Equation (5.6.3). The independence of $\sigma(M_v - M_u : s \le u \le v \le t)$ and $\mathcal{F}_s$ is thus established. ■

Remark 5.6.1 If the filtration $\mathcal{F}_t$ is not right-continuous, one can work with $\mathcal{F}_{t+}$ as the underlying filtration. For, a continuous local martingale adapted to $\mathcal{F}_t$ remains a continuous local martingale with respect to $\mathcal{F}_{t+}$. The only change in the above theorem consists in replacing the σ-field $\mathcal{F}_s$ in the statement of part (ii) to $\mathcal{F}_{s+}$.

The next application of the Itô formula is known as the stochastic integral representation of martingales. Let $W = (W_1, \ldots, W_d)$ be a d-dimensional Wiener process on $(\Omega, \mathcal{F}, P)$. Let $\mathcal{F}_t^W = \sigma(W(s) : 0 \le s \le t) \vee P$-null sets.

Theorem 5.6.2 (Martingale representation) *A square integrable martingale M_t adapted to $\mathcal{F}_t^W$ with $t \in [0, T]$, and $M_0 = 0$, has a continuous modification $\tilde{M}_t$ given by*

$$\tilde{M}_t = \sum_{j=1}^{d} \int_0^t \Phi_j(s)\, dW_j(s) \tag{5.6.4}$$

where each $\Phi_j \in \mathcal{M}_W^2[0, T]$.

Proof

Step 1 Let $f_j \in L^2[0, T]$ for each $j = 1, \ldots, d$. Define

$$X_f(t) = \exp\left[\sum_{j=1}^{d} \int_0^t f_j(s)\, dW_j(s) - \frac{1}{2} \sum_{j=1}^{d} \int_0^t f_j^2(s)\, ds\right].$$

By the Itô formula, we have

$$X_f(T) = 1 + \sum_{j=1}^{d} \int_0^T X_f(t) f_j(t)\, dW_j(t). \tag{5.6.5}$$

From the definition of $X_f(t)$,

$$X_f^2(t) = X_{2f}(t) \exp\left[\sum_{j=1}^{d} \int_0^t f_j^2(s)\, ds\right].$$

From Equation (5.6.5), $X_{2f}(t)$ is a nonnegative local martingale, and hence a supermartingale, so that $E[X_{2f}(t)] \leq 1$ for all $t \in [0, T]$. Therefore,

$$EX_f^2(T) \leq \exp\left[\sum_{j=1}^{d} \int_0^T f_j^2(s)\, ds\right] < \infty.$$

Step 2 Let $\mathcal{Z}$ denote the class of all random variables (r.v.s) Z in $L^2 = L^2(\Omega, \mathcal{F}_T^W, P)$ given by a stochastic integral $\sum_{j=1}^{d} \int_0^T \Phi_j(s)\, dW_j(s)$ where each $\Phi_j \in \mathcal{M}_W^2[0, T]$. By Equation (5.6.5), the random variable $X_f^2(T) - 1$ is in $\mathcal{Z}$. If $Y \in L^2$ with $EY = 0$ and orthogonal to $\mathcal{Z}$, then

$$E\left[YX_f(T)\right] = 0 \quad \text{for all } f = (f_1, \ldots, f_d), \text{ with each } f_j \in L^2[0, T].$$

Let π be a partition given by $0 = t_0 < t_1 \cdots < t_n \leq T$. By taking each f_j to be a step function that uses the partition π, we obtain

$$\begin{aligned} E[YX_f(T)] &= E\left\{E\left[YX_f(T) \mid W_{t_1}, W_{t_2}, \ldots, W_{t_n}\right]\right\} \\ &= E\left[X_f(T) E\left(Y \mid W_{t_1}, W_{t_2}, \ldots, W_{t_n}\right)\right]. \end{aligned}$$

Since $X_f(T)$ is positive valued, one infers that $E(Y \mid W_{t_1}, W_{t_2}, \ldots, W_{t_n})$ is zero. By the arbitrariness of $t_1, \ldots, t_{n-1}$, we get $Y = 0$ a.s. Thus, $L^2 = \mathcal{Z}$.

Step 3 Since $M_T \in L^2$, by Step 2, there exists $\Phi_j \in \mathcal{M}_W^2[0, T]$ for each $j = 1, \ldots, d$ such that

$$M_T = \sum_{j=1}^{d} \int_0^T \Phi_j(s)\, dW_j(s).$$

Using the martingale property of stochastic integrals,

$$M_t = E\left[M_T \mid \mathcal{F}_t^W\right] = \sum_{j=1}^{d} \int_0^t \Phi_j(s)\, dW_j(s)$$

a.s. for each $t \in [0, T]$. Therefore, the continuous version of the stochastic integral process on the right side is the process $\tilde{M}_t$. ■

The next application of the Itô formula is to prove an important L^p-inequality for stochastic integrals. In fact, the inequality holds for any continuous, locally square integrable martingale and is known as the Burkhölder-Davis-Gundy inequality. When $p = 2$, the inequality is known as the Doob inequality.

Let $\mathcal{F}_t$ be a filtration that satisfies the usual hypotheses and $H \in \mathcal{L}_W^2[0, t]$ for all $t > 0$. Then $M_t = \int_0^t H_s\, dW_s$ is a continuous, $\mathcal{F}_t$-adapted, local martingale with $\langle M \rangle_t = \int_0^t H_s^2\, ds$. We will use the notation M_t^* to denote $\sup_{0 \leq s \leq t} |M_s|$.

Theorem 5.6.3 (Burkhölder-Davis-Gundy inequality) *For any $0 < p < \infty$, there exist universal constants c_p and C_p such that*

$$c_p\, E\left[\langle M\rangle_t^p\right] \le E\left[(M_t^*)^{2p}\right] \le C_p\, E\left[\langle M\rangle_t^p\right]. \tag{5.6.6}$$

Proof By stopping the local martingale $\{M_t\}$ at

$$\tau_n = \inf\left\{t \ge 0 : |M_t| > n \text{ or } \langle M\rangle_t > n\right\},$$

we obtain a bounded martingale $M_{t\wedge\tau_n}$. Clearly $\tau_n \uparrow \infty$ a.s. as $n \to \infty$. If we establish the theorem for the martingale $M_{t\wedge\tau_n}$, we would obtain (5.6.6) by letting $n \to \infty$. Therefore, without loss of generality, we will assume that M_t is a bounded martingale.

It is useful to recall the Doob L^p-inequality proved in Chapter 3 for any $0 < p < \infty$:

$$E\left[(M_t^*)^{2p}\right] \le \left(\frac{2p}{2p-1}\right)^{2p} E\left(|M_t|^{2p}\right). \tag{5.6.7}$$

The theorem is easy to prove when $p = 1$. Indeed, we have $E[M_t^2] = E\langle M\rangle_t$ by the Itô isometry. Hence, by using (5.6.7), we obtain (5.6.6) with $c_p = 1$ and $C_p = 4$.

Step 1 We will prove the upper bound in (5.6.6) when $p > 1$.

Applying the Itô formula to the function $f(x) = |x|^{2p}$ where $p > 1$,

$$|M_t|^{2p} = 2p\int_0^t |M_s|^{2p-1}\mathrm{sgn}(M_s)\, dM_s + \int_0^t p(2p-1)|M_s|^{2p-2}\, d\langle M\rangle_s.$$

Upon taking expectation,

$$\begin{aligned} E(|M_t|^{2p}) &= p(2p-1)E\int_0^t |M_s|^{2p-2}\, d\langle M\rangle_s \\ &\le p(2p-1)E\left[\sup_{0\le s\le t}|M_s|^{2p-2}\,\langle M\rangle_t\right] \\ &\le p(2p-1)\,\left\{E\left[(M_t^*)^{2p}\right]\right\}^{(p-1)/p}\left\{E\left[\langle M\rangle_t^p\right]\right\}^{1/p} \end{aligned} \tag{5.6.8}$$

by the Hölder inequality. Multiply inequality (5.6.8) by $\left(\frac{2p}{2p-1}\right)^{2p}$, and use (5.6.7) to get

$$E\left[(M_t^*)^{2p}\right] \le \left(\frac{2p}{2p-1}\right)^{2p} p(2p-1)\,\left\{E\left[(M_t^*)^{2p}\right]\right\}^{(p-1)/p}\left\{E\left[\langle M\rangle_t^p\right]\right\}^{1/p}.$$

Setting $K_p = \left(\frac{2p}{2p-1}\right)^{2p} p(2p-1)$, we get

$$\left\{E\left[(M_t^*)^{2p}\right]\right\}^{1/p} \le K_p\left\{E\left[\langle M\rangle_t^p\right]\right\}^{1/p}.$$

Define $C_p = K_p^p$, which finishes this step.

Step 2 In this step, we will prove the upper bound in (5.6.6) when $0 < p < 1$. Define $N_t = \int_0^t \langle M\rangle_s^{(p-1)/2}\, dM_s$. Then N_t is a square integrable martingale since

$$E(N_t^2) = E\left[\int_0^t \langle M\rangle_s^{(p-1)}\, d\langle M\rangle_s\right] = \frac{1}{p} E\left[\langle M\rangle_t^p\right]. \tag{5.6.9}$$

By the Itô formula,

$$\begin{aligned} N_t \langle M\rangle_t^{(1-p)/2} &= \int_0^t \langle M\rangle_s^{(1-p)/2}\, dN_s + \int_0^t N_s\, d\left(\langle M\rangle_s^{(1-p)/2}\right) \\ &= M_t + \int_0^t N_s\, d\left(\langle M\rangle_s^{(1-p)/2}\right). \end{aligned}$$

Hence, $|M_t| \le 2N_t^* \langle M\rangle_t^{(1-p)/2}$, so that $M_t^* \le 2N_t^* \langle M\rangle_t^{(1-p)/2}$. By the Hölder inequality,

$$\begin{aligned} E\left[(M_t^*)^{2p}\right] &\le 2^{2p} \left\{E\left[(N_t^*)^2\right]\right\}^p \left\{E\left(\langle M\rangle_t^p\right)\right\}^{1-p} \\ &\le 2^{2p} 4^p \left\{E\left(N_t^2\right)\right\}^p \left\{E\left(\langle M\rangle_t^p\right)\right\}^{1-p}, \end{aligned}$$

where we have used the upper bound in (5.6.6) for $p = 1$, which has already been shown. Using (5.6.9),

$$\begin{aligned} E\left[(M_t^*)^{2p}\right] &\le \left(\frac{16}{p}\right)^p \left\{E\left(\langle M\rangle_t^p\right)\right\}^p \left\{E\left(\langle M\rangle_t^p\right)\right\}^{1-p} \\ &= \left(\frac{16}{p}\right)^p \left\{E\left(\langle M\rangle_t^p\right)\right\}. \end{aligned}$$

Step 3 By the Itô formula, $M_t^2 = 2\int_0^t M_s\, dM_s + \langle M\rangle_t$, so that

$$\langle M\rangle_t = M_t^2 - 2\int_0^t M_s\, dM_s.$$

Raise both sides to the power p, and then use on the right side the basic inequality

$$|a+b|^p \le \lambda_p \left\{|a|^p + |b|^p\right\}$$

for any two real numbers a and b, where $\lambda_p = 2^{p-1}$ if $p > 1$ and 1 if $0 < p < 1$. Therefore, there exists a constant k_p such that

$$\begin{aligned} E\left[\langle M\rangle_t^p\right] &\le k_p \left\{E\left[(M_t^*)^{2p}\right] + E\left|\int_0^t M_s\, dM_s\right|^p\right\} \\ &\le k_p \left\{E\left[(M_t^*)^{2p}\right] + C_p E\left[\left(\int_0^T M_s^2\, d\langle M\rangle_s\right)^{p/2}\right]\right\} \end{aligned}$$

where we have used the upper bound in (5.6.6) for the martingale $N_t := \int_0^t M_s \, dM_s$. By the Cauchy-Schwarz inequality,

$$E\left[\langle M\rangle_t^p\right] \leq k_p \left[E\left[(M_t^*)^{2p}\right] + C_p \left\{E\left[(M_t^*)^{2p}\right]\right\}^{1/2} \left\{E\left[\langle M\rangle_t^p\right]\right\}^{1/2}\right].$$

If we denote $\{E[\langle M\rangle_t^p]\}^{1/2}$ as x, and $\left\{E[(M_t^*)^{2p}]\right\}^{1/2}$ as y, the above inequality can be written as

$$x^2 - k_p C_p xy - k_p y^2 \leq 0.$$

The left side is a quadratic function of x. To satisfy the inequality, x has to be less than the largest root of the quadratic function. Hence

$$x \leq \frac{1}{2}\left(k_p C_p + \sqrt{(k_p C_p)^2 + 4k_p}\right) y$$

so that $[\frac{1}{2}(k_p^2 C_p^2 + \sqrt{k_p C_p + 4k_p})]^{-1} x \leq y$.

Thus, $c_p \, E[\langle M\rangle_t^p] \leq E[(M_t^*)^{2p}]$, with $c_p = [\frac{1}{2}(k_p C_p + \sqrt{k_p^2 C_p^2 + 4k_p})]^{-2}$. ■

Next, we present an extension of the Itô formula to convex functions of a one-dimensional, continuous semimartingale. We use the word extension since convex functions need not be differentiable. However, for any convex function f, right and left derivatives of f exist at each point x. We denote them as $f_+'(x)$ and $f_-'(x)$, respectively. The second derivative exists only as a measure μ obtained from the definition $\mu[a,b) = f_l'(b) - f_l'(a)$ for all finite $a < b$. This is enough for us, since for any continuously differentiable function g with compact support,

$$\int_{-\infty}^{\infty} g(x) \, d\mu(x) = -\int_{-\infty}^{\infty} g'(x) f_-'(x) \, dx$$

by an integration by parts.

Theorem 5.6.4 *Let $f : \mathbb{R}^1 \to \mathbb{R}^1$ be a convex function and X, a real-valued, continuous semimartingale. Then $f(X)$ is a continuous semimartingale, and*

$$f(X_t) = f(X_0) + \int_0^t f'(X_s) \, dX_s + A_t, \tag{5.6.10}$$

where f_-' is the left derivative of f, and A is an adapted, right-continuous, increasing process.

Proof Define $\tau_m = \inf\{t : |X_t| \geq m\}$ for all $m \in \mathbb{N}$. Clearly, $\tau_m \uparrow \infty$ a.s. as $m \to \infty$, and the stopped process, X^{τ_m}, is bounded by m. Therefore, without loss of generality, we can assume that X is bounded by m.

Let g be a positive C^∞ function with compact support contained in $(-\infty, 0]$ such that

$$\int_{-\infty}^{\infty} g(s) \, ds = 1.$$

Define the functions f_n, for each $n \in \mathbb{N}$, as follows:

$$f_n(t) = n \int_{-\infty}^{\infty} f(t+s)g(ns)\, ds.$$

Then, f_n is convex since f is. Also, each f_n is a C^∞ function, and pointwise, $\lim_{n\to\infty} f_n = f$ and $\lim_{n\to\infty} f_n' = f_-'$. By the Itô formula, for each n, we have

$$f_n(X_t) = f_n(X_0) + \int_0^t f_n'(X_s)\, dX_s + A_t^n, \tag{5.6.11}$$

where $A_t^n = \frac{1}{2}\int_0^t f''(X_s)d\langle X\rangle_s$, an increasing process. Since f_n' is continuous on the interval $[-m, m]$, it is bounded. Therefore, by the dominated convergence theorem for stochastic integrals,

$$\int_0^t f_n'(X_s)\, dX_s \to \int_0^t f_-'(X_s)\, dX_s$$

uniformly on compacts in probability as $n \to \infty$. Consequently, by letting $n \to \infty$ in (5.6.11), it follows that A^n converges to a process A uniformly on compacts in probability. The process A is continuous, adapted, and increasing in t. ■

Tanaka Formula

Using Theorem 5.6.4 for the convex function $f(x) = (x-a)^+$ for any fixed real number a, there exists an increasing continuous process $\{L_t^a\}$ called the **local time** of X such that

$$(X_t - a)^+ = (X_0 - a)^+ + \int_0^t 1_{\{X_s > a\}}\, dX_s + \frac{1}{2}L_t^a. \tag{5.6.12}$$

Likewise,

$$(X_t - a)^- = (X_0 - a)^- - \int_0^t 1_{\{X_s \le a\}}\, dX_s + \frac{1}{2}L_t^a \tag{5.6.13}$$

so that by combining these two equations, one can write

$$|X_t - a| = |X_0 - a| + \int_0^t \operatorname{sgn}(X_s - a)\, dX_s + L_t^a, \tag{5.6.14}$$

which is known as the **Tanaka formula** for X.

Proposition 5.6.5 *For each fixed a, the measure dL_t^a is a.s. carried by the set $\{t : X_t = a\}$.*

Proof We know that $|X_t - a|$ is a semimartingale. Therefore, by the Itô formula, we have

$$(X_t - a)^2 = (X_0 - a)^2 + 2\int_0^t |X_s - a|\, d(|X_s - a|) + \langle |X - a| \rangle_t.$$

By the Tanaka formula, we can write

$$(X_t - a)^2 = (X_0 - a)^2 + 2\int_0^t |X_s - a|\operatorname{sgn}(X_s - a)\, dX_s + 2\int_0^t |X_s - a|\, dL_s^a + \langle X \rangle_t. \tag{5.6.15}$$

On the other hand, the Itô formula applied to the semimartingale $\{(X_t - a)\}$ yields

$$(X_t - a)^2 = (X_0 - a)^2 + 2\int_0^t (X_s - a)\, dX_s + \langle X \rangle_t. \tag{5.6.16}$$

By comparing the two equations above, we conclude that

$$\int_0^t |X_s - a|\, dL_s^a = 0,$$

and the proof is over. ■

It is well known that there exists a jointly measurable, adapted process $\tilde{L}$ such that for each real number a, $\tilde{L}(a, \cdot, \cdot)$ is indistinguishable from L^a. One can find a proof in [41]. We will, from now on, assume that L^a is jointly measurable and adapted.

The Itô-Tanaka formula

If f is the difference of two convex functions and X, a continuous semimartingale, then

$$f(X_t) = f(X_0) + \int_0^t f_l'(X_s)\, dX_s + \frac{1}{2}\int_{\mathbb{R}} L_t^a f''\,(da). \tag{5.6.17}$$

Proposition 5.6.6 (Occupation Density Formula) *Let X be a continuous semimartingale. For any real-valued, bounded, Borel-measurable function,*

$$\int_0^t \phi(X_s)\, d\langle X \rangle_s = \int_{\mathbb{R}^1} \phi(a) L_t^a\, da. \tag{5.6.18}$$

Proof If ϕ can be written as f'' for some $f \in C^2(\mathbb{R})$, then upon comparing the Itô and the Itô-Tanaka formulas, the result follows readily. The equality thus holds almost surely for a countable dense set of functions $\{\phi_n\}$ with each $\phi_n \in C_0(\mathbb{R})$. A monotone class argument can be used to complete the proof. ■

Next, we present a few simple but useful results on stochastic exponentials which follow from the Itô formula.

Lemma 5.6.7 *Suppose that $\{M_t\}$ is a strictly positive, continuous $\mathcal{F}_t$-local martingale such that $M_0 = 1$ a.s. Then M_t can be represented as*

$$M_t = \exp\left\{N_t - \frac{1}{2}\langle N\rangle_t\right\}$$

where $\{N_t\}$ is a continuous $\mathcal{F}_t$-local martingale with $N_0 = 0$. The above representation for M_t uniquely determines $\{N_t\}$.

Proof For all $n \in \mathbb{N}$, define the stopping times σ_n by

$$\sigma_n = \inf\left\{t \geq 0 : M_t \leq \frac{1}{n} \text{ or } M_t \geq n\right\}$$

with the convention that infimum of the empty set is infinity. Then, for each n, the process $\{M_{t\wedge\sigma_n}\}$ is a bounded, local martingale, and hence a martingale. Moreover, $\sigma_n \uparrow \infty$ a.s., so that $\{\sigma_n\}$ is a localizing sequence for $\{M_t\}$. Let $f_n \in C_b^2(\mathbb{R})$ be a real-valued function such that

$$f_n(x) = \log x \text{ if } x \in \left[\frac{1}{n}, n\right].$$

By the Itô formula applied to the function f_n, and the process M_t, we get

$$\log M_{t\wedge\sigma_n} = \int_0^{t\wedge\sigma_n} \frac{1}{M_s}\, dM_s - \frac{1}{2}\int_0^{t\wedge\sigma_n} \frac{1}{M_s^2}\, d\langle M\rangle_s.$$

Take $N_t = \int_0^t \frac{1}{M_s}\, dM_s$. Then $\{N_t\}$ is the $\mathcal{F}_t$-local martingale which gives the desired representation.

If there exists another continuous, $\mathcal{F}_t$-local martingale $\{R_t\}$ such that

$$M_t = \exp\left\{R_t - \frac{1}{2}\langle R\rangle_t\right\},$$

then the local martingale $N_t - R_t$ is equal to a process of bounded variation given by $\frac{1}{2}(\langle N\rangle_t - \langle R\rangle_t)$. Hence, $N_t - R_t = 0$, which establishes the uniqueness of $\{N_t\}$. ∎

Remark 5.6.2 In the above lemma, if $\{M_t\}$ is not strictly positive but simply nonnegative, then the sequence of stopping times σ_n defined in the above proof doesn't increase to infinity; rather, $\sigma_n \uparrow \sigma$, where

$$\sigma = \inf\{t \geq 0 : M_t = 0\}$$

with the convention that infimum of the empty set is infinity. Proceeding exactly as in the proof of Lemma 5.6.7, we obtain that

$$M_t = \exp\left\{N_t - \frac{1}{2}\langle N\rangle_t\right\} \quad \text{if } t < \sigma$$

where $N_t = \int_0^{t\wedge\sigma_n} \frac{1}{M_s} dM_s$. Since $\{M_t\}$ is a nonnegative local martingale and hence, a nonnegative supermartingale, $M_t = 0$ for $t \geq \sigma$. Thus,

$$M_t = \begin{cases} \exp\left\{N_t - \frac{1}{2}\langle N\rangle_t\right\} & \text{if } t < \sigma; \\ 0 & \text{if } t \geq \sigma. \end{cases}$$

Lemma 5.6.8 *For all $\theta \in \mathbb{R}$, let $M_\theta(t) = \exp\left\{\theta N_t - \frac{1}{2}\theta^2 A_t\right\}$ be a continuous martingale. Suppose that on some open neighborhood I of $\theta = 0$, and for all t, the following bounds hold a.s.:*

(i) $|M_\theta(t)| \leq C_1$.

(ii) $|\frac{d}{d\theta} M_\theta(t)| \leq C_2$.

(iii) $|\frac{d^2}{d\theta^2} M_\theta(t)| \leq C_3$, *where C_j is a nonrandom constant depending on I but not on t.*

Then, $\{N_t\}$ and $\left\{N_t^2 - A_t\right\}$ are continuous $\mathcal{F}_t$-martingales.

Proof For any $0 \leq s \leq t$, and $A \in \mathcal{F}_s$,

$$\begin{aligned} \int_A E\left[\left(\frac{d}{d\theta} M_\theta(t)\right)_{\theta=0} \middle| \mathcal{F}_s\right] &= \int_A \left(\frac{d}{d\theta} M_\theta(t)\right)_{\theta=0} dP \\ &= \left(\frac{d}{d\theta}\int_A M_\theta(t)\, dP\right)_{\theta=0} \\ &= \left(\frac{d}{d\theta}\int_A M_\theta(s)\, dP\right)_{\theta=0} \\ &= \int_A \left(\frac{d}{d\theta} M_\theta(s)\right)_{\theta=0} dP \end{aligned}$$

by using the hypotheses (i) and (ii). Thus,

$$E\left[\left(\frac{d}{d\theta} M_\theta(t)\right)_{\theta=0} \middle| \mathcal{F}_s\right] = \left(\frac{d}{d\theta} M_\theta(s)\right)_{\theta=0};$$

that is, $E[N_t|\mathcal{F}_s] = N_s$. Proceeding similarly and using hypothesis (iii) as well, we obtain the second assertion, since $\left(\frac{d^2}{d\theta^2} M_\theta(t)\right)_{\theta=0} = N_t^2 - A_t$. ■

Theorem 5.6.9 *Suppose that $\{N_t\}$ is an $\mathbb{R}^d$-valued, continuous, $\mathcal{F}_t$-adapted process with $N_0 = 0$, and $\{A_t\}$ is a $d \times d$ matrix-valued, $\mathcal{F}_t$-adapted, increasing process. Then the following statements are equivalent:*

(i) *The process $\{N_t\}$ is a continuous $\mathcal{F}_t$-local martingale with $\langle N^i, N^j\rangle_t = A_t^{ij}$ for all $i, j = 1, 2, \ldots, d$, so that $\langle N\rangle_t = trA$.*

(ii) *For all $\theta \in \mathbb{R}^d$, the process $M_\theta(t) := \exp\left\{(\theta, N_t) - \frac{1}{2}(A_t\theta, \theta)\right\}$ is a continuous $\mathcal{F}_t$-local martingale.*

Proof

Step 1 A simple application of the Itô formula shows that (i) implies (ii). To show the reverse implication, let us first take the dimension $d = 1$. Define the stopping times

$$T_n = \inf\{t : |N_t| \geq n \text{ or } A_t \geq n\}$$

for each n. For any fixed neighborhood I of $\theta = 0$, we have

$$\begin{aligned}
|M_\theta(t \wedge T_n)| &\leq \exp\left\{n|\theta| + \frac{1}{2}\theta^2 n\right\} \leq C_1(n, I) \\
\left|\frac{d}{d\theta} M_\theta(t \wedge T_n)\right| &= |M_\theta(t \wedge T_n)|\, |N_{t\wedge T_n} - \theta A_{t\wedge T_n}| \\
&\leq C_1(n, I) n(1 + |\theta|) \leq C_2(n, I) \\
\left|\frac{d^2}{d\theta^2} M_\theta(t \wedge T_n)\right| &= |M_\theta\,(t \wedge T_n)|\, \left\{[N_{t\wedge T_n} - \theta A_{t\wedge T_n}]^2 - A_{t\wedge T_n}\right\} \\
&\leq C_1(n, I)\, \left[(n^2(1 + |\theta|^2) + n\right] \leq C_3(n, I).
\end{aligned}$$

Thus, the hypotheses of Lemma 5.6.8 are satisfied. Hence, the processes $\{N_{t\wedge T_n}\}$ and $\{N^2_{t\wedge T_n} - A_{t\wedge T_n}\}$ are continuous martingales. Thus, the theorem is proved when $d = 1$.

Step 2 If $d > 1$, set $X_t = (\theta, N)$, and $V_t = (A_t\theta, \theta)$, where $\theta \in \mathbb{R}^d$. These are real-valued processes. Our hypothesis, namely, statement (ii) implies that for every real α and $\theta \in \mathbb{R}^d$, $\exp\{\alpha X_t - \frac{1}{2}\alpha^2 V_t\}$ is a continuous local martingale. Therefore, Step 1 can be applied to infer that $\{X_t\}$ and $\{X_t^2 - V_t\}$ are continuous martingales.

Thus, $X_t = \sum_{j=1}^d \theta^j N_t^j$ is a continuous local martingale with

$$\left\langle \sum_j \theta^j N^j, \sum_j \theta^j N^j \right\rangle_t = \sum_j A_t^{ij}\theta^i\theta^j.$$

Therefore, by varying θ appropriately, one obtains that N is a continuous martingale with $\langle N^i, N^j \rangle_t = A_t^{ij}$. ∎

5.7 The Girsanov Theorem

The Girsanov theorem is a result on changing a given probability measure by an equivalent or an absolutely continuous probability measure. As we shall see in the next chapter, it plays an important role in the study of weak solutions, addition or removal of drift in a stochastic differential equation, and in applications to finance theory.

Let $(\Omega, \mathcal{F}, P)$ be a complete probability space. Let $(\mathcal{F}_t : t \geq 0)$ be an increasing right-continuous family of σ-fields such that $\mathcal{F}_0$ contains all P-null sets. Suppose that $(M_t, \mathcal{F}_t)$ is a continuous local martingale with $M_0 = 0$ a.s. Define

$$X_t = \exp\left\{M_t - \frac{1}{2}\langle M\rangle_t\right\}.$$

Lemma 5.7.1 *The $\mathcal{F}_t$-adapted process X_t is a nonnegative continuous local martingale and a supermartingale. It is a martingale if and only if $E(X_t) = 1$ for each $t \geq 0$.*

Proof By the Itô formula, $X_t = 1 + \int_0^t X_s\, dM_s$. Therefore, X_t is a continuous $\mathcal{F}_t$-local martingale. If τ_n is a localizing sequence of stopping times for X_t, then for each n,

$$EX_{t\wedge\tau_n} = 1,$$

and if $s \leq t$,

$$X_{s\wedge\tau_n} = E[X_{t\wedge\tau_n}|\mathcal{F}_s].$$

In the above equalities, we can let $n \to \infty$ and use the Fatou lemma, since $X_{t\wedge\tau_n} \geq 0$, and $X_{t\wedge\tau_n} \to X_t$ a.s. as $n \to \infty$. Thus, $X_t \in L^1(P)$, and $X_s \geq E[X_t|\mathcal{F}_s]$.

To prove the last assertion, assume that $EX_t = 1$ for all t. For any given $s \leq t$, define the set

$$A = \left\{\omega : X_s > E(X_t|\mathcal{F}_s)\right\}.$$

Then $A^c = \{\omega : X_s = E(X_t|\mathcal{F}_s)\}$. Since $EX_s = EX_t$, we obtain

$$\int_A X_s\, dP = \int_A E(X_t|\mathcal{F}_s)\, dP.$$

Therefore, $P(A) = 0$, so that $X_s = E(X_t|\mathcal{F}_s)$ a.s. ∎

Lemma 5.7.2 *Let P and Q be two probability measures on $(\Omega, \mathcal{F})$. Suppose that $Q << P$ relative to $\mathcal{F}_t$ for each $t \geq 0$. Denote the corresponding Radon-Nikodym derivative by*

$$dQ = L_t\; dP \quad on \quad \mathcal{F}_t.$$

Then, an $\mathcal{F}_t$-adapted, continuous process M_t is a Q-local martingale if M_tL_t is a P-local martingale.

Proof For any $s \leq t$, and $A \in \mathcal{F}_s$,

$$\int_A E[L_t|\mathcal{F}_s]\, dP = \int_A L_t\, dP = Q(A)$$
$$= \int_A L_s\, dP$$

so that L_t is a P-martingale. Without loss of generality, we may assume that we are taking the right-continuous version of L_t. Consider any arbitrary time $T > 0$, and stopping time τ satisfying $\tau \leq T$. By the optional sampling theorem,

$$\int_A L_T\,dP = \int_A L_\tau\,dP \;\;\forall\; A \in \mathcal{F}_\tau.$$

Thus, $dQ = L_\tau dP$ on $\mathcal{F}_\tau$.

We are given that $M_t L_t$ is a P-local martingale. Let $\{\tau_n\}$ denote the localizing sequence of stopping times so that $(M_{t\wedge\tau_n} L_{t\wedge\tau_n}, \mathcal{F}_t)$ is a P-martingale for each n. Then, for any $s \le t$, and $A \in \mathcal{F}_{s\wedge\tau_n}$,

$$\int_A M_{t\wedge\tau_n} L_{t\wedge\tau_n}\,dP = \int_A M_{s\wedge\tau_n} L_{s\wedge\tau_n}\,dP;$$

that is,

$$\int_A M_{t\wedge\tau_n}\,dQ = \int_A M_{s\wedge\tau_n}\,dQ.$$

In other words, $(M_{t\wedge\tau_n}, \mathcal{F}_{t\wedge\tau_n})$ is a Q-martingale for each n, so that $(M_{t\wedge\tau_n}, \mathcal{F}_t)$ is also a Q-martingale. Furthermore, $\tau_n \uparrow \infty$ Q-a.s. so that $(M_t, \mathcal{F}_t)$ is a Q-local martingale. ■

Remark 5.7.1 Suppose that a sequence of stopping times $\{T_n\}$ with $T_n \uparrow \infty$ a.s. and a sequence of $\mathcal{F}_{T_n\wedge t}$-adapted continuous processes $\{M_n\}$ are given. Following the notation of Lemma 5.7.2, assume that, for each n, $M_n(t)L_{T_n\wedge t}$ is a continuous, $\mathcal{F}_t$-adapted martingale with respect to the measure P. Arguing as in the proof of Lemma 5.7.2, for any $s \le t$, and $A \in \mathcal{F}_{s\wedge T_n}$,

$$\int_A M^n(t) L_{t\wedge T_n}\,dP = \int_A M^n(s) L_{s\wedge T_n}\,dP;$$

that is,

$$\int_A M^n(t)\,dQ = \int_A M^n(s)\,dQ.$$

Thus, for each n, $\{M^n\}$ is a $\mathcal{F}_{t\wedge T_n}$-adapted Q-martingale and hence a $\mathcal{F}_t$-adapted Q martingale.

Theorem 5.7.3 *Suppose that $M = (M_t^1, \ldots, M_t^d)$ is a d-dimensional continuous local martingale defined on $(\Omega, \mathcal{F}, (\mathcal{F}_t), P)$ with $M_0 = 0$ and $\mathcal{F}_0 = \{\Omega, \Phi\}$. Let Q be a probability measure such that $Q \equiv P$ relative to $\mathcal{F}_T$ for some $T > 0$. On $\mathcal{F}_t$, let us denote $\frac{dQ}{dP}$ by L_t, which is assumed to be continuous on $[0, T]$. Write L as the stochastic exponential of a continuous martingale N, a representation guaranteed by Lemma 5.6.7. Define*

$$V_t = (V^1, \ldots, V^d) \quad \text{where} \quad V_t^j = \langle M^j, N\rangle_t.$$

Define

$$Z_t = M_t - V_t. \tag{5.7.1}$$

Then $\{Z_t\}$ is a d-dimensional, continuous, $\mathcal{F}_t$-local martingale with respect to the measure Q, and

$$\langle Z^i, Z^j\rangle = \langle M^i, M^j\rangle. \tag{5.7.2}$$

Proof By the hypothesis on $\mathcal{F}_0$, we have $L_0 = 1$. Since $Q \equiv P$ on $\mathcal{F}_T$, it follows that $L_t > 0$ P a.s. Therefore, by Lemma 5.6.7,

$$L_t = \exp\left\{N_t - \frac{1}{2}\langle N\rangle_t\right\}.$$

Define $A_t^{ij} = \langle M^i, M^j\rangle_t$. By Theorem 5.6.9, it suffices to show that for all $\theta \in \mathbb{R}^d$, the process $\exp\{(\theta, Z_t) - \frac{1}{2}(A_t\theta, \theta)\}$ is a continuous $(\mathcal{F}_t, Q)$ local martingale. By Lemma 5.7.2, this follows if we can show that

$$\exp\left\{(\theta, Z_t) - \frac{1}{2}(A_t\theta, \theta)\right\} \exp\left\{N_t - \frac{1}{2}\langle N\rangle_t\right\} \tag{5.7.3}$$

is a continuous $(\mathcal{F}_t, P)$ local martingale. Note that

$$(A_t\theta, \theta) = \sum_{i,j} \theta_i\theta_j\langle M^i, M^j\rangle_t = \langle(\theta, M)\rangle_t,$$

and, by definition of V,

$$(\theta, V)_t = \sum_j \theta_j V_t^j = \sum_j \theta_j\langle M^j, N\rangle_t = \langle(\theta, M), N\rangle_t.$$

Using this, we can evaluate the exponent in Equation 5.7.3 as follows:

$$\begin{aligned}
&(\theta, Z_t) - \frac{1}{2}(A_t\theta, \theta) + N_t - \frac{1}{2}\langle N\rangle_t \\
&= (\theta, M_t) + N_t - \frac{1}{2}\langle(\theta, M)\rangle_t - \frac{1}{2}\langle N\rangle_t - (\theta, V)_t \\
&= (\theta, M_t) + N_t - \frac{1}{2}[\langle(\theta, M)\rangle_t + 2\langle(\theta, M), N\rangle_t + \langle N\rangle_t] \\
&= (\theta, M_t) + N_t - \frac{1}{2}\langle(\theta, M) + N\rangle_t.
\end{aligned}$$

Hence, Equation 5.7.3 can be written as $\exp\{(\theta, M_t) + N_t - \frac{1}{2}\langle(\theta, M) + N\rangle_t\}$, which is a continuous P-local martingale by Lemma 5.7.1. ■

Remark 5.7.2 (i) In the above theorem, we have assumed that $\mathcal{F}_0 = \{\Omega, \Phi\}$ to ensure that $L_0 = 1$. The condition that $M_0 = 0$ is not crucial and is made for convenience.

(ii) Using the Itô formula, we can write V_t^j as $\int_0^t \frac{1}{L_s} d\langle M^j, L\rangle_s$ Q-a.s. Indeed,

$$\log L_t = \int_0^t \frac{1}{L_s}\, dL_s - \frac{1}{2}\int_0^t \frac{1}{L_s^2}\, d\langle L\rangle_s,$$

and from the definition of N_t, we have $N_t = \log L_t + \frac{1}{2}\langle N\rangle_t$. Hence,

$$\langle M, N\rangle_t = \left\langle M, \int_0^t \frac{1}{L_s}\, dL_s \right\rangle_t = \int_0^t \frac{1}{L_s}\, d\langle M, L\rangle_s.$$

From Theorem 5.7.3, it follows that, under the probability measure Q,

$$Z_t = M_t - \int_0^t \frac{1}{L_s}\, d\langle M, L\rangle_s \tag{5.7.4}$$

is a continuous, $\mathcal{F}_t$-local martingale with respect to Q.

Next, we consider the case when Q is absolutely continuous with respect to P in the place of equivalence of P and Q. In the next theorem, we show that, even under this weaker assumption, the conclusion of Theorem 5.7.3 holds.

Theorem 5.7.4 *In the setup of Theorem 5.7.3, suppose that $Q << P$ on $\mathcal{F}_T$ instead of $Q \equiv P$. Then,*

$$V_t = \int_0^t \frac{1}{L_s}\, d\langle M, L\rangle_s$$

exists Q-a.s., and

$$Z_t = M_t - \int_0^t \frac{1}{L_s}\, d\langle M, L\rangle_s$$

is a Q-local martingale.

Proof Define $T_n = \inf\{t \geq 0 : L_t \leq \frac{1}{n}\}$ with the convention that infimum of the empty set is infinity. Suppose that the increasing sequence of stopping times T_n converges to τ a.s. with respect to P. Define $L_\tau = L_T$ on $\{\tau = \infty\}$. Since $L_\tau = 0$ on the set $\tau \leq T$,

$$\int_{\{\tau \leq T\}} L_\tau\, dP = 0.$$

On the other hand, $\frac{dQ}{dP} = L_{T\wedge\tau}$ relative to $\mathcal{F}_{T\wedge\tau}$, so that

$$\int_{\{\tau \leq T\}} L_\tau\, dP = \int_{\{\tau \leq T\}} L_{T\wedge\tau}\, dP = Q\{\tau \leq T\}.$$

Thus, $Q\{\tau \leq T\} = 0$, so that $T_n \uparrow \infty$ Q-a.s. Therefore, $\int_0^t \frac{1}{L_s}\, d\langle M, L\rangle_s$ exists Q-a.s.

The process $L_t > 0$ on $[0, T_n]$ for each n. Moreover, $\int_0^{t\wedge T_n} \frac{1}{L_s}\, d\langle M, L\rangle_s$ exists Q-a.s. Hence, by Equation (5.7.4), the process $M_{t\wedge T_n} - \int_0^{t\wedge T_n} \frac{1}{L_s}\, d\langle M, L\rangle_s$ is a Q-local martingale for each n. The proof is over, since a local local martingale is a local martingale. ∎

The above theorems are generalizations of a famous result by Girsanov which is given below. Girsanov's theorem is a deep and important result with many uses in stochastic analysis.

Theorem 5.7.5 *Let $\{W_t\}$ be a d-dimensional Wiener martingale defined on $(\Omega, \mathcal{F}, (\mathcal{F}_t), P)$, and $\{f(t)\}$ be a d-dimensional $\mathcal{F}_t$-adapted process such that*

$$\int_0^T |f(t)|^2 dt < \infty \quad P\text{-a.s.}$$

Define

$$L_t = \exp\left\{\int_0^t (f(s), dW_s) - \frac{1}{2}\int_0^t |f(s)|^2\, ds\right\}$$

where $\int_0^t (f(s), dW_s) = \sum_{j=1}^d \int_0^t f_j(s)\, dW_j(s)$. Assume that $EL_T = 1$. Define the probability measure Q by

$$dQ = L_T\, dP \quad \text{on } \mathcal{F}_T.$$

If $Z_t = W_t - \int_0^t f(s)\, ds$, then $\{Z_t\}$ is a d-dimensional $\mathcal{F}_t$-Wiener martingale with respect to Q.

Proof We are given that $EL_T = 1$. Therefore we can invoke Lemma 5.7.1 to conclude that L_t is a martingale . All the conditions of Theorem 5.7.4 hold. Identifying W_t with M_t, and N_t with $\int_0^t (f(s), dW_s)$, we get

$$V_t^j = \langle M^j, N\rangle_t = \int_0^t f_j(s)\, ds,$$

so that $Z_t = W_t - \int_0^t f(s)\, ds$ is a $\mathcal{F}_t$-local martingale with respect to Q since $\langle Z\rangle_t = tI$, where I is the $d \times d$ identity matrix by the Lévy characterization of Brownian motion given in Theorem 5.6.1. ■

The next result is an extension of the Girsanov theorem and is a consequence of Theorem 5.7.4. In Theorem 5.7.4, the process L_t can be written as

$$L_t = 1 + \int_0^t L_s\, dN_s$$

where N_t is a local martingale. In the following theorem, we consider a special case,

$$L_t = 1 + \int_0^t b_s\, dW_s$$

where b_t is an adapted, square integrable process on $[0, T]$.

Theorem 5.7.6 *Let $\{W_t\}$ be a d-dimensional Wiener martingale defined on $(\Omega, \mathcal{F}, (\mathcal{F}_t), P)$. Let Q be a probability measure on $\mathcal{F}_T$ such that*

$$dQ = L_T\, dP.$$

Assume that $L_t = 1 + \int_0^t b_s\, dW_s$ *and* $\{b_t\}$ *is a d-dimensional, measurable,* $\mathcal{F}_t$*-adapted process such that*

$$\int_0^T |b_s|^2\, ds < \infty \quad P\text{-a.s.}$$

Then, on the probability space $(\Omega, \mathcal{F}, (\mathcal{F}_t), Q)$, *the process*

$$Z_t := W_t - \int_0^t \frac{b_s}{L_s}\, ds \tag{5.7.5}$$

is a Wiener martingale.

Proof The proof essentially follows from that of Theorem 5.7.4. Since the stopping times T_n introduced in the proof of Theorem 5.7.4 increase to ∞, we get $L_s > 0$ a.s. with respect to Q for all $s \leq T$. Hence, $A := \inf\{L_s : 0 \leq s \leq T\} > 0$, so that

$$\int_0^t \frac{|b_s|}{L_s}\, ds \leq \frac{1}{A} \int_0^t |b_s|\, ds < \infty \quad Q\text{-a.s.}$$

Thus, Z_t given by Equation (5.7.5) exists Q-a.s. and is a Wiener martingale by the Lévy characterization of Brownian motion. ■

In the Girsanov theorem and its generalizations, it was assumed that $dQ = L_T\, dP$. To ensure that Q is a *probability* measure, we required $E^P L_T = 1$. In applications, the continuous local martingale N_t is given, and hence we can only conclude that L_t is a supermartingale by Lemma 5.7.1. However, we need L_t to be a martingale to guarantee that $EL_T = 1$. A sufficient condition for L_t to be a martingale is presented below. The condition is easily verifiable and is known as the **Novikov criterion**. First, we need the following basic estimate.

Lemma 5.7.7 *Let N be a continuous local martingale. Then,*

$$E\left[\exp\left\{\frac{1}{2}N_t\right\}\right] \leq \left[E\exp\left\{\frac{1}{2}\langle N\rangle_t\right\}\right]^{1/2}.$$

Proof Let L denote the stochastic exponential of N. Then,

$$\exp\{N_t\} = L_t\left[\exp\left\{\frac{1}{2}\langle N\rangle_t\right\}\right].$$

Upon taking square root on both sides,

$$\exp\left\{\frac{1}{2}N_t\right\} = L_t^{1/2}\left[\exp\left\{\frac{1}{2}\langle N\rangle_t\right\}\right]^{1/2}.$$

By the Cauchy-Schwarz inequality,

$$E\left[\exp\left\{\frac{1}{2}N_t\right\}\right] \leq (EL_t)^{1/2}\left[E\exp\left\{\frac{1}{2}\langle N\rangle_t\right\}\right]^{1/2}.$$

The proof is over upon noting that $EL_t \leq 1$. ■

Theorem 5.7.8 (The Novikov criterion) *Let N_t be a continuous local martingale, and*

$$E\left[\exp\left\{\frac{1}{2}\langle N\rangle_T\right\}\right] < \infty. \tag{5.7.6}$$

Then, L_t is a martingale. In particular, $EN_T = 1$.

Proof

Step 1 The condition (5.7.6) implies that $E\langle N\rangle_T < \infty$. Therefore, N is L^2-bounded so that N is a uniformly integrable martingale. In addition, the above lemma implies that $\exp\left\{\frac{1}{2}N_T\right\} \in L^1(P)$, which allows us to conclude that $\exp\left\{\frac{1}{2}N\right\}$ is a submartingale.

In fact, $\exp\left\{\frac{1}{2}N\right\}$ is a uniformly integrable submartingale. Indeed, for any $c > 0$,

$$\begin{aligned}\int_{\{N_t\geq c\}} \exp\left\{\frac{1}{2}N_t\right\} dP &\leq \int_{\{N_t\geq c\}} E\left[\exp\left\{\frac{1}{2}N_T\right\} \,\middle|\, \mathcal{F}_t\right] dP \\ &\leq \int_{\{N_t\geq c\}} \exp\left\{\frac{1}{2}N_T\right\} dP \\ &\leq \int_{\left\{\sup_{0\leq t\leq T} N_t\geq c\right\}} \exp\left\{\frac{1}{2}N_T\right\} dP\end{aligned}$$

so that

$$\sup_{0\leq t\leq T}\int_{\{N_t\geq c\}} \exp\left\{\frac{1}{2}N_t\right\} dP \leq \int_{\left\{\sup_{0\leq t\leq T} N_t\geq c\right\}} \exp\left\{\frac{1}{2}N_T\right\} dP.$$

Since N is a continuous process and $N_T \in L^1(P)$, the family of events $\left\{\sup_{0\leq t\leq T} N_t \geq c\right\}$ decreases to a P-null set as $c \uparrow \infty$. Thus,

$$\lim_{c\to\infty}\int_{\left\{\sup_{0\leq t\leq T} N_t\geq c\right\}} \exp\left\{\frac{1}{2}N_T\right\} dP = 0,$$

and the uniform integrability of the process $\exp\{\frac{1}{2}N\}$ follows.

Step 2 Pick any $a \in (0, 1)$. Let $L_a(t)$ denote the stochastic exponential of aN. We will show that $L_a(t)$ is a uniformly integrable martingale.

We can write $L_a(t)$ as

$$L_a(t) = (L_t)^{a^2}\left(\exp\left\{\frac{a}{1+a}N_t\right\}\right)^{1-a^2}. \tag{5.7.7}$$

Arguing as in Step 1, the process $\exp\{\frac{a}{1+a}N\}$ is a uniformly integrable submartingale, since $\frac{a}{1+a} < \frac{1}{2}$.

Fix any set $A \in \mathcal{F}$, and $t \leq T$. Using Hölder's inequality (with $\frac{1}{p} = a^2$ and $\frac{1}{q} = 1 - a^2$) in Equation (5.7.7),

$$E[1_A L_a(t)] \leq [EL_t]^{a^2} \left[E\left(1_A \exp\left\{\frac{a}{1+a}N_t\right\}\right)\right]^{1-a^2}$$
$$\leq \left[E\left(1_A \exp\left\{\frac{a}{1+a}N_t\right\}\right)\right]^{1-a^2}$$

since $EL_t \leq 1$. From this, we can conclude that $\{L_a(t)\}$ is a uniformly integrable martingale, since $\exp\left\{\frac{a}{1+a}N\right\}$ is uniformly integrable.

Step 3 From Step 2, we get $EL_a(t) = 1$. Using Equation (5.7.7) and the Hölder inequality as before,

$$1 = EL_a(t) \leq [EL_t]^{a^2} \left[E\left(\exp\left\{\frac{a}{1+a}N_t\right\}\right)\right]^{1-a^2}$$
$$\leq [EL_t]^{a^2} \left[E\left(\exp\left\{\frac{1}{2}N_t\right\}\right)\right]^{2a(1-a)}.$$

The last line uses the Jensen inequality. Thus,

$$1 \leq [EL_t]^{a^2} \left[E\left(\exp\left\{\frac{1}{2}N_t\right\}\right)\right]^{2a(1-a)}.$$

Allow $a \uparrow 1$ to get $1 \leq EL_t$ for all t. Since EL_t also satisfies $EL_t \leq 1$, we obtain $EL_t = 1$. By Lemma 5.7.1, L is a martingale. ■

There are only two well-known criteria to answer the question: If N is a local martingale, under what conditions is L a martingale? Often, the Novikov criterion is easy to verify. The other is known as the Kazamaki criterion, which stipulates that $\exp\left\{\frac{1}{2}N\right\}$ is a uniformly integrable submartingale. A nice discussion and comparison of the two criteria can be found in the book by Revuz and Yor [62] as well as in Protter [60].

Exercises

1. Write $P\{\int_0^3 f(s)\,dW(s) > 3\}$ explicitly as an integral in each of the following cases:

(i)
$$f(s) = \begin{cases} -1, & \text{if } 0 \leq s < 1; \\ W^2(1), & \text{if } 1 \leq s < 3. \end{cases}$$

(ii)
$$f(s) = \begin{cases} 1, & \text{if } 0 \leq s < 1; \\ 1_{\{W(1/2)>0\}}, & \text{if } 1 \leq s \leq 3. \end{cases}$$

2. Evaluate $\int_0^1 W^2(t)\,dW(t)$ directly from the definition of a stochastic integral. Do the same for $\int_0^1 e^{-s}\,dW(s)$.

3. Use the Itô formula to show that

$$X_t = (W_t + t)\exp\left\{-W_t - \frac{1}{2}t\right\}$$

is a martingale.

4. Let H be in $\mathcal{M}_W$, and X, a bounded $\mathcal{F}_s$-measurable random variable for a fixed s. Show that

$$X\int_s^t H_u\,dW_u = \int_s^t XH_u\,dW_u.$$

5. Let W be a standard one-dimensional Wiener process and H, an adapted-right, continuous, bounded process. Prove that, for fixed t,

$$\lim_{h\to 0}(W(t+h) - W(t))^{-1}\int_t^{t+h} H(s)\,dW(s) = H(t)$$

in probability. The result holds for unbounded H as well, provided that it is continuous.

6. Show that the process $\{W^4(t) - 6tW^2(t) + 3t^2\}$ is a martingale where W is a standard Wiener process.

7. Let $X(t) = W(t) + t$ for all $t \geq 0$ where W is a standard Wiener process.

   (i) Find a function f such that $f(X(t)$ is a martingale.

   (ii) Let $\tau = \inf\{t : X(t) = -a \text{ or } b\}$. Find $P\{X(\tau) = b\}$.

   What is $E(\tau)$?

8. Let f be a convex function from $\mathbf{R}^1 \to \mathbf{R}^1$. Show that f is continuous and has finite right and left derivatives at all $x \in \mathbb{R}$.

9. (Continuation of Problem 8) Suppose that D^+f denotes the right derivative, and D^-f, the left derivative. Show that

$$D^+f(x) \leq D^-f(y) \leq D^+f(y)$$

for all $x < y$. Deduce that $D^-f(\cdot)$ and $D^+f(\cdot)$ are both non-decreasing functions on $\mathbb{R}$.

10. (Continuation of Problem 9) Show that D^-f is left continuous and D^+f is right continuous on $\mathbb{R}$.

6 Stochastic Differential Equations

The mathematical models that arise in sciences such as biology, physics, and economics are stochastic rather than deterministic. The accumulation of errors inherent in such phenomena leads to the formulation of stochastic differential equations. The theory of stochastic differential equations provides us with a probabilistic method to study a class of stochastic processes known as diffusion processes. Solutions to a wide class of parabolic partial differential equations can be studied by means of stochastic differential equations.

6.1 Introduction

Let $(\Omega, \mathcal{F}, P)$ be a complete probability space on which the following are defined:

(i) $W = \{W_t : 0 \leq t \leq T\}$, a standard k-dimensional Wiener process.

(ii) ξ, *ad*-dimensional random vector.

We assume that ξ is independent of the Wiener process W. For each t, define the σ-field

$$\mathcal{F}_t = \sigma\,(\xi,\ W_s\ 0 \leq s \leq t) \vee (\text{all } P\text{-null sets in } \mathcal{F}).$$

It is clear that the filtration $(\mathcal{F}_t)$ satisfies the usual conditions and W is an $\mathcal{F}_t$-adapted Wiener process.

Let $b = (b_1, \ldots, b_d)$ be an d-dimensional vector of functions where each coordinate satisfies the following:

(i) $b_j : [0, T] \times \mathbb{R}^d \times \Omega \to \mathbb{R}$.

(ii) b_j is $\mathcal{B}[0, T] \times \mathcal{B}(\mathbb{R}^d) \times \mathcal{F}$ measurable.

(iii) For each t, $b_j(t, \cdot, \cdot)$ is measurable with respect to $\mathcal{B}(\mathbb{R}^d) \times \mathcal{F}_t$.

Let σ be a $d \times k$-dimensional matrix of functions with each entry σ_{ij} satisfying the same conditions listed above for b_j.

Definition 6.1.1 *A d-dimensional process $X = \{X_t\}$, $t \in [0,T]$, defined on $(\Omega, \mathcal{F}, P)$, is called a* **strong solution** *of the stochastic differential equation*

$$dX_t = b(t, X_t)\,dt + \sigma(t, X_t)\,dW_t \tag{6.1.1}$$

with initial condition $X_0 = \xi$ if the following assertions hold:

1. *X is $\mathcal{F}_t$ adapted with continuous sample paths.*
2. *$\int_0^T (|b(t,X_t)| + |\sigma(t,X_t)|^2)\,dt < \infty$ a.s. where $|\cdot|$ is used to denote both the d-dimensional norm and the norm of a matrix.*
3. *For each $t \in [0,T]$,*

$$X_t = \xi + \int_0^t b(s, X_s)\,ds + \int_0^t \sigma(s, X_s)\,dW_s \text{ a.s.} \tag{6.1.2}$$

Thus, the stochastic differential equation (6.1.1) is interpreted as a stochastic integral equation given by (6.1.2). The stochastic integral on the right side of (6.1.2) is taken in the sense of Itô.

Definition 6.1.2 *The stochastic differential equation (6.1.1) with initial condition $X_0 = \xi$ has a unique strong solution if, for any two strong solutions $X = \{X_t\}$ and $Y = \{Y_t\}$ on $(\Omega, \mathcal{F}, P)$, one has*

$$P\left\{\omega : X_t(\omega) = Y_t(\omega)\ \forall\ t \in [0,T]\right\} = 1.$$

The notion of uniqueness given above is called **strong uniqueness** of solutions. In order to establish the existence and uniqueness of strong solutions of Equation (6.1.1), we need to impose additional restrictions on b and σ:

Hypotheses H

For all $t \in [0,T]$, and $x, y \in \mathbb{R}^d$,

H.1 $|b(t,x)|^2 + |\sigma(t,x)|^2 \le K(1 + |x|^2)$ a.s.

H.2 $|b(t,x) - b(t,y)|^2 + |\sigma(t,x) - \sigma(t,y)|^2 \le K|x-y|^2$ a.s.,

where K is a positive constant independent of t and x.

The above conditions are known as linear growth and Lipschitz continuity of b and σ respectively. Other sets of hypotheses on b and σ to prove existence and uniqueness of solutions will be presented later. Before proving the main result, a useful inequality due to Gronwall is recalled below:

Lemma 6.1.1 *Let f and g be two functions in $L^1[a,b]$. Suppose that*

$$f(t) \le g(t) + C\int_a^t f(s)\,ds \;\;\forall\; t \in [a,b] \tag{6.1.3}$$

for a constant $C > 0$. Then, for all $t \in [a,b]$,

$$f(t) \le g(t) + C\int_a^t e^{C(t-s)}\,g(s)\,ds. \tag{6.1.4}$$

In particular, if $g(t) \equiv G$, a constant, then the conclusion becomes

$$f(t) \le Ge^{C(t-a)}.$$

Proof To prove (6.1.4), we need to show that $\int_a^t f(s)\,ds \le \int_a^t e^{C(t-s)}\,g(s)\,ds$. Consider the inequality (6.1.3) at time s instead of t. Multiply both sides of (6.1.3) by e^{-Cs} and rearrange the terms to obtain

$$e^{-Cs}f(s) - Ce^{-Cs}\int_a^s f(u)\,du \le e^{-Cs}g(s). \tag{6.1.5}$$

Integrating both sides of (6.1.5) from a to t, we get

$$\int_a^t e^{-Cs}f(s)\,ds - C\int_a^t e^{-Cs}\int_a^s f(u)\,du\,ds \le \int_a^t e^{-Cs}g(s)\,ds. \tag{6.1.6}$$

Clearly,

$$\begin{aligned} C\int_a^t e^{-Cs}\int_a^s f(u)\,du\,ds &= C\int_a^t f(u)\int_u^t e^{-Cs}\,ds\,du \\ &= \int_a^t e^{-Cu}f(u)\,du - \left(\int_a^t f(u)\,du\right)e^{-Ct} \end{aligned}$$

by interchanging the order of integration. Using this in inequality (6.1.6), one obtains

$$\left(\int_a^t f(u)\,du\right)e^{-Ct} \le \int_a^t e^{-Cs}g(s)\,ds.$$

We have thus shown that $\int_a^t f(s)\,ds \le \int_a^t e^{C(t-s)}\,g(s)\,ds$, which completes the proof. ■

Though the above lemma is quite useful, it is worthwhile to mention the following extension, which is important in applications. In fact, we presented the proof of Lemma 6.1.1 in a manner such that it can be used with minor modifications to prove the next result.

Lemma 6.1.2 *Let μ be a Borel measure on an interval $[a, b]$. Let f and g be two functions in $L^1(\mu)$ satisfying*

$$f(t) \leq g(t) + \int_{[a,t)} f(s)\mu(ds) \ \forall \ t \in [a, b]. \tag{6.1.7}$$

Then, for all $t \in [a, b]$,

$$f(t) \leq g(t) + \int_{[a,t)} e^{\mu[s,t)} g(s) \, \mu(ds).$$

In particular, if $g(t) \equiv G$, a constant, then the conclusion becomes

$$f(t) \leq Ge^{\mu[a,t)}.$$

Proof The proof is analogous to that of Lemma 6.1.1. The only change needed is to multiply the inequality (6.1.7) by $e^{-\mu[a,s)}$ instead of e^{-Cs}. ∎

6.2 Existence and Uniqueness of Solutions

Theorem 6.2.1 *Suppose that $(\Omega, \mathcal{F}, (\mathcal{F}_t), P)$, W, ξ, b, and σ are as given in Section 6.1. Let b and σ satisfy the Hypotheses H. Then the stochastic differential equation (6.1.1) with initial condition ξ has a unique strong solution $X = \{X_t\}$.*

Proof We will first assume that $E|\xi|^2 < \infty$. The proof of existence of solutions uses the Picard iteration scheme. We break the proof into several steps.

Step 1 Define $X_t^0 \equiv \xi$, and

$$X_t^1 = \xi + \int_0^t b(s, X_s^0) \, ds + \int_0^t \sigma(s, X_s^0) \, dW_s.$$

Upon squaring both sides, one obtains

$$|X_t^1|^2 \leq 3 \left[|\xi|^2 + \left| \int_0^t b\left(s, X_s^0\right) \, ds \right|^2 + \left| \int_0^t \sigma\left(s, X_s^0\right) \, dW_s \right|^2 \right] \tag{6.2.1}$$

where

$$\left| \int_0^t \sigma\left(s, X_s^0\right) \, dW_s \right|^2 \quad \text{stands for} \quad \sum_{i=1}^d \left(\sum_{j=1}^k \int_0^t \sigma_{ij}\left(s, X_s^0\right) \, dW_j(s) \right)^2.$$

In the inequality (6.2.1), first take supremum over t in $[0, T]$, and then expectation to obtain

$$E \sup_{0\leq t\leq T} |X_t^1|^2 \leq 3\left[E|\xi|^2 + TE\int_0^T \left(|b(s, X_s^0)|^2 + 4\sum_{i=1}^{d}\sum_{j=1}^{k} \sigma_{ij}^2(s, X_s^0)\right) ds\right]$$

by using the Doob inequality. By hypothesis (H.1), we have

$$\begin{aligned}
E\left(\sup_{0\leq t\leq T} |X_t^1|^2\right) &\leq 3\left[E|\xi|^2 + (4\vee T)E\int_0^T \left(|b(s, X_s^0)|^2 + |\sigma(s, X_s^0)|^2\right) ds\right] \\
&\leq 3\left[E|\xi|^2 + (4\vee T)\, KE\int_0^T \left(1 + |X_s^0|^2\right) ds\right] \\
&\leq 3\left[E|\xi|^2 + (4\vee T)\, KT\left(1 + E\sup_{0\leq s\leq T} |X_s^0|^2\right)\right] \\
&< \infty
\end{aligned}$$

since $X_s^0 \equiv \xi \in L^2(P)$. From the definition of the process X^1, we have that $\{X_t^1\}$ is an $\mathcal{F}_t$-adapted process with continuous paths.

Step 2 Induction: Suppose that the processes $X^k = \{X_t^k\}$ for $0 \leq k \leq m$ are to be $\mathcal{F}_t$-adapted with continuous paths, and having the following properties:

$$E\left[\sup_{0\leq s\leq T} |X_t^k|^2\right] < \infty \tag{6.2.2}$$

and, for $0 < k \leq m$,

$$X_t^k = \xi + \int_0^t b(s, X_s^{k-1})\, ds + \int_0^t \sigma(s, X_s^{k-1})\, dW_s. \tag{6.2.3}$$

Define

$$X_t^{m+1} = \xi + \int_0^t b(s, X_s^m)\, ds + \int_0^t \sigma(s, X_s^m)\, dW_s. \tag{6.2.4}$$

Then we claim that X^{m+1} is an $\mathcal{F}_t$-adapted process with continuous paths, and $E\left(\sup_{0\leq s\leq T} |X_t^{m+1}|^2\right) < \infty$. Indeed, by the induction hypothesis on X^m, and the condition (H.1), the right side of (6.2.4) is well defined. Hence, X^{m+1} is continuous and $\mathcal{F}_t$-adapted. The proof of $E[\sup_{0\leq s\leq T} |X_t^{m+1}|^2] < \infty$ follows exactly as in Step 1.

Step 3 We thus have a sequence of continuous $\mathcal{F}_t$-adapted processes $\{X_t^k\}$ for $k \geq 1$, satisfying Equation (6.2.3) and the requirement (6.2.2). We will show that X^k converges uniformly on $[0, T]$ to a continuous adapted process $X = \{X_t\}$.

Define $\rho_1(t) = E[\sup_{0\leq s\leq t} |X_s^1 - \xi|^2]$, and in general

$$\rho_{m+1}(t) = E \sup_{0\leq s\leq t} |X_s^{m+1} - X_s^m|^2.$$

Then by the Doob inequality and hypotheses (H.1),

$$\begin{aligned}\rho_1(t) &\leq 2\left(tE\int_0^t |b(s,\xi)|^2\, ds + 4E\int_0^t |\sigma(s,\xi)|^2\, ds\right)\\ &\leq 2(T+4)TK(1+E|\xi|^2)\end{aligned}$$

Let us denote the above bound by R.
By using hypothesis (H.2), we obtain that

$$\begin{aligned}\rho_{m+1}(t) &\leq 2tE\int_0^t \left|b(s,X_s^m) - b\left(s,X_s^{m-1}\right)\right|^2 ds + 8E\int_0^t \left|\sigma\left(s,X_s^m\right) - \sigma\left(s,X_s^{m-1}\right)\right|^2 ds\\ &\leq 2(T+4)K\int_0^t E\left[\sup_{0\leq r\leq s} \left|X_r^m - X_r^{m-1}\right|^2\right] ds; \quad \text{that is,}\end{aligned}$$

$$\rho_{m+1}(t) \leq 2(T+4)K\int_0^t \rho_m(s)\, ds. \tag{6.2.5}$$

Let us denote the constant $2(T+4)K$ as C. By iterating the inequality (6.2.5), we get

$$\begin{aligned}\rho_{m+1}(t) &\leq C^m \int_0^t \int_0^{t_{m-1}} \cdots \int_0^{t_1} \rho_1(s)\, ds\, dt_1 \cdots dt_{m-1}\\ &\leq \frac{RC^m t^m}{m!}\end{aligned}$$

for each m, and $t \in [0,T]$. Using this estimate, we have

$$\begin{aligned}\sum_{m=0}^{\infty} P\left\{\sup_{0\leq t\leq T} \left|X_t^{m+1} - X_t^m\right| > \frac{1}{(m+1)^2}\right\} &\leq \sum_{m=0}^{\infty}(m+1)^4 E\left[\sup_{0\leq t\leq T} \left|X_t^{m+1} - X_t^m\right|^2\right]\\ &\leq R\sum_{m=0}^{\infty} \frac{(m+1)^4 (CT)^m}{m!} < \infty.\end{aligned}$$

Hence, we can conclude by the Borel-Cantelli lemma that

$$\sum_{m=0}^{\infty} \sup_{0\leq t\leq T} \left|X_t^{m+1} - X_t^m\right| < \infty \text{ a.s.}$$

Thus, $X_t^0 + \sum_{m=0}^{\infty}(X_t^{m+1} - X_t^m)$ converges uniformly on $[0,T]$, P-a.s., to a continuous that $\mathcal{F}_t$-adapted process as we denote $X = \{X_t\}$.

Step 4 In this step, we prove that $E[\sup_{0\le t\le T}|X_t|^2]$ is finite. Indeed, for any $n > m$, consider

$$\begin{aligned}|X_t^n - X_t^m|^2 &= \left|\sum_{j=m}^{n-1}\left(X_t^{j+1} - X_t^j\right)\right|^2 \\ &\le \left(\sum_{j=m}^{n-1} 2^{-j}\right)\sum_{j=m}^{n-1} 2^j \left|X_t^{j+1} - X_t^j\right|^2 \\ &\le 2^{-(m-1)}\sum_{j=0}^{\infty} 2^j \left|X_t^{j+1} - X_t^j\right|^2.\end{aligned}$$

Upon taking supremum over t on both sides,

$$\sup_{0\le t\le T}\left|X_t^n - X_t^m\right|^2 \le 2^{-(m-1)}\sup_{0\le t\le T}\sum_{j=0}^{\infty} 2^j \left|X_t^{j+1} - X_t^j\right|^2. \tag{6.2.6}$$

For each fixed m, $\sup_{0\le t\le T}|X_t^n - X_t^m|^2 \to \sup_{0\le t\le T}|X_t - X_t^m|^2$ a.s. as $n \to \infty$, by Step 3. Consider

$$\begin{aligned}E\sum_{j=0}^{\infty} 2^j \sup_{0\le t\le T}\left|X_t^{j+1} - X_t^j\right|^2 &\le \sum_{j=0}^{\infty} 2^j \rho_{j+1}(T) \\ &\le R\sum_{j=0}^{\infty}\frac{2^j(CT)^j}{j!}.\end{aligned}$$

We thus conclude, by (6.2.6), that for any d, $\sup_{0\le t\le T}|X_t^n - X_t^m|^2$ is dominated by an integrable random variable. Hence, by the dominated convergence theorem,

$$E\left[\sup_{0\le t\le T}|X_t - X_t^m|^2\right] \le 2^{-(m-1)}Re^{2CT}. \tag{6.2.7}$$

In particular, $E[\sup_{0\le t\le T}|X_t|^2] < \infty$, and $X_t^m \to X_t$ uniformly in t, in $L^2(P)$ as $m \to \infty$.

Step 5 Identification of the limit: First of all, by hypothesis (H.1), it is clear that

$$\begin{aligned}E\int_0^T \left|b(s, X_s)\right|^2 ds + E\int_0^T \left|\sigma(s, X_s)\right|^2 ds &\le KT\left[1 + E\sup_{0\le t\le T}|X_t|^2\right]) \\ &< \infty.\end{aligned}$$

For any $t \in [0, T]$, by hypothesis (H.2),

$$E\left[\int_0^t |b(s, X_s) - b(s, X_s^m)|^2 \, ds + E\int_0^t |\sigma(s, X_s) - \sigma(s, X_s^m)|^2 \, ds\right]$$
$$\leq KE\int_0^t |X_s - X_s^m|^2 \, ds$$
$$\leq KT\, E \sup_{0 \leq t \leq T} |X_t - X_t^m|^2.$$

Using the estimate (6.2.7), it follows that

$$\int_0^t b(s, X_s^m) \, ds \to \int_0^t b(s, X_s) \, ds \quad \text{and}$$

$$\int_0^t \sigma(s, X_s^m) \, dW_s \to \int_0^t \sigma(s, X_s) \, dW_s$$

in $L^2(P)$ as $m \to \infty$. Taking the $L^2(P)$-limit as $m \to \infty$ in (6.2.4) we obtain

$$X_t = \xi + \int_0^t b(s, X_s) \, ds + \int_0^t \sigma(s, X_s) \, dW_s \quad \text{a.s.} \tag{6.2.8}$$

Thus, $X = \{X_t\}$ is shown to be a strong solution *provided* that $E|\xi|^2 < \infty$.

Step 6 We will show the existence of a strong solution without the restriction that $E|\xi|^2 < \infty$. Given any random variable ξ independent of W, define the probability measure Q by

$$Q(F) = \frac{1}{C}\int_F \frac{1}{1 + |\xi|^2} \, dP \quad \text{for all } F \in \mathcal{F} \tag{6.2.9}$$

where C is the normalizing constant.

It is clear that P and Q are equivalent measures, and $E^Q |\xi|^2 < \infty$. If $F \in \mathcal{F}_T^W$, then

$$Q(F) = \frac{1}{C}\int_\Omega \frac{1_F}{1 + |\xi|^2} dP = P(F)$$

by using the independence of ξ and W. Therefore, W is a Wiener process on $(\Omega, \mathcal{F}, Q)$.

Besides, for any $A \in \sigma(\xi)$ and $F \in \mathcal{F}_T^W$,

$$Q(A \cap F) = \frac{1}{C}\int_\Omega \frac{1_A 1_F}{1 + |\xi|^2} \, dP$$
$$= E\left(\frac{1_A}{C(1 + |\xi|^2)}\right) P(F) \quad \text{by independence of } \xi \text{ and } W$$
$$= Q(A)P(F)$$
$$= Q(A)Q(F)$$

so that ξ and W are independent in $(\Omega, \mathcal{F}, Q)$. Therefore, our arguments in Steps 1 to 5 can be used in the space $(\Omega, \mathcal{F}, (\mathcal{F}_t), Q)$ to obtain a strong solution, denoted by $X = \{X_t\}$. Thus, Q-a.s., we have

$$\int_0^T \left(|b(t, X_t)| + |\sigma(t, X_t)|^2\right) dt < \infty,$$

and

$$X_t = \xi + \int_0^t b(s, X_s)\, ds + \int_0^t \sigma(s, X_s)\, dW_s.$$

Therefore, they remain valid P-a.s. as well due to the equivalence of the measures P and Q. We have thus shown that X is a strong solution in the probability space $(\Omega, \mathcal{F}, (\mathcal{F}_t), P)$. It is worthwhile to note that the process $\{\int_0^t \sigma(s, X_s)\, dW_s : 0 \le t \le T\}$ is a P-*local* martingale, adapted to $\mathcal{F}_t$.

Step 7 Uniqueness: Suppose that X and Y are two strong solutions on $(\Omega, \mathcal{F}, (\mathcal{F}_t), P)$. Then

$$X_t - Y_t = \int_0^t \left[b(s, X_s) - b(s, Y_s)\right] ds + \int_0^t \left[\sigma(s, X_s) - \sigma(s, Y_s)\right] dW_s. \tag{6.2.10}$$

By equivalence of P and Q where Q is the measure defined by Equation (6.2.9), it suffices to show that $Q\{X_t = Y_t \ \forall\ t \in [0, T]\} = 1$. Define for each $n \in \mathbb{N}$

$$\tau_n = \inf\left\{t \ge 0 : |X_t| \vee |Y_t| \ge n\right\}$$

with the convention that infimum of the empty set is infinity. By the path continuity of solutions, $\tau_n \uparrow \infty$ a.s. For any $t \in [0, T]$, we have

$$\begin{aligned} \sup_{0\le s\le t} \left|X_{s\wedge\tau_n} - Y_{s\wedge\tau_n}\right|^2 \le 2 &\left[\sup_{0\le s\le t} \int_0^{s\wedge\tau_n} [b(u, X_u) - b(u, Y_u)]\, du|^2 \right. \\ &\left. + \sup_{0\le s\le t} \left|\int_0^{s\wedge\tau_n} [\sigma(u, X_u) - \sigma(u, Y_u)]\, dW_u\right|^2\right]. \end{aligned}$$

Taking expectation,

$$\begin{aligned} E^Q\left(\sup_{0\le s\le t\wedge\tau_n} |X_s - Y_s|^2\right) &\le 2TE^Q \int_0^{t\wedge\tau_n} \left|b(u, X_u) - b(u, Y_u)\right|^2 du \\ &\quad + 8E^Q \int_0^{t\wedge\tau_n} \left|\sigma(u, X_u) - \sigma(u, Y_u)\right|^2 du \\ &\le 2(T+4)K \int_0^t E^Q\left(\sup_{0\le r\le u\wedge\tau_n} |X_r - Y_r|^2\right) du \end{aligned}$$

where we have used the Doob inequality and hypothesis H.2. Applying the Gronwall lemma with $f(t) = E^Q\left(\sup_{0\leq s\leq t\wedge\tau_n} |X_s - Y_s|^2\right)$, and $g(t) \equiv 0$, we conclude that $f(t) \equiv 0$. Hence, $Q\{X_t = Y_t \ \forall\, t \in [0, T]\} = 1$ by letting $n \to \infty$. ■

Next, we define adapted functionals so that the coefficients b and σ in the stochastic differential equation (6.1.2) can, at time t, depend, not just on X_t, but on the path of the solution X up to time t. Let $C_d = C([0, T], \mathbb{R}^d)$ denote the space of all continuous functions on $[0, T]$ taking values in $\mathbb{R}^d$. Let $\mathcal{B}_t(C_d)$ be the σ-field generated by the finite-dimensional cylinder sets up to time t. Let $\mathcal{B}(C_d) := \mathcal{B}_T(C_d)$.

Definition 6.2.1 *The functional $f : [0, T] \times C_d \times \Omega \to \mathbb{R}$ is called an* **adapted functional** *if the following conditions are satisfied:*

(i) f is $\mathcal{B}[0, T] \times \mathcal{B}(C_d) \times \mathcal{F}$-measurable.
(ii) For each t, $f(t, \cdot, \cdot)$ is measurable with respect to $\mathcal{B}_{t+}(C_d) \times \mathcal{F}_t$.

A d-dimensional functional $b = (b_1, \ldots, b_d)$ is an adapted functional if each b_j is adapted in the above sense. Likewise, a matrix-valued functional $\sigma = (\sigma_{ij})$ where $i = 1, \ldots, d$ and $j = 1, \ldots, k$ is an adapted functional if each σ_{ij} is. Suppose that b is an d-adapted functional and σ is an $d \times k$ matrix-valued adapted functional. Let b and σ satisfy the following hypotheses:

Hypotheses G

For all $t \in [0, T]$, and $f, g \in C_d$, there exists a constant $K > 0$ such that

G.1 $|b(t,f)|^2 + |\sigma(t,f)|^2 \leq K(1 + \|f\|_t^2)$ a.s.
G.2 $|b(t,f) - b(t,g)|^2 + |\sigma(t,f) - \sigma(t,g)|^2 \leq K\|f - g\|_t^2$ a.s.,

where $\|f\|_t$ is the supremum norm of f over $[0, t]$.

An extension of Theorem 6.2.1 is stated below. It can be proved exactly along the same lines as the proof of Theorem 6.2.1.

Theorem 6.2.2 *Suppose that $(\Omega, \mathcal{F}, (\mathcal{F}_t), P)$, W, ξ are as in Section 6.1. Let b and σ be adapted functionals satisfying Hypotheses G. Then the stochastic differential equation*

$$dX_t = b(t, X)\, dt + \sigma(t, X)\, dW_t \tag{6.2.11}$$

with initial condition ξ has a unique strong solution $X = \{X_t\}$.

6.3 Linear Stochastic Differential Equations

In particular instances such as linear stochastic differential equations, the strong solution can be explicitly obtained by using integrating factors, stochastic exponentials, and the Itô formula. We develop the linear theory by the following series of examples. The setup given in Section 6.1 is assumed. By Theorem 6.2.1, a unique strong solution exists in each of the following examples.

Example 6.3.1 *Consider the stochastic differential equation*

$$dX_t = bX_t\, dt + dW_t$$

with $X_0 = \xi$, and b, any constant. Consider the discounted process

$$Y_t = e^{-bt} X_t = f(t, X_t)$$

where $f(t, x) = e^{-bt} x$. By the Itô formula applied to f,

$$Y_t = \xi + \int_0^t e^{-bs}\, dW_s.$$

Therefore,

$$X_t = \xi e^{bt} + \int_0^t e^{b(t-s)}\, dW_s. \tag{6.3.1}$$

The above solution is exactly what one would obtain by viewing the stochastic differential equation ω-wise as an ordinary differential equation.

Example 6.3.2 *Consider the stochastic differential equation*

$$dX_t = \sigma X_t\, dW_t$$

with initial condition $X_0 = \xi$, and σ, any constant. From Chapter 6, we know that by the Itô lemma, $X_t = \xi\, \mathcal{E}(\sigma W)_t$ where E denotes the stochastic exponential given by

$$E(\sigma W_t) = \exp\left\{\sigma W_t - \frac{\sigma^2}{2} t\right\}.$$

The appearance of $-\frac{\sigma^2}{2}t$ in the exponent is a surprise. It illustrates the important fact that stochastic differential equations cannot be solved by interpreting them as ordinary differential equations.

Example 6.3.3 *Consider the stochastic differential equation*

$$dX_t = (\sigma_1 X_t + \sigma_2)\, dW_t$$

with $X_0 = \xi$, *and* σ_1, σ_2, *any constants. Analogous to the method employed in Example 6.3.1, define*

$$Y_t = \frac{X_t}{E(\sigma_1 W)_t}$$

where the stochastic exponential is used since the linear term in the differential equation is stochastic.

Clearly, $Y_t = f(t, X_t, W_t)$ *where* $f(t,x,w) = x \exp\left\{-\sigma_1 w + \frac{\sigma_1^2}{2} t\right\}$. *Applying the Itô formula to* f*, one obtains*

$$\begin{aligned}
Y_t &= \xi + \int_0^t f_s'(t, X_s, W_s)\, ds + \int_0^t f_x'(s, X_s, W_s)\, dX_s + \int_0^t f_w'(s, X_s, W_s)\, dW_s \\
&\quad + \frac{1}{2} \int_0^t f_{ww}''(s, X_s, W_s)\, ds + \int_0^t f_{xw}''(s, X_s, W_s)\, d\langle X, W\rangle_s \\
&= \xi + \frac{\sigma_1^2}{2} \int_0^t Y_s\, ds + \int_0^t \frac{1}{E(\sigma_1 W)_s} (\sigma_1 X_s + \sigma_2)\, dW_s - \sigma_1 \int_0^t Y_s\, dW_s \\
&\quad + \frac{1}{2}\sigma_1^2 \int_0^t Y_s\, ds - \sigma_1 \int_0^t \frac{1}{E(\sigma_1 W)_s} (\sigma_1 X_s + \sigma_2)\, ds \\
&= \xi + \sigma_2 \int_0^t \frac{1}{E(\sigma_1 W)_s}\, dW_s - \sigma_1 \sigma_2 \int_0^t \frac{1}{E(\sigma_1 W)_s}\, ds
\end{aligned}$$

upon simplification. Thus, from the definition of Y_t*, we have*

$$X_t = E(\sigma_1 W)_t \left[\xi + \sigma_2 \int_0^t \frac{1}{E(\sigma_1 W)_s}\, dW_s - \sigma_1\sigma_2 \int_0^t \frac{1}{E(\sigma_1 W)_s}\, ds\right]. \tag{6.3.2}$$

Whereas the first two terms on the right side of Equation (6.3.2) are analogous to the terms in Equation (6.3.1), the last term is a surprise. It arises from the covariation process between $\{X_t\}$ *and* $\{W_t\}$.

Example 6.3.4 *Consider*

$$dX_t = bX_t\, dt + \sigma X_t\, dW_t$$

where b *and* σ *are constants, and* $X_0 = \xi$. *As in Example 6.3.1, let* $Y_t = e^{-bt} X_t$. *Then by the Itô formula,*

$$Y_t = \xi + \int_0^t \sigma Y_s\, dW_s.$$

By using Example 6.3.2, it follows that $Y_t = \xi E(\sigma W)_t$. *Hence,*

$$X_t = \xi e^{bt} E(\sigma W)_t.$$

Example 6.3.5 *Consider*

$$dX_t = (b_1 X_t + b_2)\,dt + (\sigma_1 X_t + \sigma_2)\,dW_t$$

where $b_1, b_2, \sigma_1, \sigma_2$ are constants, and $X_0 = \xi$. Since linearity is present in both terms on the right side of the equation, the correct discounted process can be guessed from the previous example, and is given by

$$Y_t = \frac{X_t e^{-b_1 t}}{E(\sigma_1 W)_t}.$$

Since Y_t is a function of t, X_t, and W_t, one may use the Itô lemma as in Example 6.3.3 to obtain

$$Y_t = \xi + b_2 \int_0^t \frac{e^{-b_1 s}}{E(\sigma_1 W)_s}\,ds + \sigma_2 \int_0^t \frac{e^{-b_1 s}}{E(\sigma_1 W)_s}\,dW_s - \sigma_1\sigma_2 \frac{e^{-b_1 s}}{E(\sigma_1 W)_s}\,ds.$$

Thus,

$$X_t = E(\sigma_1 W)_t e^{b_1 t}\left[\xi + b_2 \int_0^t \frac{e^{-b_1 s}}{E(\sigma_1 W)_s}\,ds \right.$$
$$\left. + \sigma_2 \int_0^t \frac{e^{-b_1 s}}{E(\sigma_1 W)_s}\,dW_s - \sigma_1\sigma_2 \int_0^t \frac{e^{-b_1 s}}{E(\sigma_1 W)_s}\,ds\right].$$

The above expression for X_t is an explicit function of the Wiener process and time. We give the expression in more detail and generality in the following remark.

Remark 6.3.1 Suppose the constants b_1, b_2, σ_1, and σ_2 are replaced by deterministic continuous functions of time in Example 6.3.5. Then the solution is given by X_t

$$= \exp\left[\int_0^t \left(b_1(u) - \frac{1}{2}\sigma_1^2(u)\right) du + \int_0^t \sigma_1(u)\,dW_u\right]\left\{\xi + \right.$$
$$\int_0^t \exp\left[-\int_0^s \left(b_1(u) - \frac{1}{2}\sigma_1^2(u)\right) du - \int_0^s \sigma_1(u)\,dW_u\right](b_2(s) - \sigma_1(s)\sigma_2(s))\,ds$$
$$\left. + \int_0^t \exp\left[-\int_0^s \left(b_1(u) - \frac{1}{2}\sigma_1^2(u)\right) du - \int_0^s \sigma_1(u)\,dW_u\right]\sigma_2(s)\,dW_s\right\}.$$

Other examples of stochastic differential equations that admit explicit solutions will be given later.

6.4 Weak Solutions

In several applications, approximate solutions of a given stochastic differential equation can readily be obtained, and they converge in distribution to a limit. It is natural to expect that the limit is a solution of the stochastic differential equation. However, the proof of identification of the limit as a solution necessitates a change of the underlying probability space. For instance, such a change may occur by a use of the Skorohod representation theorem. One may also change the given probability measure by the Girsanov transformation. Whenever a stochastic differential equation is solved in a probability space other than the given space, we are led to distinguish such solutions and call them weak solutions.

It is worthwhile to mention that weak solutions to a stochastic differential equation exist under milder conditions on the coefficients. To define a weak solution, we assume that the functions (or adapted functionals) b and σ are as in Section 6.1. They are *not* assumed to satisfy the Hypotheses H (or Hypotheses G), which are too strong to show the existence of a weak solution.

In this section, we will prove the important result of Yamada and Watanabe which says that a pathwise unique weak solution of a stochastic differential equation is a unique strong solution.

Definition 6.4.1 *Suppose that, on some probability space $(\Omega, \mathcal{F}, P)$, there exists an increasing family $(\mathcal{G}_t)$ of sub σ-fields of $\mathcal{F}$, a d-dimensional random variable ξ with a given distribution μ, and continuous $\mathcal{G}_t$-adapted processes $W = \{W_t\}$ and $X = \{X_t\}$ such that*

(i) $(W_t, \mathcal{G}_t, P)$ *is a k-dimensional Wiener martingale that is independent of* ξ.

(ii) $P\left\{\omega : \int_0^T [|b(t, X_t(\omega))| + |\sigma(t, X_t(\omega))|^2] < \infty\right\} = 1.$

(iii) $X_t(\omega) = \xi(\omega) + \int_0^t b(s, X_s(\omega))\, ds + \int_0^T \sigma(s, X_s)\, dW_s(\omega)$ *P-a.s. for all t.*

Then the family $(\Omega, \mathcal{F}, (\mathcal{G}_t), P, \xi, \{W_t\}, \{X_t\})$ is called a **weak solution** *of the stochastic differential equation* (6.1.1).

First of all, it is clear that a strong solution is also a weak solution. Next, we define a notion of uniqueness which is well suited for weak solutions.

Definition 6.4.2 *A weak solution of the stochastic differential equation* (6.1.1) *is said to be* **unique in law** *if, for any two weak solutions given by $(\Omega_i, \mathcal{F}_i, (\mathcal{G}_t^i), P_i, \xi^i, (W_t^i), (X_t^i))$, $i = 1, 2$, the two processes $\{X_t^1\}$ and $\{X_t^2\}$ have the same probability distribution or law. In other words,*

$$P_1 \left\{\omega : X^1(\cdot, \omega) \in B\right\} = P_2 \left\{\omega : X^2(\cdot, \omega) \in B\right\}$$

for all finite-dimensional cylinder set B in $C([0, T] : \mathbb{R}^d)$.

In particular, note that $P_i\{X^i(0) \in B\} = \mu(B)$ where μ is the distribution of ξ, and B, any Borel set in $\mathbb{R}^d$.

A related notion of uniqueness is given below, where the underlying probability space remains unaltered, but the filtration is changed.

Definition 6.4.3 *A weak solution of the stochastic differential equation* (6.1.1) *is said to be* **pathwise unique** *if, for any two weak solutions given by $(\Omega, \mathcal{F}, (\mathcal{G}_t^i), P, \xi^i, (W_t), (X_t^i))$, $i = 1, 2$, the following holds:*

$$P\left\{X^1(t) = X^2(t) \ \ \forall t \geq 0\right\} = 1.$$

First, we prove a result of Yamada and Watanabe that states that pathwise uniqueness implies uniqueness in law for weak solutions. For this purpose and for later use, we recall the basic concept **regular conditional probability distribution** (rcpd) of X given Y where (X, Y) is a given random vector. We assume that (X, Y) takes values in a complete, separable metric space E. Let $\mathcal{E}$ denote the Borel σ-field of E.

From Chapter 1, we know that one can leave the original probability space on which the random variables (and processes) were defined and construct them (canonically), without changing their probability distributions, on the representation space, that is, the space that contains the values of the random variable or the process. Therefore, we can denote E itself as Ω, and $\mathcal{E}$ as $\mathcal{F}$.

Definition 6.4.4 *Suppose that $(\Omega, \mathcal{F})$ is the measurable space described above. Let (X, Y) be a random vector defined on it. Let P be the probability distribution of (X, Y). Then a function $Q : \Omega \times \sigma(X) \to [0, 1]$ is called an rcpd for X given Y if the following conditions are satisfied:*

(i) For any fixed $\omega \in \Omega$, $Q(\omega, \cdot)$ is a probability measure on $(\Omega, \sigma(X))$.

(ii) For any fixed $A \in \sigma(X)$, $Q(\cdot, A)$ is a $\sigma(Y)$-measurable function.

(iii) For each $A \in \sigma(X)$, $Q(\omega, A) = P\{A|Y\}(\omega)$ P-a.s.

It is well known that a unique regular conditional probability exists if Ω is a complete, separable metric space, and $\mathcal{F}$, its Borel σ-field. Consider any $\omega \in \{Y = y\}$ for an arbitrary but fixed y. Let us denote the common value of $Q(\omega, A)$ by $Q(y, A)$ for each $A \in \sigma(X)$. Then, we have

$$Q(y, A) = P\{A \mid Y = y\} \quad P\text{-a.s.}$$

where $P\{A \mid Y = y\}$ stands for the common value of $P\{A \mid \sigma(Y)\}(\omega)$ for any ω that satisfies $Y(\omega) = y$.

We return to the context of the Yamada-Watanabe result. Suppose we have two weak solutions, $(\Omega_i, \mathcal{F}_i, (\mathcal{G}_t^i), P_i, \xi^i, (W_t^i), (X_t^i))$ with $i = 1, 2$, of Equation (6.1.1). Recall that X_0^i has a common distribution denoted by μ. Let us denote the distribution of $(X_0^i, W^i, X^i - X_0^i)$ by ν_i. Let us define the process

$$Y^i = X^i - X_0^i \quad \text{for} \quad i = 1, 2.$$

Then, ν_i is a probability measure on the space

$$S := \mathbb{R}^d \times C([0, \infty); \mathbb{R}^k) \times C([0, \infty); \mathbb{R}^d)$$

equipped with its Borel σ-field denoted by $\mathcal{S}$. For all Borel subsets Γ,

$$\nu_i(\Gamma) = P_i\left\{(X_0^i, W^i, Y^i) \in \Gamma\right\}.$$

Consider the probability space $(S, \mathcal{S}, \nu_i)$. Since S is a complete, separable metric space, the regular conditional distribution of Y^i given (X_0^i, W^i), denoted by Q_i, exists. Let η

denote the distribution of the k-dimensional Wiener process W^i. If x, w denotes the generic values of X_0^i, W^i, we obtain

$$\nu_i(B \times A) = \int_B Q_i(x, w : A)\mu(dx)\eta(dw)$$

for all $B \in \mathcal{B}(\mathbb{R}^d) \times \mathcal{B}(C([0,\infty) : \mathbb{R}^k))$ and $A \in \mathcal{B}(C([0,\infty) : \mathbb{R}^d))$. By the independence of X_0^i and W^i, the joint distribution of (X_0^i, W^i) is the product of μ and η.

Define the probability space $\Omega = S \times C([0,\infty); \mathbb{R}^d)$ with σ-field $\mathcal{F}$ given by the completion of $\mathcal{S} \times \mathcal{B}(C[0,\infty) : \mathbb{R}^d)$ with respect to the measure

$$P(d\omega) = Q_1\left(x, w; dy^1\right) Q_2\left(x, w; dy^2\right) \mu(dx)\eta(dw) \tag{6.4.1}$$

where $\omega = (x, w, y^1, y^2)$. Define the filtration $(\mathcal{G}_t)$ by

$$\mathcal{G}_t = \sigma\left(x, w(s), y^1(s), y^2(s) : 0 \le s \le t\right) \vee P\text{-null sets of } \mathcal{F}.$$

Let $\mathcal{F}_t = \mathcal{G}_{t+}$. Then the filtration $(\mathcal{F}_t)$ satisfies the usual conditions. From the definition of the measure P, it is clear that

$$P\left\{\omega : \left(x, w, y^i\right) \in B\right\} = \nu_i(B) = P_i\left\{\left(X_0^i, W^i, Y^i\right) \in B\right\}$$

for all $B \in \mathcal{S}$. Therefore, the distribution of $(w, x + y^i)$ under P is the same as that of (W^i, X^i) under P_i.

Theorem 6.4.1 *If weak solutions of Equation* (6.1.1) *are pathwise unique, then they are also unique in law.*

Proof Suppose that $(\Omega_i, \mathcal{F}_i, (\mathcal{G}_t^i), P_i, \xi, (W_t^i), (X_t^i))$ for $i = 1, 2$ are two weak solutions of Equation (6.1.1). Then, by the construction described prior to the statement of this theorem, the two solutions can be represented in a single probability space. Indeed, $(w, x + y^i)$ for $i = 1, 2$ on $(\Omega, \mathcal{F}, (\mathcal{F}_t), P)$ are weak solutions such that the distribution of (W^i, X^i) under P_i coincides with the distribution under P of $(w, x + y^i)$.

By pathwise uniqueness, $P\left\{x + y^1(t) = x + y^2(t) \;\; \forall \; t\right\} = 1$. Therefore,

$$P\left\{\omega : y^1 = y^2\right\} = 1.$$

Hence, for any $B \in \mathcal{S}$,

$$\begin{aligned}
\nu_1(B) &= P\left\{\omega : \left(x, w, y^1\right) \in B\right\} \\
&= P\left\{\omega : \left(x, w, y^1\right) \in B, \; y^1 = y^2\right\} \\
&= P\left\{\omega : \left(x, w, y^2\right) \in B, \; y^1 = y^2\right\} \\
&= P\left\{\omega : \left(x, w, y^2\right) \in B\right\} \\
&= \nu_2(B).
\end{aligned}$$

In other words, the solutions have the same distribution. ∎

We illustrate the ideas of weak and strong solutions and the two notions of uniqueness in the following example.

Example 6.4.2 *Let B_t be a Wiener process on a probability space $(\Omega, \mathcal{F}, P)$ and let $\mathcal{F}_t = \mathcal{F}_t^B$. Define*

$$\sigma(x) = \begin{cases} 1 & \text{if } x > 0, \\ -1 & \text{if } x \leq 0. \end{cases}$$

Consider the stochastic differential equation

$$dX_t = \sigma(X_t)\, dB_t \quad \text{with} \quad X_0 = 0. \tag{6.4.2}$$

First, a weak solution exists.

Proof Toward this, define

$$W_t = \int_0^t \sigma(B_s)\, dB_s.$$

Then, W is a continuous martingale and $\langle W \rangle_t = \int_0^t \sigma^2(B_s)\, ds = t$. Therefore, W_t is a Wiener process. Besides,

$$\int_0^t \sigma(B_s)\, dW_s = \int_0^t \sigma^2(B_s)\, dB_s = B_t.$$

Thus, $B_t = \int_0^t \sigma(B_s)\, dW_s$ with $B_0 = 0$. In other words, the given Wiener process B_t is a solution of (6.4.2) if we change the driving Wiener process to W_t. In other words, $(\Omega, \mathcal{F}, (\mathcal{F}_t^B), P, 0, W_t, B_t)$ is a *weak* solution of Equation (6.4.2).

In addition, note that

$$\int_0^t \sigma(-B_s)\, dW_s = \int_0^t \sigma(-B_s)\sigma(B_s)\, dB_s = -B_t.$$

Thus $-B_t$ is also a weak solution of (6.4.2) on $(\Omega, \mathcal{F}, (\mathcal{F}_t^B), P)$, with W_t as the driving Wiener process. Hence, we do not have pathwise uniqueness of solutions.

However, uniqueness in law holds for the weak solutions. Indeed, let X_t be a weak solution of (6.4.2) with W_t as a Wiener process on some probability space. Then, X_t is a continuous martingale, and $\langle X \rangle_t = t$. Hence, X_t is a Wiener process. The law of any weak solution is thus the Wiener measure. ■

Next, we prove an important result of Yamada and Watanabe that states that weak existence and pathwise uniqueness of solutions imply the existence of a unique strong solution. First, let us remark that, under Hypotheses H, the strong solution X_t of Equation (6.1.1) can be written as a Borel-measurable functional of ξ and $\{W_s : 0 \leq s \leq t\}$ since

$$\begin{aligned} &\sigma(X_s : 0 \leq s \leq t) \vee (\text{all } P\text{-null sets in } \mathcal{F}) \\ &\subseteq \mathcal{F}_t = \sigma(\xi,\ W_s\ 0 \leq s \leq t) \vee (\text{all } P\text{-null sets in } \mathcal{F}). \end{aligned}$$

However, in the context of weak solutions, the σ-field $\mathcal{G}_t$ may be larger than $\mathcal{F}_t$. Hence a weak solution X_t may not admit a representation as a Borel-measurable functional of ξ and $\{W_s : 0 \le s \le t\}$.

We need the following two lemmas. From Chapter 1, recall the notation $\mathcal{B}_t(C([0,\infty) : \mathbb{R}^d))$ for the σ-field in $C([0,\infty) : \mathbb{R}^d)$ generated by the collection of all finite-dimensional cylinder sets based on $0 \le t_1 < t_2 \cdots < t_n)$ with $t_n \le t$.

Using the notation built before, let us define the σ-field

$$\mathcal{H}_t := \left[\mathcal{B}\left(\mathbb{R}^d\right) \times \mathcal{B}_t\left(C\left([0,\infty) : \mathbb{R}^k\right)\right)\right] \vee \left\{(\mu \times \eta)\text{-null sets}\right\}.$$

Lemma 6.4.3 *For any fixed $A \in \mathcal{B}_t(C([0,\infty) : \mathbb{R}^d))$, consider $Q_i(x, w, A)$ as functions of x, w for $i = 1, 2$. $Q_i(x, w, A)$ are $\mathcal{H}_t$ measurable.*

Proof Let $i = 1$. Consider the rcpd of Y^1 restricted to the interval $[0, t]$ given X_0^1, and W^1 restricted to the interval $[0, t]$. We can call it the rcpd of $\mathcal{B}_t(C([0,\infty) : \mathbb{R}^d))$ given $\mathcal{B}(\mathbb{R}^d) \times \mathcal{B}_t(C([0,\infty) : \mathbb{R}^k))$. Let us denote it as Q_t^1. Then, $Q_t^1(x, w, A)$ is $\mathcal{B}(\mathbb{R}^d) \times \mathcal{B}_t(C([0,\infty) : \mathbb{R}^k))$-measurable, and

$$P_1(B \times A) = \int_B Q_t^1(x, w, A)\mu(dx)\eta(dw)$$

for all $B \in \mathcal{B}(\mathbb{R}^d) \times \mathcal{B}_t(C([0,\infty) : \mathbb{R}^k))$.

Let us now take $B = B_1 \times (B_2 \cap B_3)$ where $B_1 \in \mathcal{B}(\mathbb{R}^d)$, $B_2 \in \mathcal{B}_t(C([0,\infty) : \mathbb{R}^k))$, and B_3 to be of the form

$$B_3 = \left\{w : w(t+s) - w(t) \in F\right\}$$

for some $s > 0$ and $F \in \mathcal{B}(C([0,\infty) : \mathbb{R}^k))$. Then,

$$\begin{aligned}
\int_B Q_t^1(x, w, A)\mu(dx)\eta(dw) &= \int_{B_1 \times B_2} Q_t^1(x, w, A)\mu(dx)\eta(dw)\eta(B_3) \\
&= \int_{B_1 \times B_2} Q_t^1(x, w, A)\mu(dx)\eta(dw)\eta(B_3) \\
&= P_1(B_1 \times B_2 \times A)\eta(B_3) \\
&= P_1(B_1 \times B_2 \times A)P_1\left(\mathbb{R}^d \times B_3 \times C\left([0,\infty) : \mathbb{R}^d\right)\right) \\
&= P_1(B \times A).
\end{aligned}$$

Hence, we can conclude that $Q_t^1(x, w, A)$ coincides with $Q^1(x, w, A)$, $\mu \times \eta$ almost surely. The $\mathcal{H}_t$ measurability of $Q^1(\cdot, \cdot, A)$ is thus established. ■

Lemma 6.4.4 *There exists a Borel-measurable function $f : \mathbb{R}^d \times C([0,\infty) : \mathbb{R}^k) \to C([0,\infty) : \mathbb{R}^d)$ such that for almost every (x, w) with respect to $\mu \times \eta$, we have*

$$Q^1(x, w, \{f(x, w)\}) = Q^2(x, w, \{f(x, w)\}) = 1. \tag{6.4.3}$$

Proof For each fixed x, w, define a measure Q on $\mathcal{B}(C([0,\infty) : \mathbb{R}^d)^{\times 2})$ by

$$Q(x, w, dy_1, dy_2) = Q^1(x, w, dy_1)Q^2(x, w, dy_2).$$

By definition of the measure P in (6.4.1), we have

$$P(dx, dw, dy_1, dy_2) = Q(x, w, dy_1, dy_2)\mu(dx)\eta(dw).$$

Let us set $A = \left\{(y_1, y_2) \in C([0,\infty) : \mathbb{R}^d)^{\times 2} : y_1 = y_2\right\}$. Then, by pathwise uniqueness,

$$Q(x, w, A) = 1$$

for almost every (x, w) with respect to $\mu \times \eta$. By definition of Q and Fubini's theorem, it follows that, for each such (x, w), there exists a corresponding y such that, for $i = 1, 2$,

$$Q^i(x, w, \{y\}) = 1.$$

Therefore, we can write this y as $f(x, w)$. The Borel measurability of f is a consequence of that of Q^i. ■

Remark 6.4.1 From Lemma 6.4.3, it follows that $f^{-1}(A) \in \mathcal{H}_t$ for any $A \in \mathcal{B}_t(C([0,\infty), \mathbb{R}^d))$ for all t. In other words, f is $\mathcal{H}_t \;/\; \mathcal{B}_t(C([0,\infty), \mathbb{R}^d))$ measurable.

Theorem 6.4.5 *Suppose that the stochastic differential equation* (6.1.1) *has a weak solution* $(\Omega, \mathcal{F}, (\mathcal{G}_t), P, \xi, \{W_t\}, \{X_t\})$ *which is pathwise unique. Then there exists a Borel-measurable function*

$$g : \mathbb{R}^d \times C\left([0,\infty); \mathbb{R}^k\right) \to C\left([0,\infty) : \mathbb{R}^d\right)$$

such that

$$X = g(\xi, W) \text{ } P\text{-a.s.} \tag{6.4.4}$$

The function g *has the property that* g *is* $\mathcal{H}_t \;/\; \mathcal{B}_t(C([0,\infty), \mathbb{R}^d))$ *measurable.*

Proof Set $g(x, w) = x + f(x, w)$ where f is as in Lemma 6.4.4. Since

$$P\left\{(x, w, y_1, y_2) \in \Omega : y_1 = y_2 = f(x, w)\right\} = 1,$$

we obtain (6.4.4). The measurability properties of g follow from Lemma 6.4.3. ■

Suppose that we are given a probability space $(\Omega, \mathcal{F}, P)$ with a filtration $\{\mathcal{F}_t\}$, an initial random variable α with distribution μ, and an independent Wiener process $\{B_t\}$ adapted to $\mathcal{F}_t$ and defined on it. Then Theorem 6.4.5 implies that the process $g(\alpha, W)$ is a *strong* solution of the stochastic differential equation (6.1.1) on this space, with α as the initial data and $\{B_t\}$ as the driving Wiener process. Thus, we can state

Theorem 6.4.6 *Suppose that the stochastic differential equation (6.1.1) has a weak solution X which is pathwise unique. Then, there exists a unique strong solution of (6.1.1) on any probability space, which is rich enough to support the initial random variable and an independent Wiener process.*

We can obtain weak solutions of stochastic differential equations even with a measurable drift coefficient by removal of drift using the Girsanov theorem.

Theorem 6.4.7 *Consider the d-dimensional stochastic differential equation*

$$X_t = X_0 + \int_0^t b\,(s, X_s)\, ds + B_t \tag{6.4.5}$$

on the interval $[0, T]$, where B is a d-dimensional Brownian motion, and X_0 is a random variable, independent of W, with its distribution denoted by μ. Assume that b is a bounded Borel-measurable $\mathbb{R}^d$-valued function on $[0, T] \times \mathbb{R}^d$. Then, there exists a weak solution of Equation (6.4.5) for any any given initial distribution μ.

Proof Consider a Wiener martingale $\{W_t, \mathcal{F}_t : 0 \le t \le T\}$ with distribution of W_0 given by μ on a probability space $(\Omega, \mathcal{F}, P)$. Then,

$$M_t = \exp\left\{\sum_{j=1}^{d} \int_0^t b_j(s, W_s)\, ds - \frac{1}{2}\int_0^t |b(s, W_s)|^2\, ds\right\}$$

is a martingale. Define the probability measure Q defined by $\frac{dQ}{dP} = M_T$. By the Girsanov theorem, under Q the process

$$B_t = W_t - W_0 - \int_0^t b(s, W_s)\, ds$$

for $0 \le t \le T$ is a standard d-dimensional Brownian motion on $(\Omega, \mathcal{F}, Q)$. Thus,

$$W_t = W_0 + \int_0^t b(s, W_s)\, ds + B_t.$$

If we rename W as X, then X is a weak solution of (6.4.5) on the space $(\Omega, \mathcal{F}, Q)$. ■

6.5 Markov Property

Let $(\Omega, \mathcal{F}, P)$ be a probability space with a filtration $(\mathcal{F}_t)$. An $\mathcal{F}_t$-adapted process $\{X_t\}$ with values in $\mathbb{R}^d$ is called a **Markov process** if the following property is satisfied:

For all $s \le t$, $E[f(X_t)|\mathcal{F}_s] = E[f(X_t) \,|\, X_s]$ a.s. for every bounded Borel-measurable function f with values in $\mathbb{R}$.

The above statement is the Markov property. If $f = 1_B$, where B is a d-dimensional Borel set, the property reduces to one about conditional probability. If the time s denotes the present, and t denotes a future time, the Markov property can be stated simply as follows. The probability of a future event given information on the past and present is equal to the probability of the event given information on the present. Thus, information on the past is ignored. An equivalent statement of the Markov property is given below:

For any $0 \leq s_1 \leq s_2 \leq \ldots \leq s_n \leq s \leq t$, and for every real, bounded Borel function f on $\mathbb{R}^d$,

$$E\{f(X_t)|X_{s_1}, X_{s_2}, \ldots, X_{s_n}, X_s\} = E\left[f(X_t)|X_s\right] \quad \text{a.s.}$$

We will describe the ideas that lead to the proof of the Markov property of solutions of stochastic differential equations. Fix any positive s, T such that $T > s$. For $t \in [s, T]$, let $X(x, t, \omega)$ be the strong solution of

$$dX_t = b(t, X_t)\, dt + \sigma(t, X_t)\, dW_t \tag{6.5.1}$$

and

$$X_s = x$$

where $x \in \mathbb{R}^d$, and b and σ satisfy the conditions listed in Section 6.1 so that a unique strong solution exists. Since the initial time is fixed as s, the solution is adapted to the filtration

$$\mathcal{F}_s^{W,t} = \sigma\{W_v - W_u : s \leq u \leq v \leq t\} \vee \{\text{all } P\text{-null sets}\}.$$

Lemma 6.5.1 *Suppose that $f : \mathbb{R}^d \times \Omega \to \mathbb{R}$ is a bounded $\mathcal{B}(\mathbb{R}^d) \times \mathcal{F}_s^{W,t}$-measurable function. If Z is a d-dimensional, $\mathcal{F}_s$-measurable random variable, then*

$$E[f(Z, \omega)|\mathcal{F}_s] = g(Z) \tag{6.5.2}$$

where $g(x) = E[f(x, \omega)]$.

Proof Define the class of sets

$$\mathcal{S} = \left\{S \in \mathcal{B}\left(\mathbb{R}^d\right) \times \mathcal{F}_s^{W,t} : \text{Equation (6.5.2) holds for } f = 1_S\right\}.$$

We claim that $\mathcal{S}$ is a Dynkin class. For, it is clear that $\mathbb{R}^d \times \Omega \in \mathcal{S}$. If $\{S_n\}$ is an increasing sequence of sets in $\mathcal{S}$ with S as the limit, then by the monotone convergence theorem,

$$E\left\{1_S(Z, \omega)|\mathcal{F}_s\right\} = E\left\{1_{S_n}(Z, \omega)|\mathcal{F}_s\right\} = \lim_{n\to\infty} g_n(Z)$$

where

$$g_n(x) = \int_\Omega 1_{S_n}(x, \omega) P(d\omega) \to \int_\Omega 1_S(x, \omega) P(d\omega) = g(x).$$

Thus, $\lim_{n\to\infty} g_n(Z) = g(Z)$ so that $S \in \mathcal{S}$.

If $A \subseteq B$, and $A, B \in \mathcal{S}$, then by the linearity of conditional expectation,

$$E\left\{1_{B\setminus A}(Z,\omega)|\mathcal{F}_s\right\} = E\left\{1_B(Z,\omega)|\mathcal{F}_s\right\} - E\left\{1_A(Z,\omega)|\mathcal{F}_s\right\} = g_1(Z) - g_2(Z)$$

where

$$g_1(x) - g_2(x) = \int_\Omega 1_B(x,\omega)p(d\omega) - \int_\Omega 1_A(x,\omega)p(d\omega) = \int_\Omega 1_{B\setminus A}(x,\omega)P(d\omega).$$

Thus, $B \setminus A \in \mathcal{S}$.

Next, define the class of sets

$$\mathcal{R} = \left\{R = B \times F \text{ where } B \in \mathcal{B}\left(\mathbb{R}^d\right); \text{and } F \in \mathcal{F}_s^{W,t}\right\}.$$

Then, $\mathcal{R} \subseteq \mathcal{S}$. Indeed,

$$E\left\{1_{B\times F}(Z,\omega)|\mathcal{F}_s\right\} = 1_B(Z)E\left\{1_F(\omega) \mid \mathcal{F}_s\right\} = 1_B(Z)E(1_F)$$

by independence of $\mathcal{F}_s$ and $\mathcal{F}_s^{W,t}$. Define $g(x) = E[1_R(x,\omega)]$. Then, $g(x) = 1_B(x)E(1_F)$ so that $E\left\{1_{B\times F}(Z,\omega)|\mathcal{F}_s\right\} = g(Z)$.

It is quite simple to observe that $\mathcal{R}$ is closed under intersection so that it is a π-system. By the Dynkin class theorem, $\sigma(\mathcal{R}) \subseteq \mathcal{S}$. Thus, we can conclude that $\mathcal{B}(\mathbb{R}^d) \times F_s^{W,t} = \mathcal{S}$. It is quite routine to extend the equality (6.5.2) to bounded simple functions, and hence to any bounded $\mathcal{B}(\mathbb{R}^d) \times F_s^{W,t}$-measurable function. ■

Next, we need certain measurability and continuity properties of the solution $X(x,t,\omega)$ of the stochastic differential equation (6.5.1). We will show that $X(x,t,\omega)$ is $\mathcal{B}(\mathbb{R}^d) \times \mathcal{F}_s^{W,t}$-measurable as a function of (x,ω) for each fixed $t \geq s$. Let X_t^x denote $X(x,t)$ in what follows.

Lemma 6.5.2 *There exists a constant C depending on K and T such that for any x and y in* $\mathbb{R}^d$, *and* $s \in [0,T]$,

$$E\left\{\sup_{s\leq t\leq T} |X_t^x - X_t^y|^2\right\} \leq C|x-y|^2. \tag{6.5.3}$$

Proof For each $N \geq 1$, define the stopping time

$$\tau_N = T \wedge \inf\left\{t \geq s : \left|X_t^x\right|^2 + \left|Y_t^y\right|^2 \geq N\right\}.$$

By the path continuity of strong solutions, $\lim_{N\to\infty} \tau_N = T$ a.s. For any $0 \leq s \leq T$, we have

$$X_{u\wedge\tau_N}^x - X_{u\wedge\tau_N}^y = x - y + \int_s^{u\wedge\tau_N} \left[b\left(r, X_r^x\right) - b\left(r, X_r^y\right)\right] dr$$
$$+ \int_s^{u\wedge\tau_N} \left[\sigma\left(r, X_r^x\right) - \sigma\left(r, X_r^y\right)\right] dW_r,$$

and hence

$$\left|X^x_{u\wedge\tau_N} - X^y_{u\wedge\tau_N}\right|^2 \leq 3\left[|x-y|^2 + \left|\int_s^{u\wedge\tau_N} \left[b\left(r,X^x_r\right) - b\left(r,X^y_r\right)\right] dr\right|^2 + \left|\int_s^{u\wedge\tau_N} \left[\sigma\left(r,X^x_r\right) - \sigma\left(r,X^y_r\right)\right] dW_r\right|^2\right]. \tag{6.5.4}$$

Let us denote $E[\sup_{0\leq u\leq t} |X^x_{u\wedge\tau_N} - X^y_{u\wedge\tau_N}|^2]$ as $f(t)$. In (6.5.4), take supremum over $0 \leq u \leq t$, and then take expectation termwise to get

$$\begin{aligned} f(t) &\leq 3\left[|x-y|^2 + E\sup_{0\leq u\leq t}\left|\int_s^{u\wedge\tau_N} \left[b\left(r,X^x_r\right) - b\left(r,X^y_r\right)\right] dr\right|^2 + E\sup_{s\leq u\leq t}\left|\int_0^{u\wedge\tau_N} \left[\sigma\left(r,X^x_r\right) - \sigma\left(r,X^y_r\right)\right] dW_r\right|^2\right] \\ &\leq 3|x-y|^2 + 3TE\int_0^{t\wedge\tau_N} \left|b\left(r,X^x_r\right) - b\left(r,X^y_r\right)\right|^2 dr + 12E\int_s^{t\wedge\tau_N} \left|\sigma\left(r,X^x_r\right) - \sigma\left(r,X^y_r\right)\right|^2 dr \\ &\leq 3|x-y|^2 + 3K(T+4)\int_s^t f(r)\, dr \end{aligned}$$

where we have used the Doob inequality and hypothesis H.2. Applying the Gronwall lemma, we conclude that

$$f(t) \leq 3|x-y|^2 e^{3K(T+4)(t-s)}.$$

Thus, $C = 3\exp\left\{3K(T+4)T\right\}$ in the estimate (6.5.3). ■

The proof of the next lemma is quite instructive and useful in establishing regularity of processes with respect to a parameter. We are interested in showing the Borel measurability of X^x_t with respect to the initial condition x.

Lemma 6.5.3 *There exists a function $Y(x,t,\omega)$ defined on $\mathbb{R}^d \times [s,T] \times \Omega$ with values in $\mathbb{R}^d$ with the following properties:*

(i) For each $x \in \mathbb{R}^d$, the process $Y(x,t)$ is a solution of (6.5.1) with continuous paths.

(ii) For each $t \in [s,T]$, the restriction of $Y(x,u,\omega)$ to $\mathbb{R}^d \times [s,t] \times \Omega$ is $\mathcal{B}(\mathbb{R}^d) \times \mathcal{B}[s,t] \times \mathcal{F}^{W,t}_s$-measurable.

Proof Consider any rational number in $\mathbb{R}^d$ of the form

$$a_{m,k} = \left(k_1 2^{-m}, \ldots, k_d 2^{-m}\right)$$

where each k_j is an integer. Let $X(a_{m,k}, t)$ be the solution of (6.5.1) with initial condition $a_{m,k}$. Since the solution is adapted to $\mathcal{F}_s^{W,t}$ and has continuous paths, we get by Chapter 1 that the solution is progressively measurable; that is, $X(a_{m,k}, u)$ restricted to $[s, t] \times \Omega$ is $\mathcal{B}[s, t] \times \mathcal{F}_s^{W,t}$-measurable.

For $x \in \mathbb{R}^d$, let $a_{m,k}(x)$ be such that $k_j 2^{-m} \leq x_j \leq k_{j+1} 2^{-m}$ for all $j = 1, \ldots, n$, where $x = (x_1, \ldots, x_n)$. Define

$$Y^m(x, t, \omega) = X(a_{m,k}(x), t, \omega).$$

The process $Y^m(x, t)$ has continuous paths. The map $(x, t, \omega) \to Y^m(x, t, \omega)$ is $\mathcal{B}(\mathbb{R}^d) \times \mathcal{B}[s, t] \times \mathcal{F}_s^{W,t}$ measurable. Define

$$Y(x, t, \omega) = \limsup_{m \to \infty} Y^m(x, t, \omega). \tag{6.5.5}$$

Let $X(x, t)$ be the unique strong solution of (6.5.1). Then by lemma 6.5.2,

$$\begin{aligned} E\left\{\sup_{s \leq t \leq T} \left|X(x, t) - Y^m(x, t)\right|^2\right\} &\leq C|x - a_{m,k}(x)|^2 \\ &\leq Cd2^{-2m}. \end{aligned}$$

From the Borel-Cantelli lemma it follows that

$$P\left\{\sup_{s \leq t \leq T} |X(x, t) - Y^m(x, t)| > 1/m \text{ infinitely often}\right\} = 0.$$

Hence for each x,

$$P\left\{\omega : Y^m(x, t, \omega) \to X(x, t, \omega) \text{ uniformly on } [s, T]\right\} = 1.$$

We can thus conclude from (6.5.5) that, for almost all ω,

$$Y(x, t, \omega) = X(x, t, \omega)$$

for all $t \in [s, T]$ and for each x. Therefore, $Y(x, t)$ is a solution of (6.5.1) with initial time s and condition x. It has continuous paths. The required measurability property follows from (6.5.5). ■

A direct consequence of Lemma 6.5.3 is the following:

Lemma 6.5.4 *The stochastic differential equation (6.5.1) has a unique solution $X(x, t)$ which is $\mathcal{B}(\mathbb{R}^d) \times \mathcal{F}_s^{W,t}$-measurable as a function of (x, ω).*

We introduce the following important notation. For fixed s, x, and t with $t \geq s$, define the probability measure

$$P(s, x, t, B) = P\left\{\omega : X(x, t, \omega) \in B\right\}.$$

Since $P(s, x, t, B) = \int I_B(X(x, t, \omega))P(d\omega)$, it follows from Lemma 6.5.3 and the Fubini theorem that $P(s, \cdot, t, B)$ is a Borel-measurable function of x for each fixed s, t, and B.

Let us now consider

$$\begin{aligned} dZ_t &= b(t, Z_t)\,dt + \sigma(t, Z_t)\,dW_t \quad t \geq s \\ Z_s &= \eta, \end{aligned} \tag{6.5.6}$$

where η is an $\mathcal{F}_s$-measurable random variable. Note that $\mathcal{F}_s$ denotes $\mathcal{F}_0^s$.

Lemma 6.5.5 *The unique $\mathcal{F}_t$-adapted strong solution of (6.5.6) is given by $X(\eta(\omega), t, \omega)$, where $X(x, t, \omega)$ is the solution given by Lemma 6.5.3.*

Proof

Step 1 First we show that the integral $\int_s^t \sigma(u, X(\eta, u))\,dW_u$ is a local martingale. Define $\bar{\sigma}(t, \omega) = \sigma(t, X(\eta(\omega), t, \omega))$. From Lemma 6.5.3, $\bar{\sigma}(t, \omega)$ is jointly measurable and $\mathcal{F}_t$-adapted. By the linear growth condition,

$$|\bar{\sigma}(t, \omega)|^2 \leq K(1 + |X(\eta(\omega), t, \omega)|^2)$$

so that

$$P\left\{\omega : \int_s^T |X(\eta(\omega), t, \omega)|^2\,dt < \infty\right\} \leq P\left\{\omega : \int_s^T |\bar{\sigma}(t, \omega)|^2\,dt < \infty\right\}.$$

Let A denote the set $\{\omega : \int_s^T |X(\eta(\omega), t, \omega)|^2\,dt < \infty\}$. We will show that $P(A) = 1$. Indeed, for any x, $P\{\omega : \int_s^T |X(x, t, \omega)|^2\,dt < \infty\} = 1$. Call this event A_x. Let Q denote the regular conditional probability of $X(\eta, t)$ given η, and μ, the probability distribution of η. Then,

$$P(A) = \int Q(x; A)\,\mu(dx) = \int Q(x, A_x)\,\mu(dx) = \int P(A_x)\,\mu(dx) = 1$$

where $Q(x, A_x) = P(A_x)$ by the independence of η and $X(x, t)$. This proves that $\int_s^t \sigma(u, X(\eta, u))\,dW_u$ is a local martingale for $t \in [s, T]$.

Step 2 Define the event

$$B_\eta = \left\{\omega : X(\eta, t) = \eta + \int_s^t b(u, X(\eta, u))\,du + \int_s^t \sigma(u, X(\eta, u))\,dW_u\right\}.$$

Then,

$$\begin{aligned}
P(B_\eta) &= \int Q(x, B_\eta)\, \mu(dx) \\
&= \int P\{B_\eta | \eta = x\}\, \mu(dx) \\
&= \int P\{B_x \mid \eta = x\}\, \mu(dx) \\
&= \int P(B_x)\, \mu(dx) \\
&= 1,
\end{aligned}$$

which completes the proof. ■

So far, the initial time s was arbitrary but fixed. Now, we need to vary s. Therefore, let us introduce the notation

$$X_{s,x}(t) := X(x,t) \quad \text{and} \quad X_{s,\eta}(t) := X(\eta, t).$$

Consider the equation introduced in Section 6.1:

$$dX_t = b(t, X_t)\, dt + \sigma\,(t, X_t)\; dW_t \text{ with } \quad X_0 = \xi. \tag{6.5.7}$$

Here, ξ is a random variable which is independent of the Wiener process W. The coefficients b and σ satisfy the Hypotheses H given in Section 6.1. We will prove the Markov property of the unique solution X_t obtained in Section 6.1.

Let η be any $\mathcal{F}_s$-measurable random variable where $\mathcal{F}_s = \sigma(\xi) \vee \mathcal{F}_s^W$. We know that $X_{s,\eta}(t)$ is the solution of (6.5.6) by Lemma 6.5.5. Then, from Lemma 6.5.1, we have

$$E\left[\phi(X_{s,\eta}(t) \mid \mathcal{F}_s\right] = g(\eta). \tag{6.5.8}$$

where $g(x) = E[\phi(X_{s,x}(t)) \mid \mathcal{F}_s] = E(\phi(X_{s,x}(t))$. If we take $\eta = X_s$, then $X_{s,\eta}(t) = X_t$ by Lemma 6.5.5. Therefore, (6.5.8) yields

$$E\left[\phi(X_t) \mid \mathcal{F}_s\right] = g(X_s). \tag{6.5.9}$$

Hence, $E[\phi(X_t) \mid \mathcal{F}_s]$ is $\sigma(X_s)$ measurable. In other words,

$$E\left[\phi(X_t) \mid \mathcal{F}_s\right] = E\left[\phi(X_t) \mid X_s\right], \tag{6.5.10}$$

which is the Markov property of X_t with respect to $\mathcal{F}_t$.

Definition 6.5.1 *Let $\{X_t\}$ be a Markov process with values in $\mathbb{R}^d$. The* **transition probability function** *of the Markov process is a function $P(s, x, t, B)$ where $s < t$, $x \in \mathbb{R}^d$, and $B \in \mathcal{B}(\mathbb{R}^d)$ with the following properties:*

(a) *For all* s, x, t, $P(s, x, t, \cdot)$ *is a probability measure on* $\mathcal{B}(\mathbb{R}^d)$.

(b) *For each* s, t, B, $P(s, \cdot, t, B)$ *is* $\mathcal{B}(R^n)$ *measurable.*

(c) $P[X_t \in B \mid X_s] = P(s, X_s, t, B)$ *a.s.*

(d) *For all* $s < u < t$, *and* $x \in \mathbb{R}^d$, $B \in \mathcal{B}(\mathbb{R}^d)$,

$$P(s, x, t, B) = \int_{\mathbb{R}^d} P(s, x, u, dy)P(u, y, t, B)$$

The first three properties in the above definition are equivalent to the requirement that $P(s, x, t, B)$ is a regular conditional probability distribution of X_t given $\mathcal{F}_s^X$. The property (d) in the definition is known as the **Chapman-Kolmogorov equation** for $\{X_t\}$.

Theorem 6.5.6 *The unique strong solution* $\{X_t : 0 \le t \le T\}$ *of the stochastic differential equation* (6.5.7) *is a continuous Markov process such that*

$$P[X_t \in B \mid \mathcal{F}_s] = P[X_t \in B \mid X_s] = P(s, X_s, t, B)$$

where $P(s, x, t, B)$ *is the transition probability function given by* $P(s, x, t, B) = P\left\{X_{s,x}(t) \in B\right\}$.

Proof The Markov property of X_t has already been shown. By independence of $\mathcal{F}_s$ and $\mathcal{F}_s^{W,t}$,

$$P(s, x, t, B) = E\big[1_B\left(X_{s,x}(t)\right)\big] = E\big[1_B\left(X_{s,x}(t)\right) \mid \mathcal{F}_s\big].$$

In the notation used in Equation (6.5.8), if we take $\phi = 1_B$, then $g(x) = P(s, x, t, B)$. Therefore, by Equation (6.5.9),

$$E\left\{\phi(X_t) \mid \mathcal{F}_s\right\} = g(X_s) = P(s, X_s, t, B);$$

that is, $P\left\{X_t \in B \mid \mathcal{F}_s\right\} = P(s, X_s, t, B)$.

It remains to verify the Chapman-Kolmogorov equation for $P(s, x, t, B)$. Let $s < u$. Since $\int_{\mathbb{R}^d} f(y)\, P(s, x, u, dy) = E[f(X_{s,x}(u))]$ for any bounded, Borel-measurable function f defined on $\mathbb{R}^d$, we have

$$\begin{aligned}
\int_{\mathbb{R}^d} P(s, x, u, dy)\, P(u, y, t, B) &= E\big[P(u, X_{s,x}(u), t, B)\big] \\
&= E\big[P\left\{X_{s,x}(t) \in B \mid X_{s,x}(u)\right\}\big] \\
&= E\big[E\left\{I_B(X_{s,x}(t)) \mid X_{s,x}(u)\right\}\big] \\
&= E(I_B(X_{s,x}(t))) = P\left\{X_{s,x}(t) \in B\right\} \\
&= P(s, x, t, B).
\end{aligned}$$

Thus, $P(s, x, t, B)$ is a transition probability function. ■

6.6 Generators and Diffusion Processes

In the last section, we showed that the solution X_t of the stochastic differential equation (6.5.7) has a transition probability function which satisfies

$$P(s, x, t, B) = P[X_{s,x}(t) \in B].$$

For any $0 \leq s \leq t \leq T$, define an operator T_s^t on the space of Borel-measurable functions $f : \mathbb{R}^d \to \mathbb{R}$ by

$$T_s^s f(x) = f(x) \quad \text{and} \quad T_s^t f(x) = \int_{\mathbb{R}^d} f(y)\, P(s, x, t, dy) \tag{6.6.1}$$

if the integral exists. Then T_s^t defines a family of operators on the space of bounded, Borel-measurable functions, and has the semigroup property:

$$T_s^t f = T_s^u T_u^t f \quad \text{for all } 0 \leq s \leq u \leq t \leq T.$$

This is simply a restatement of the Chapman-Kolmogorov equation. The family $\{T_s^t\}$ is known as the *semigroup for the Markov process* X_t.

Theorem 6.6.1 *Let X_t be the solution of the stochastic differential equation* (6.5.7). *Let L_t be the operator defined on C_b^2 by*

$$L_t f(x) = \frac{1}{2} \sum_{i,j} a_{ij}(t, x) \frac{\partial^2 f}{\partial x_i \partial x_j}(x) + \sum_i b_i(t, x) \frac{\partial f}{\partial x_i}(x) \tag{6.6.2}$$

where $a_{ij} = (\sigma\sigma^*)_{ij}$. Then, $\int_{\mathbb{R}^d} |(L_t f)(y)|\, P(s, x, t, dy) < \infty$, and

$$T_s^t f(x) = f(x) + \int_s^t T_s^u L_u f(x) du \tag{6.6.3}$$

for all $x \in \mathbb{R}^d$, and $0 \leq s < t \leq T$.

Proof By the definition of $X_{s,x}(t)$,

$$X_{s,x}(t) = x + \int_0^t b\,(u, X_{s,x}(u))\; du + \int_s^t \sigma\,(u, X_{s,x}(u))\; dW_u.$$

By the Itô formula, $f(X_{s,x}(t))$ equals

$$f(x) + \int_s^t L_u f(X_{s,x}(u))\; du + \sum_{i,j} \int_s^t \sigma_{ij}\,(u, X_{s,x}(u))\; \frac{\partial f}{\partial x_i}(X_{s,x}(u))\; dW_j(u).$$

The stochastic integral is a square integrable martingale. Hence, upon taking expectation on both sides,

$$E[f\,(X_{s,x}(t))] = f(x) + \int_s^t E[L_u f\,(X_{s,x}(u))]\; du.$$

In other words,

$$T_s^t f(x) = f(x) + \int_s^t T_s^u L_u f(x) du. \qquad \blacksquare$$

In view of Equation (6.6.3), we say that the family of operators $\{L_t\}$ defined on C_b^2 is the **extended generator** of the process X_t. If the coefficients b and σ are, in addition to the Hypotheses H, continuous functions, then L_t satisfies

$$\lim_{t \to s} \frac{T_s^t f(x) - f(x)}{t - s} = L_s f(x) \tag{6.6.4}$$

for all $f \in C_b^2$, and all $x \in \mathbb{R}^d$. To show this, note that $T_s^u L_u f(x)$ is a continuous function of $u \in [s, T]$. Hence, by using (6.6.3), one obtains (6.6.4).

A family of operators $\{L_s\}$ that satisfies Equation (6.6.4) is known as the **weak infinitesimal generator** for the semigroup $\{T_s^t\}$. The adjective "weak" is used since a pointwise evaluation is made in Equation (6.6.4). From the above paragraph, if b and σ are continuous in t, then the notions of extended generator and weak infinitesimal generator coincide on C_b^2.

The Markov process X_t has stationary transition probabilities if

$$P(s, x, t, B) = P(0, x, t - s, B).$$

If the coefficients b and σ are autonomous, that is,

$$b(t, x) = b(x) \quad \text{and} \quad \sigma(t, x) = \sigma(x),$$

then X_t has stationary transition probabilities. In this situation, $T_s^t = T_0^{t-s}$, which we will denote as S_{t-s}. The extended generator L doesn't depend on time. Hence, we can write (6.6.3) more simply as

$$S_t f(x) = f(x) + \int_0^t S_u L f(x) du.$$

Next, we start with a definition of diffusion processes. There is no standard definition of diffusions, though the underlying idea is more or less the same. Indeed, a diffusion process is a Markov process with continuous paths whose infinitesimal generator is specified by a differential operator on a suitable class of smooth functions. Our aim in introducing diffusions at this juncture is to further elaborate on infinitesimal generators for solutions of stochastic differential equations.

The term "diffusion" arises from modeling the motion and spread of a chemical concentration in a medium by the random motion of their molecules. Diffusion processes are quite useful in studying various biological phenomena such as the growth of a particular biological species or the prevalence of a particular genotype in a population. In such studies, diffusions arise as limits of discrete-time models.

Definition 6.6.1 *A Markov process X_t with transition function $P(s,x,t,B)$ is called a* **diffusion process** *if it has continuous paths and if there exists an $\mathbb{R}^d$-valued function $b(t,x)$ and an $d \times d$-matrix-valued function $a(t,x)$ that satisfy the following conditions: For every bounded, open neighborhood U_x of x,*

(i) $\lim_{h\downarrow 0} \frac{1}{h} \int_{U_x^c} P(t,x,t+h,dy) = 0.$

(ii) $\lim_{h\downarrow 0} \frac{1}{h} \int_{U_x} (y_i - x_i)\, P(t,x,t+h,dy) = b_i(t,x)$ *for all* $1 \le i \le d$.

(iii) $\lim_{h\downarrow 0} \frac{1}{h} \int_{U_x} (y_i - x_i)\,(y_j - x_j)\, P(t,x,t+h,dy) = a_{ij}(t,x)$ *for all* $1 \le i,j \le d$.

The vector $b(t,x)$ is called the **drift**, and the matrix $a(t,x)$, the **diffusion *coefficient*** of the process.

Theorem 6.6.2 *Let X_t be the solution of*

$$dX_t = b(t,X_t)\,dt + \sigma(t,X_t)\,dW_t \text{ with } X_0 = \xi$$

where the coefficients b and σ satisfy the Hypotheses H given in Section 6.1, and ξ is a random variable that is independent of the Wiener process W. In addition, let b and σ be continuous functions. Then, X_t is a diffusion process with drift b and diffusion coefficient $a = \sigma\sigma^*$.

Proof Suppose U_x is any bounded, open set containing x. Let V_x be an open set containing x such that its closure $\overline{V}_x \subset U_x$. Choose a C_b^2 function f taking values in $[0,1]$ such that $f = 0$ on V_x, and $f = 1$ on U_x^c.

By Equation (6.6.4), it follows that

$$\lim_{h\to 0} \frac{T_s^{s+h} f(x) - f(x)}{h} = L_s f(x)$$

where $L_s f(x)$ is given by Equation (6.6.2). By our choice of f, $f(x)$ as well as $L_s f(x)$ are both zero. Hence, $\lim_{h\to 0} \frac{T_s^{s+h} f(x)}{h} = 0$; that is,

$$\lim_{h\to 0} \frac{1}{h} \int f(y)\, P(s,x,s+h,dy) = 0.$$

The verification of the requirement (i) in the definition of a diffusion process is completed upon noting that

$$\int f(y)\, P(s,x,s+h,dy) \ge \int_{U_x^c} f(y)\, P(s,x,s+h,dy) = \int_{U_x^c} P(s,x,s+h,dy).$$

The verification of (ii) is quite similar to that of (iii). So, we will verify the requirement (iii). Suppose V_x is a bounded open set containing $\overline{U}_x$. Let g be a C_b^2 function with $0 \le g \le 1$, $g = 1$ on $\overline{U}_x$, and $g = 0$ on V_x^c. For $i,j = 1,\ldots,n$, define

$$f^{ij}(y) = (y_i - x_i)(y_j - x_j)g(y).$$

Clearly,

$$\frac{\partial f^{ij}(y)}{\partial y_i}\Big|_{y=x} = \frac{\partial f^{ij}(y)}{\partial y_j}\Big|_{y=x} = 0.$$

Likewise, $\frac{\partial f^{ij}(y)}{\partial y_i \partial y_j}\big|_{y=x} = 1$. All other second-order partial derivatives of f^{ij} evaluated at $y = x$ are zero. Hence, $L_s f^{ij}(x) = a_{ij}(s, x)$. By Equation (6.6.4),

$$\lim_{h \to 0} \frac{1}{h}\left[\int f^{ij}(y)\, P(s, x, s + h, dy) - f^{ij}(x)\right] = a_{ij}(s, x).$$

Since $f^{ij}(x) = 0$,

$$\lim_{h \to 0} \frac{1}{h}\int f^{ij}(y)\, P(s, x, s + h, dy) = a_{ij}(s, x).$$

Since V_x is bounded, there exists a K such that $|y_i - x_i| \leq K$ for all $y \in V_x$. It follows that

$$\begin{aligned}
&\left|\frac{1}{h}\int f^{ij}(y)\, P(s, x, s + h, dy) - \frac{1}{h}\int_{U_x} f^{ij}(y)\, P(s, x, s + h, dy)\right| \\
&\quad \leq \frac{K^2}{h}\int_{V_x \setminus U_x} P(s, x, s + h, dy) \\
&\quad \leq \frac{K^2}{h}\int_{U_x^c} P(s, x, s + h, dy),
\end{aligned}$$

which tends to 0 as $h \to 0$. Therefore,

$$\lim_{h \to 0} \frac{1}{h}\int_{U_x} (y_i - x_i)(y_j - x_j)\, P(s, x, s + h, dy) = a_{ij}(x).$$

The proof is over. ■

Exercises

1. Consider the one-dimensional stochastic differential equation

$$dX_t = X_t\, dt + dW_t, \quad \text{with} \quad X_0 = x.$$

Show that

$$P\{\lim_{t \to \infty} X_t = \infty\} = \frac{1}{\sqrt{\pi}}\int_{-\infty}^{x} e^{-s^2}\, ds.$$

2. Show that for $0 \leq t < 1$,

$$X_t = (t-1)\int_0^t \frac{1}{s-1}\, dW_s$$

is the solution of the equation

$$dX_t = dW_t + \frac{1}{t-1} X_t\, dt$$

with initial condition $X_0 = 0$.

3. Solve explicitly the one-dimensional stochastic differential equation:

$$dX(t) = \left(\sqrt{1+X^2(t)} + \frac{1}{2}X(t)\right) dt + \sqrt{1+X^2(t)}\, dW(t)$$

with $X(0) = x \in \mathbf{R}^1$.

4. Show that the stochastic differential equation

$$X(t) = 3\int_0^t X^{1/3}(s)\, ds + 3\int_0^t X^{2/3}(s)\, dW(s)$$

has uncountably many solutions of the form

$$X_t^a = \begin{cases} 0 & \text{if } 0 \leq t \leq T_a, \\ W^3(t) & \text{if } T_a \leq t < \infty \end{cases}$$

where $0 \leq a \leq \infty$ and $T_a = \inf\{s \geq a : W(s) = 0\}$.

5. Prove that the stochastic differential equation

$$dX(t) = dW(t) + X^{1/3}(t)\, dt$$

with $X(0) = x$ has a unique strong solution.

6. For all $t \geq 0$, consider the equation $X_t = \int_0^t |X_s| ds + W_t$.
   (i) Show that X is a submartingale.
   (ii) Prove that $\liminf_{t \to 0} \dfrac{\int_0^t E|X_s|\, ds}{t^{3/2}}$ is nonzero.

7 The Martingale Problem

Suppose that $X := \{X_t\}$ is a continuous, time-homogeneous Markov process with infinitesimal generator L. Then one can show that

$$M^f(t) = f(X_t) - \int_0^t Lf(X_s)\, ds$$

is a martingale for each $f \in \mathcal{D}(L)$. In other words, $\{X_t\}$ is a solution of the martingale problem for L. It is easier to determine an operator, such as the weak infinitesimal generator of X, or simply, a differential operator associated with X than the full generator. The differential operator is not a closed operator with its domain as, for instance, $C_b^2(\mathbb{R}^d)$. Hence it cannot be the full generator of a Markov process. The differential operator admits a closed extension, though such extensions are not unique. To bypass these operator-theoretic hurdles, Stroock and Varadhan initiated a martingale approach to study diffusion processes that is known as the martingale problem.

The martingale problem provides us with a new concept for the solution of a stochastic differential equation. Using this notion, existence and uniqueness of solutions of stochastic differential equations can be proved under milder conditions on the coefficients. Besides, such solutions are weak solutions of stochastic differential equations.

The statement of the martingale problem posed by a stochastic differential equation, and of the martingale problem equivalent formulations can be found in Section 7.1. Existence results for solutions of martingale problems are presented in Section 7.2. Uniqueness of solutions under suitable conditions is studied in Section 7.4 after a brief discussion of the required background results in analysis. The Markov property, and further results on uniqueness of solutions are proved in the last two sections.

7.1 Introduction

Let $X := \{X_t\}$ be a continuous, time-homogeneous, $\mathbb{R}^d$-valued Markov process defined on a probability space $(\Omega, \mathcal{F}, P)$ with infinitesimal generator L. Let $\{S_t\}$ denote the semigroup generated by L. We will denote the domain of the semigroup by B, a Banach space contained in the space of bounded Borel-measurable functions, and

$$L : D(L) \subset B \to B.$$

One may recall that the connection between X and $\{S_t\}$ is made by

$$E\left[f(X_{t+s}) \,|\mathcal{F}_s^X\right] = S_t f(X_s) = \int_{\mathbb{R}^d} f(y)\, P(t, X_s, dy)$$

for all $s, t \geq 0$ and $f \in B$. Here, $P(t, x, dy)$ denotes the transition probability measure for the process X.

The first proposition states an important property enjoyed by such processes, which led Stroock and Varadhan to formulate the martingale problem.

Proposition 7.1.1 *Let $X := \{X_t\}$ be a time-homogeneous Markov process as described above with generator L. Then, for any f in the domain of L, the process*

$$M_t = f(X_t) - \int_0^t Lf(X_r)\, dr$$

is an $\mathcal{F}_t^X$ martingale.

Proof It is clear that M_t is $\mathcal{F}_t^X$ measurable. The finiteness of $E|M_t|$ follows readily, since functions in the domain of L are bounded. For any $t \geq s \geq 0$, consider

$$\begin{aligned}
E(M_t \mid \mathcal{F}_s^X) &= \int f(y)\, P(t-s, X_s, dy) - \int_s^t \int Lf(y)\, P(r-s, X_s, dy)\, dr \\
&\quad - \int_0^s Lf(X_r)\, dr \\
&= S_{t-s} f(X_s) - \int_s^t S_{r-s} Lf(X_s)\, dr - \int_0^s Lf(X_r)\, dr \\
&= S_{t-s} f(X_s) - \int_0^{t-s} S_u Lf(X_s)\, du - \int_0^s Lf(X_r)\, dr \\
&= f(X_s) - \int_0^s Lf(X_r)\, dr
\end{aligned}$$

by recalling from Theorem 6.6.1 that

$$S_{t-s} f(X_s) = f(X_s) + \int_0^{t-s} S_u Lf(X_s)\, du.$$

Thus, $E(M_t \mid \mathcal{F}_s^X) = M_s$. ■

Before we define martingale problems, we need the following notation. Let $B(\mathbb{R}^d)$ denote the space of all bounded, Borel-measurable functions on $\mathbb{R}^d$. $C(\mathbb{R}^d)$ will denote the space of real-valued, continuous functions on $\mathbb{R}^d$, and $C_b(\mathbb{R}^d)$, the space of bounded, real-valued, continuous functions on $\mathbb{R}^d$. The spaces $B(\mathbb{R}^d)$ and $C_b(\mathbb{R}^d)$ are Banach spaces under the supremum norm: $\|f\| = \sup_{x\in\mathbb{R}^d} |f(x)|$.

Let A be an operator in $C_b(\mathbb{R}^d)$ with domain D.

Definition 7.1.1 *A process $X = \{X_t\}$ with continuous paths defined on a probability space $(\Omega, \mathcal{F}, P)$ is called a solution to the* **martingale problem** *for the initial distribution μ and the operator A if the following hold:*

1. *The distribution of X_0 is μ.*
2. *For any $f \in D$, the process $M_f(t) := f(X_t) - \int_0^t Af(X_s)\,ds$ is a $\mathcal{F}_t^X$-martingale.*

In the definition, note that the probability space is not fixed in advance. Frequently, one uses the path space to be Ω with $\mathcal{F}$ given by the Borel σ-field of Ω. If X_t is taken to be the canonical process $X_t(\omega) = \omega(t)$, then it becomes clear that the solution of a martingale problem is a statement about the existence of a probability measure P under which the above two conditions hold.

The next result provides us with a simple way to check that a process X satisfies the second requirement in the above definition. Such an X will be called a solution to the martingale problem for the operator A, since there is no reference to the initial distribution.

Proposition 7.1.2 *The following are equivalent:*

(i) X is a solution to the martingale problem for the operator A.

(ii)
$$E\left[\left(f\left(X_{t_{n+1}}\right) - f\left(X_{t_n}\right) - \int_{t_n}^{t_{n+1}} Af(X_s)\,ds\right)\prod_{j=1}^{n} h_j\left(X_{t_j}\right)\right] = 0 \tag{7.1.1}$$

for all $f \in D(A)$, $0 \le t_1 < t_2 \cdots < t_{n+1}$, $h_1, h_2, \ldots, h_n \in B(\mathbb{R}^d)$, and $n \ge 1$.

Proof If X is a solution to the martingale problem for A, then

$$E\left[M_f(t_{n+1})\prod_{j=1}^{n} h_j\left(X_{t_j}\right) \mid \mathcal{F}_{t_n}^X\right] = \prod_{j=1}^{n} h_j(X_{t_j})M_f(t_n)$$

since $\prod_{j=1}^n h_j(X_{t_j})$ is $\mathcal{F}_{t_n}^X$-measurable, and M_f is a $\mathcal{F}_t^X$-martingale. Thus, (7.1.1) holds.

To prove the converse, we need to show that, for all t_n, t_{n+1} such that $t_{n+1} > t_n$, and any $F \in \mathcal{F}_{t_n}^X$,

$$E\left[\left(M_f(t_{n+1}) - M_f(t_n)\right) 1_F\right] = 0.$$

Consider all functions of the form $\prod_{j=1}^n h_j(X_{t_j})$ with $h_1, h_2, \ldots, h_n \in B(\mathbb{R}^d)$, and $0 \le t_1 < t_2 < \cdots < t_n$, $n \ge 1$. They generate the σ-field $\mathcal{F}_{t_n}^X$, and hence there exists a sequence of such functions that converge a.s. to 1_F. Taking the limit in (7.1.1) along such a sequence and using the bounded convergence theorem allows us to conclude that statement (ii) implies (i). ■

The next result is a nice application of the previous proposition. It shows that in the context of martingale problems, the finite-dimensional distributions of a solution of the martingale problem are determined by its one-dimensional distributions.

Proposition 7.1.3 *Suppose that for each initial distribution μ, any two solutions X and Y of the martingale problem for (μ, A) have the same one-dimensional distributions. Then, X and Y have the same finite-dimensional distributions.*

Proof Suppose that X and Y are defined on $(\Omega_1, \mathcal{F}_1, P_1)$ and $(\Omega_2, \mathcal{F}_2, P_2)$, respectively. For any $t_0 \geq 0$, we have, for any strictly positive bounded Borel-measurable function g defined on $\mathbb{R}^d$,

$$E^{P_1}\left[g(X_{t_0})\right] = E^{P_2}\left[g(Y_{t_0})\right] \tag{7.1.2}$$

by the hypothesis. Define the measures

$$Q_1(B) = \frac{E^{P_1}\left[1_B g(X_{t_0})\right]}{E^{P_1}\left[g(X_{t_0})\right]} \quad \text{for any } B \in \mathcal{F}_1$$

and

$$Q_2(C) = \frac{E^{P_2}\left[1_C g(Y_{t_0})\right]}{E^{P_2}\left[g(Y_{t_0})\right]} \quad \text{for any } C \in \mathcal{F}_2.$$

Take any $f \in D(A)$, and $0 \leq t_1 < t_2 \ldots < t_{n+1}$ for any $n \geq 1$. Let $h_1, h_2, \ldots, h_n$ be any bounded, Borel-measurable function on $\mathbb{R}^d$. We will use the following notation:

$$\eta(X) = \left[\left(f(X_{t_{n+1}}) - f(X_{t_n}) - \int_{t_n}^{t_{n+1}} Af(X_s)\, ds\right)\prod_{j=1}^{n} h_j(X_{t_j})\right].$$

Define the processes

$$\hat{X}(t) = X_{t_0+t} \quad \text{and} \quad \hat{Y}(t) = Y_{t_0+t}$$

for all $t \geq 0$. Then,

$$E^{P_1}[\eta(\hat{X})g(X_{t_0})] = 0$$

by Proposition 7.1.2, so that

$$E^{Q_1}\left[\eta(\hat{X})\right] = \frac{E^{P_1}\left[\eta(X(t_0 + \cdot))g(X_{t_0})\right]}{E^{P_1}\left[g(X_{t_0})\right]} = 0.$$

Likewise, $E^{Q_2}[\eta(\hat{Y})] = 0$. Thus $\hat{X}$ and $\hat{Y}$ are two solutions of the martingale problem for A. They have a common initial distribution. Indeed, for any function ϕ bounded and Borel measurable on $\mathbb{R}^d$,

$$\begin{aligned} E^{Q_1}\left[\phi(\hat{X}(0))\right] &= \frac{E^{P_1}\left[\phi(X_{t_0})g(X_{t_0})\right]}{E^{P_1}\left[g(X_{t_0})\right]} \\ &= \frac{E^{P_2}\left[\phi(Y_{t_0})g(Y_{t_0})\right]}{E^{P_2}\left[g(Y_{t_0})\right]} \quad \text{by Equation (7.1.2)} \\ &= E^{Q_2}\left[\phi(\hat{Y}(0))\right]. \end{aligned}$$

Therefore, the hypothesis of the theorem yields

$$E^{Q_1}\left[f(\hat{X}(t))\right] = E^{Q_2}\left[f(\hat{Y}(t))\right]$$

for all $f \in B(\mathbb{R}^d)$, and any $t \geq 0$; that is,

$$E^{P_1}\left[f(X_{t_0+t})g(X_{t_0})\right] = E^{P_2}\left[f(Y_{t_0+t})g(Y_{t_0})\right].$$

Since t_0 and t are arbitrary, and f, g are any functions in $B(\mathbb{R}^d)$, we have shown that X and Y have the same two-dimensional distributions.

Repeating the above procedure, and using induction, the proof is completed. ■

The definition of martingale problems allows us to construct the solution X on any probability space. Since the process X has continuous paths, we can take $\Omega = C([0,\infty); \mathbb{R}^d)$ as the space of all continuous $\mathbb{R}^d$-valued functions defined on $[0,\infty)$. Define $X_t(\omega) = \omega(t)$ for all $\omega \in \Omega$. Equipped with the topology of uniform convergence on compact subsets of $[0,\infty)$, the space Ω is a complete separable metric space. Let $\mathcal{F}$ equal the Borel σ-field of Ω, and $\mathcal{F}_{s,t} = \sigma(X(r) : s \leq r \leq t)$ for any $0 \leq s \leq t$. When $s = 0$, $\mathcal{F}_{0,t}$ will be simply written as $\mathcal{F}_t$. With such a canonical choice of $(\Omega, \mathcal{F})$, and the process X, the definition of the martingale problem for (μ, A) can be recast as follows:

Definition 7.1.2 *A probability measure P on $(\Omega, \mathcal{F})$ is called a solution of the martingale problem for (μ, A) if the following hold:*

1. *$P\{\omega : \omega(0) \in B\} = \mu(B)$ for all Borel sets B in $\mathbb{R}^d$.*
2. *For any $f \in D(A)$ the process $M_f(t)(\omega) := f(\omega(t)) - f(\omega(0)) - \int_0^t Af(\omega(s))\, ds$ is an $\mathcal{F}_t$ martingale with respect to P.*

So far, we have considered time-homogenous martingale problems. The definition of time-inhomogeneous martingale problems can analogously be defined as follows. Let $\{A_t : t \geq 0\}$ be a family of operators defined on a common domain D, a subset of $C_b(\mathbb{R}^d)$.

Definition 7.1.3 *A probability measure P on $(\Omega, \mathcal{F})$ is called a solution of the martingale problem for $(\mu, (A_t))$ if the following hold:*

1. *$P\{\omega : X_0(\omega) \in B\} = \mu(B)$ for all Borel sets B in $\mathbb{R}^d$.*
2. *For any $f \in D$, the process $M_f(t) := f(X_t) - f(X_0) - \int_0^t A_s f(X_s)\, ds$ is an $\mathcal{F}_t$-martingale with respect to P.*

Next, we proceed to write equivalent formulations of martingale problems for which we recall the following result.

Lemma 7.1.4 *Let $(\Omega, \mathcal{F}, P)$ be any probability space equipped with a filtration $(\mathcal{F}_t)$. Let V be an adapted, nonnegative, continuous, increasing process with $V_0 = 0$ and $E(V_t) < \infty$ for all t. Let M be an adapted, real-valued, continuous process with $M_0 = 0$. The following statements are equivalent:*

(i) The process M is a local martingale with $\langle M\rangle_t = V_t$ for all $t \geq 0$.

(ii) For any $\theta \in \mathbb{R}$, $X_\theta(t) := \exp\{\theta M_t - \frac{\theta^2}{2} V_t\}$ is a local martingale.

If X_θ is a martingale with $E(e^{\theta M_t}) < \infty$ for all θ and $t \geq 0$, M is a martingale with increasing process given by V_t.

From now on, let us consider the canonical space $(\Omega, \mathcal{F}, (\mathcal{F}_t))$. Let $a : [0, \infty) \times \Omega \to \mathbb{R}^d \times \mathbb{R}^d$ be an $\mathcal{F}_t$-adapted function which is symmetric, bounded, continuous, and nonnegative definite. The last requirement means that, for all $\alpha = (\alpha_1, \alpha_2, \ldots, \alpha_d)$, there exists a constant K such that

$$0 \leq \sum_{i,j=1}^{d} a_{ij}(t, \omega)\alpha_i\alpha_j \leq K|\alpha|^2.$$

Let $b : [0, \infty) \times \Omega \to \mathbb{R}^d$ be an $\mathcal{F}_t$-adapted function which is bounded, and continuous. Define the operator A on $C_b^2(\mathbb{R}^d)$ by

$$A_t = \frac{1}{2}\sum_{i,j=1}^{d} a_{ij}(t, \omega)\frac{\partial^2}{\partial x_i \partial x_j} + \sum_{i=1}^{d} b_i(t, \omega)\frac{\partial}{\partial x_i}.$$

where $a(t, \omega) = ((a_{ij}(t, \omega)))$. Let $X(t, \omega) = \omega(t)$ for all $t \geq 0$. We have the following equivalence:

Theorem 7.1.5 *Let P be a probability on the canonical space such that $P(X_0 = x) = 1$. The following are equivalent:*

(i) For all $f \in C_b^2(\mathbb{R}^d)$, $M_f(t) = f(X_t) - f(X_0) - \int_0^t A_s f(X_s)\, ds$ is a P-martingale.

(ii) For all $\alpha \in \mathbb{R}^d$, $N_\alpha(t) = \langle \alpha, X_t - X_0 - \int_0^t b(s)\, ds\rangle$ is a P-martingale with $V_\alpha(t) = \int_0^t \langle \alpha, a(s)\alpha\rangle\, ds$ as the quadratic variation process.

(iii) For all $\alpha \in \mathbb{R}^d$, $X_\alpha(t) = \exp\{\langle \alpha, X_t - X_0 - \int_0^t b(s)\, ds\rangle - \frac{1}{2}\int_0^t \langle \alpha, a(s)\alpha\rangle\, ds\}$ is a P-martingale.

Proof

Step 1 (i) implies (ii): Consider a function $f \in C_b^2(\mathbb{R}^d)$ such that $f(x) = \langle \alpha, x\rangle$ if $|x| \leq n$. Then, by (i), $N_\alpha(t)$ is a martingale till time $\tau_n = \inf\{t \geq 0 : |X_t| \geq n\}$. By path continuity of X_t, the stopping times τ_n increase to ∞ as $n \to \infty$. Thus, N_α is a local martingale.

Next, consider a function $g \in C_b^2(\mathbb{R}^d)$ such that $g(x) = \langle \alpha, x \rangle^2$ if $|x| \le n$. Then

$$M_g(t) = \langle \alpha, X_t \rangle^2 - \langle \alpha, X_0 \rangle^2 - 2\int_0^t \langle \alpha, X_s \rangle \left\langle \alpha, b(s) \right\rangle ds - \int_0^t \left\langle \alpha, a(s)\alpha \right\rangle ds \tag{7.1.3}$$

is a local martingale. Next, we show that $N_\alpha^2(t) - Y_t$ is a local martingale where

$$Y_t = \langle \alpha, X_t \rangle^2 - \langle \alpha, X_0 \rangle^2 - 2\int_0^t \langle \alpha, X_s \rangle \left\langle \alpha, b(s) \right\rangle ds.$$

Equivalently, we show that $(N_\alpha(t) + \langle \alpha, X_0 \rangle)^2 - (Y_t + \langle \alpha, X_0 \rangle^2)$ is a local martingale. From the definition of $N_\alpha(t)$, we have

$$(N_\alpha(t) + \langle \alpha, X_0 \rangle)^2 = \langle \alpha, X_t \rangle^2 + \left(\int_0^t \left\langle \alpha, b(s) \right\rangle ds \right)^2 - 2\langle \alpha, X_t \rangle \int_0^t \left\langle \alpha, b(s) \right\rangle ds \tag{7.1.4}$$

Integrate by parts to obtain

$$\begin{aligned} \langle \alpha, X_t \rangle \int_0^t \langle \alpha, b(s) \rangle ds &= \int_0^t \langle \alpha, X_s \rangle \left\langle \alpha, b(s) \right\rangle ds + \int_0^t \int_0^s \left\langle \alpha, b(u) \right\rangle du \, d\langle \alpha, X_s \rangle \\ &= \int_0^t \langle \alpha, X_s \rangle \left\langle \alpha, b(s) \right\rangle ds + \int_0^t \left[\int_0^s \left\langle \alpha, b(u) \right\rangle du \right] dN_\alpha(s) \\ &\quad + \int_0^t \left[\int_0^s \left\langle \alpha, b(u) \right\rangle du \right] \left\langle \alpha, b(s) \right\rangle ds. \end{aligned}$$

This along with the simple observation that

$$\int_0^t \left[\int_0^s \left\langle \alpha, b(u) \right\rangle du \right] \left\langle \alpha, b(s) \right\rangle ds = \frac{1}{2} \left(\int_0^t \left\langle \alpha, b(s) \right\rangle ds \right)^2$$

allows us to write Equation (7.1.4) as

$$(N_\alpha(t) + \langle \alpha, X_0 \rangle)^2 = Y_t + \langle \alpha, X_0 \rangle^2 - 2\int_0^t \left[\int_0^s \left\langle \alpha, b(u) \right\rangle du \right] dN_\alpha(s).$$

By Equation (7.1.3), it follows that the increasing process for the local martingale N_α is given by $V_\alpha(t) = \int_0^t \langle \alpha, a(s)\alpha \rangle \, ds$. The boundedness of the function $a(s)$ implies that, for any t, $EV_\alpha(t) < \infty$. Therefore, N_α is not just a local martingale but a L^2-martingale with increasing process V_α.

Step 2 Given the statement (ii), it follows from Lemma 7.1.4 that the process X_α (defined in statement (iii)) is a local martingale. By the boundedness of the function $a(s)$, it follows that $E\exp\{V_\alpha(t)\} < \infty$. Hence, X_α is a P-martingale, which is statement (iii). It remains to show that (iii) implies (i). We will first show that, for the function $f(x) = e^{\langle\alpha,x\rangle}$, statement (i) holds. Define a finite variation process $B_\alpha(t)$ by

$$B_\alpha(t) = \exp\left\{\int_0^t \langle\alpha, b(s)\rangle\, ds + \frac{1}{2}\int_0^t \langle\alpha, a(s)\alpha\rangle\, ds\right\}.$$

Using the hypothesis and integration by parts, $X_\alpha(t)B_\alpha(t) - \int_0^t X_\alpha(s)\, dB_\alpha(s)$ is a local martingale. Therefore, $\{X_\alpha(t)B_\alpha(t) - \int_0^t X_\alpha(s)\, dB_\alpha(s)\}\exp\langle\alpha, X_0\rangle$ is a local martingale. When expanded and written in full, we get that

$$\exp\langle\alpha, X_t\rangle - \int_0^t \exp\langle\alpha, X_s\rangle \left\{\langle\alpha, b(s)\rangle + \frac{1}{2}\langle\alpha, a(s)\alpha\rangle\right\} ds$$

is a local martingale. Hence,

$$\exp\langle\alpha, X_t\rangle - \exp\langle\alpha, X_0\rangle - \int_0^t \exp\langle\alpha, X_s\rangle \left\{\langle\alpha, b(s)\rangle + \frac{1}{2}\langle\alpha, a(s)\alpha\rangle\right\} ds$$

is a local martingale. Note that for the particular function $f(x) = e^{\langle\alpha,x\rangle}$ on hand,

$$A_t f(x) = \left\{\langle\alpha, b(t)\rangle + \frac{1}{2}\langle\alpha, a(t)\alpha\rangle\right\}\exp\langle\alpha, x\rangle.$$

We have thus shown that the process $M_f(t)$ is a local martingale when $f(x) = e^{\langle\alpha,x\rangle}$. Therefore, $M_f(t)$ is a local martingale when f is a linear combination of exponential functions.

Step 3 Consider any function $f \in C_b^2(\mathbb{R}^d)$. Let B_n denote the closed ball of radius $n \in \mathbb{N}$ in $\mathbb{R}^d$. By the Stone-Weierstrass theorem, there exists a function f_n on B_n such that f_n is a linear combination of exponential functions, and f_n approximates f and its derivatives of order up to two uniformly on B_n to within $\frac{1}{n}$; that is,

$$\sup_{x\in B_n}|f(x) - f_n(x)| + \max_{1\le j\le d}\sup_{x\in B_n}\left|\frac{\partial f}{\partial x_j}(x) - \frac{\partial f_n}{\partial x_j}(x)\right|$$
$$+ \max_{1\le i,j\le d}\sup_{x\in B_n}\left|\frac{\partial f}{\partial x_i\partial x_j}(x) - \frac{\partial f_n}{\partial x_i\partial x_j}(x)\right| < \frac{1}{n}.$$

Thus, we obtain a sequence of functions $\{f_n\}$ that converges to f uniformly on compacts. The derivatives of f_n up to order two also converge uniformly on compacts to the corresponding derivatives of f. By Step 2, we know that M_{f_n} is a local martingale for each n. Therefore, $\{M_f(t \wedge \tau_R)\}$ is also a local martingale where $\tau_R = \inf\{t : |X_t| > R\}$. Being bounded, it is actually a martingale, so that for any $0 \leq s \leq t$,

$$E(M_{f_n}(t \wedge \tau_R)|\mathcal{F}_s) = M_{f_n}(s \wedge \tau_R).$$

Letting $n \to \infty$, we get $E(M_f(t \wedge \tau_R)|\mathcal{F}_s) = M_f(s \wedge \tau_R)$. As $R \to \infty$, using the bounded convergence theorem, we obtain

$$E(M_f(t)|\mathcal{F}_s) = M_f(s).$$

■

Remark 7.1.1 In the proof of Theorem 7.1.5, we showed that, for any $f \in C_0^2(\mathbb{R}^d)$, $f(X_t) - f(X_0) - \int_0^t A_t f(X_s)\,ds$ is a P-martingale. Hence, $C_0^2(\mathbb{R}^d) \subseteq D(A)$.

By the equivalence of statements given in Theorem 7.1.5, one can define martingale problems as follows.

Definition 7.1.4 *A probability measure P on the canonical space $(\Omega, \mathcal{F}, (\mathcal{F}_t))$ is a solution to the martingale problem for A_t, and initial position x if*

1. $P(X_0 = x) = 1$, *and*
2. $X_\theta(t) = \exp\{\langle\theta, X_t - X_0\rangle - \int_0^t \langle\theta, b(s)\rangle\,ds - \frac{1}{2}\int_0^t \langle\theta, a(s)\theta\rangle\,ds\}$ *is a P-martingale for all $\theta \in \mathbb{R}^d$, where X is the canonical process.*

This is the definition of martingale problems that was originally given by Stroock and Varadhan.

7.2 Existence of Solutions

Let P be a probability measure on the canonical space $(\Omega, \mathcal{F})$, and Y_t, any $\mathbb{R}^d$-valued adapted process. Let $a : [0, \infty) \times \Omega \to \mathbb{R}^d \times \mathbb{R}^d$ be jointly measurable and adapted to $\mathcal{F}_t$. Let us assume the following:

1. There exist a positive constant K such that

$$0 \leq \sum_{i,j=1}^{d} a_{ij}(s)\theta_i\theta_j \leq K|\theta|^2 \quad \text{for all } \theta \in \mathbb{R}^d.$$

2. $X_\theta(t) = \exp\{\langle\theta, Y_t - Y_0\rangle - \frac{1}{2}\int_0^t \langle\theta, a(s)\theta\rangle\,ds\}$ is a P-martingale for all $\theta \in \mathbb{R}^d$.

Lemma 7.2.1 *Under the hypotheses given above, for any $T > 0$ and $\lambda > 0$,*

$$P\left\{\sup_{0\le t\le T}\langle\theta, Y_t - Y_0\rangle \ge \lambda\right\} \le \exp\left\{-\frac{\lambda^2}{2K||\theta||^2 T}\right\} \tag{7.2.1}$$

for all $\theta \in \mathbb{R}^d$.

Proof Using the second hypothesis, one can write

$$P\left\{\sup_{0\le t\le T}\langle\theta, Y_t - Y_0\rangle \ge \lambda\right\} \le P\left\{\sup_{0\le t\le T} X_\theta(t) \ge \exp\left\{\lambda - \frac{||\theta||^2}{2}KT\right\}\right\}$$
$$\le \exp\left\{-\lambda + \frac{||\theta||^2}{2}KT\right\}$$

by the basic martingale inequality and the Chebyshev inequality. By replacing λ by $||\theta||\lambda$, one obtains the upper bound $\exp\{-||\theta||\lambda + \frac{||\theta||^2}{2}KT\}$. The bound is a function of $||\theta||$ and is minimized when $||\theta|| = \frac{\lambda}{KT}$. Hence,

$$P\left\{\sup_{0\le t\le T}\langle\theta, Y_t - Y_0\rangle \ge ||\theta||\lambda\right\} \le \exp\left\{-\frac{\lambda^2}{2KT}\right\}$$

for such θ. In fact, the inequality holds for all nonzero $\theta \in \mathbb{R}^d$ since the left side depends only on $\frac{\theta}{||\theta||}$. The proof is over upon replacing λ by $\frac{\lambda}{||\theta||}$. ■

Lemma 7.2.2 *Let the above assumptions hold. If $\alpha : [0,\infty) \times \Omega \to \mathbb{R}^d$ is a bounded, adapted, $\mathbb{R}^d$-valued process, then,*

$$X_\alpha(t) = \exp\left\{\int_0^t \langle\alpha(s), dY_s\rangle - \frac{1}{2}\int_0^t \langle\alpha(s), a(s)\alpha(s)\rangle\, ds\right\} \tag{7.2.2}$$

is a P-martingale.

Proof Define the sequence of functions $\{\alpha_n\}$ by $\alpha_n(t) = \alpha(\frac{[nt]}{n})$, where the notation $[x]$ denotes the greatest integer function of x. Then, $\alpha_n(t)$ is a simple $\mathcal{F}_t$-adapted function which converges to $\alpha(t)$ as $n \to \infty$, for each $t > 0$. If $||\alpha||$ denotes the bound for α, then the functions α_n are also bounded by $||\alpha||$. By using the second condition in the hypotheses, X_{α_n} is a P-martingale for each n.

By Lemma 7.2.1, for any $T > 0$ and $\lambda > 0$,

$$P\left\{\sup_{0\le t\le T}\int_0^t \langle\alpha_n(s), dY_s\rangle \ge \lambda\right\} \le \exp\left\{-\frac{\lambda^2}{2K||\alpha_n||^2 T}\right\}.$$

Therefore, for all $x > 1$,

$$P\left\{X_{\alpha_n}(t) \geq x\right\} \leq P\left\{\int_0^t \langle \alpha_n(s), dY_s\rangle \geq \log x\right\}$$
$$\leq \exp\left\{-\frac{(\log x)^2}{2K||\alpha_n||^2 t}\right\}$$
$$\leq \exp\left\{-\frac{(\log x)^2}{2K||\alpha||^2 t}\right\}.$$

Hence, for each $t > 0$, the sequence $\{X_{\alpha_n}(t)\}$ is bounded in $L^2(P)$ and is therefore uniformly integrable. Hence, taking the limit as $n \to \infty$, X_α inherits the martingale property. ■

Lemma 7.2.3 *Let P, Y, and a be as above. Let $\sigma : [0, \infty) \times \Omega \to \mathbb{R}^d \times \mathbb{R}^d$ be an adapted function such that*

$$\sum_{i,j=1}^{d} (\sigma\sigma^*)_{ij}(s, \omega)\theta_i\theta_j \leq C|\theta|^2 \quad \textit{for all} \;\; \theta \in \mathbb{R}^d$$

where C is a constant. Then, we have the following:

1. *The stochastic integral $N_t = \int_0^t \sigma(s)\, dY_s$ exists, and the process*

$$Y_\theta(t) = \exp\left\{\langle\theta, N_t\rangle - \frac{1}{2}\int_0^t \left\langle\theta, \sigma(s)a(s)\sigma^*(s)\theta\right\rangle\, ds\right\}$$

is a P-martingale for all $\theta \in \mathbb{R}^d$, and

2. *If $\alpha : [0, \infty) \times \Omega \to \mathbb{R}^d$ is an adapted function such that $E^P\{\int_0^t |\alpha(s)|^2 ds\}$ is finite, then $\int_0^t \langle \alpha(s), dN_s\rangle$ is well defined, and*

$$\int_0^t \langle\alpha(s), dN_s\rangle = \int_0^t \langle\sigma^*(s)\alpha(s), dY_s\rangle \;\; P\text{-a.s.} \tag{7.2.3}$$

Proof By the hypotheses, Y is a P-martingale. For each i, if σ_i denotes the ith row of the matrix σ, then $\int_0^t \sigma_i(s)\, dY_s$ exists as an Itô integral, since $\int_0^t \langle\sigma_i(s), \sigma_i(s)a(s)\rangle\, ds$ is finite. Hence N_t exists for all $t \geq 0$. Therefore, Y_θ is a P-martingale.

The integral $\int_0^t \langle\alpha(s), dN_s\rangle$ exists and is a martingale since $E^P\{\int_0^t |\alpha(s)|^2\, ds\}$ is finite. Equation (7.2.3) is easy to verify if α is a *simple* adapted process. The general statement follows since one can approximate α by a sequence of simple adapted processes. ■

Theorem 7.2.4 *Let P, Y, and a be as above. Suppose that a is symmetric and satisfies*

$$k|\theta|^2 \leq \sum_{i,j=1}^{d} |a_{ij}(s, \omega)|\theta_i\theta_j \leq K|\theta|^2 \quad \textit{for all} \;\; \theta \in \mathbb{R}^d$$

where $k > 0$ is a constant. Let σ be the positive definite, symmetric square root of a. Then,

$$Z_t := \exp\left\{\int_0^t \langle\sigma^{-1}(s)\theta, dY_s\rangle - \frac{1}{2}|\theta|^2 t\right\}$$

is a P-martingale for all $\theta \in \mathbb{R}^d$. The process

$$B_t := \int_0^t \sigma^{-1}(s)\, dY_s \tag{7.2.4}$$

is a d-dimensional, standard Wiener process with respect to P. Besides, a.s.,

$$Y_t = Y_0 + \int_0^t \sigma(s)\, dB_s. \tag{7.2.5}$$

Proof In Equation (7.2.2), take $\alpha(s) = \sigma^{-1}(s)\theta$ to obtain that the process

$$\exp\left\{\int_0^t \langle\sigma^{-1}(s)\theta, dY_s\rangle - \frac{1}{2}\int_0^t \langle\sigma^{-1}(s)\theta, a(s)\sigma^{-1}(s)\theta\rangle\, ds\right\}$$

is a P-martingale. Thus, Z is a P-martingale. Equation (7.2.3) and the Lévy characterization of a Brownian motion imply the second assertion.

Equation (7.2.5) is shown by observing that

$$\begin{aligned}\langle\theta, Y_t - Y_0\rangle &= \int_0^t \langle\theta, dY_s\rangle = \int_0^t \langle\sigma^{-1}(s)\sigma(s)\theta, dY_s\rangle \\ &= \int_0^t \langle\sigma(s)\theta, dB_s\rangle \quad \text{by (7.2.3) and (7.2.4)} \\ &= \left\langle\theta, \int_0^t \sigma(s)\, dB_s\right\rangle \quad \text{by using (7.2.3).}\end{aligned}$$

■

Corollary 7.2.5 *Let P and a be as in Theorem 7.2.4, and X, the canonical process $X_t(\omega) = \omega(t)$. Let $b : [0,\infty) \times \Omega \to \mathbb{R}^d$ be a bounded, adapted function. Then, the following are equivalent:*

1. *$X_\theta(t) = \exp\{\langle\theta, X_t - X_0\rangle - \int_0^t \langle\theta, b(s)\rangle\, ds - \frac{1}{2}\int_0^t \langle\theta, a(s)\theta\rangle\, ds\}$ is a P-martingale for each $\theta \in \mathbb{R}^d$.*
2. *There exists a Brownian motion B with respect to P such that*

$$X_t = X_0 + \int_0^t b(s)\, ds + \int_0^t \sigma(s)\, dB_s \quad P\text{-a.s.},$$

where $\sigma(s)$ is the symmetric, positive, definite square root of $a(s)$.

Proof Statement (1) implies (2): To show this, define $Y_t = X_t - \int_0^t b(s)\, ds$ for all $t \geq 0$. Then it follows that $\langle\theta, Y_t - Y_0\rangle$ is a martingale with increasing process $\int_0^t \langle\theta, a(s)\theta\rangle\, ds$. By Theorem 7.2.4, the proof is complete.

The converse is a straightforward application of the Itô formula. ■

The next result is on patching of measures, a useful method in the study of martingale problems.

Lemma 7.2.6 *On the canonical space $(\Omega, \mathcal{F}, (\mathcal{F}_t))$, let $X_t(\omega) = \omega(t)$ and τ, a given stopping time. Let P be a probability measure on $(\Omega, \mathcal{F}_\tau)$. For each $\omega \in \Omega$, let P_ω be a probability measure on $(\Omega, \mathcal{F}_{\tau(\omega),\infty})$ such that*

1. *$P_\omega\{\omega' : X_{\tau(\omega)}(\omega') = X_{\tau(\omega)}(\omega)\} = 1$ for each $\omega \in \Omega$, and*
2. *the map $\omega \to P_\omega(B)$ is $\mathcal{F}_t$ measurable on $\{\tau \leq t\}$ for any $t \geq 0$ and $B \in \mathcal{F}_{t,\infty}$.*

Then there exists a unique probability measure Q on $(\Omega, \mathcal{F})$ such that $Q = P$ on $\mathcal{F}_\tau$, and the rcpd of Q given $\mathcal{F}_\tau$ coincides with P_ω on $\mathcal{F}_{\tau(\omega),\infty}$ a.s. with respect to P.

Proof For each $\omega \in \Omega$, define a measure Q_ω on $(\Omega, \mathcal{F})$ such that its specification on cylinder sets is given by

$Q_\omega\{X_{s_1} \in B_1, \ldots, X_{s_n} \in B_n\} =$

$$\begin{cases} 1_{B_1}(X_{s_1}(\omega)), \ldots, 1_{B_k}(X_{s_k}(\omega))P_\omega\{X_{s_{k+1}} \in B_{k+1}, \ldots, X_{s_n} \in B_n\} & \text{if } \tau(\omega) \in I_k, \\ 1_{B_1}(X_{s_1}(\omega)), \ldots, 1_{B_n}(X_{s_n}(\omega)) & \text{if } s_n \leq \tau(\omega) \end{cases}$$

where $s_1 < s_2 \cdots < s_n$, $n \geq 1$, and $B_1, \ldots, B_n$ are Borel sets in $\mathbb{R}^d$, and $I_k = [s_k, s_{k+1})$. For each such cylinder set F, and $t \geq 0$, consider the event

$$\{\omega : Q_\omega(F) \in B\} \cap \{\tau \leq t\}$$

where B is any Borel set of $\mathbb{R}^1$. It is quite simple to check that

$$\{\omega : Q_\omega(F) \in B\} \cap \{\tau \leq t\}$$

is $\mathcal{F}_t$-measurable. Therefore, as a function of ω, $Q_\omega(F)$ is $\mathcal{F}_\tau$-measurable for all cylinder sets F. The class of sets

$$\{F \in \mathcal{F} : \text{the function } Q_\omega(F) \text{ is } \mathcal{F}_\tau\text{-measurable}\}$$

is a σ-field, and hence coincides with $\mathcal{F}$. It is clear from the definition of Q_ω that $Q_\omega(F) = 1_F(\omega)$ for all cylinder sets F in $\mathcal{F}_\tau$ and hence, the equality holds for all $F \in \mathcal{F}_\tau$.

Define Q on $(\Omega, \mathcal{F})$ by $Q(F) = E^P[Q_\omega(F)]$ for all $F \in \mathcal{F}$. Then,

$$Q(F) = P(F) \quad \text{for all } F \in \mathcal{F}_\tau.$$

If $F \in \mathcal{F}_\tau$ and $G \in \mathcal{F}$, then

$$Q(F \cap G) = E^P\{1_F Q_\omega(G)\} = \int_F Q_\omega(G)\, dP(\omega) = \int_F Q_\omega(G)\, dQ(\omega)$$

since P and Q coincide on $\mathcal{F}_\tau$. Thus Q_ω is an rcpd of Q given $\mathcal{F}_\tau$. It should be noted that Q_ω is determined a.s. with respect to Q. Equivalently, one can say that Q_ω is determined P-a.s.. Let $\tau(\omega) \leq t$, and F be a cylinder set in $\mathcal{F}_{t,\infty}$. One can easily show that $Q_\omega(F) = P_\omega(F)$ P-a.s. Indeed, if $\tau(\omega) < t$, then $Q_\omega(F) = P_\omega(F)$ follows from the definition of Q_ω. If $\tau(\omega) = t$, and $F = \{X_t \in B_1, X_{s_2} \in B_2, \ldots, X_{s_n} \in B_n\}$, with $t < s_2 < \cdots < s_n$, then by using the property (1) (stated in the lemma) for the measure P_ω,

$$
\begin{aligned}
P_\omega(F) &= P_\omega\{X_t(\omega) \in B_1, X_{s_2} \in B_2, \ldots, X_{s_n} \in B_n\} \\
&= I_{B_1}(X_t(\omega))P_\omega\{X_{s_2} \in B_2, \ldots, X_{s_n} \in B_n\} \\
&= Q_\omega\{X_{s_1} \in B_1, X_{s_2} \in B_2, \ldots, X_{s_n} \in B_n\}.
\end{aligned}
$$

Thus Q satisfies all the requirements of the lemma.

Since the values of Q on $\mathcal{F}_\tau$, and its rcpd given $\mathcal{F}_\tau$ are specified by the lemma, the measure Q is unique. ■

In the remainder of this section, let $a : [0,\infty) \times \Omega \to \mathbb{R}^d \times \mathbb{R}^d$ be an adapted function which is symmetric and satisfies

$$0 \leq \langle \theta, a(s)\theta \rangle \leq K|\theta|^2. \tag{7.2.6}$$

We will prove the existence of a solution P to the martingale problem for a, started at x. In other words, P satisfies

(i) $P\{X_0 = x\} = 1$, and
(ii) $X_\theta(t) = \exp\{\langle \theta, X_t - x\rangle - \frac{1}{2}\int_0^t \langle \theta, a(s)\theta\rangle\, ds\}$ is a P-martingale for all $\theta \in \mathbb{R}^d$.

Let μ_{t_0} denote the standard Wiener measure on $(\Omega, \mathcal{F}_{t_0,\infty})$ started at time t_0. In other words, $\{X_t : t \geq t_0\}$ is a standard $\mathbb{R}^d$-valued Brownian motion with $X_{t_0} = 0$. Given a symmetric, nonnegative, definite matrix a, let σ be the symmetric, nonnegative, definite square root of a. For any $x \in \mathbb{R}^d$, define the probability measure

$$\mu_{(t_0,x,a)}\{X_{t_1} \in B_1, \ldots, X_{t_n} \in B_n\} = \mu_{t_0}\{\sigma(X_{t_1} + x) \in B_1, \ldots, \sigma(X_{t_n} + x) \in B_n\}$$

for all $t_0 < t_1 < \cdots < t_n$, $n \geq 1$, and any Borel sets $B_1, \ldots, B_n$ in $\mathbb{R}^d$. Then the process $\exp\{\langle \theta, X_t - x\rangle - \frac{1}{2}(t - t_0)\langle \theta, a\theta\rangle\}$, for $t \geq t_0$, is a $\mu_{(t_0,x,a)}$-martingale for all $\theta \in \mathbb{R}^d$.

We will first take a to be an adapted, simple function of the form given by $a(t,\omega) = a_k(\omega)$ if $\frac{k}{m} \leq t < \frac{k+1}{m}$. We will call such functions m-simple.

Lemma 7.2.7 *If $a : [0,\infty) \times \Omega \to \mathbb{R}^d \times \mathbb{R}^d$ is an adapted function which is symmetric, m-simple, and satisfies (7.2.6), then there exists a probability measure P on $(\Omega, \mathcal{F})$ which solves the martingale problem for the matrix a started at any x in $\mathbb{R}^d$.*

Proof Define a sequence of measures on $(\Omega, \mathcal{F})$ by induction as follows. Let the measure $Q_1 = \mu_{0,x,a*}$ where $a* = a_0(\omega_0)$ for any ω_0 satisfying $X_0(\omega_0) = x$. Since a_0 is $\mathcal{F}_0$-measurable, it is a Borel-measurable function of X_0, and hence $a*$ is uniquely determined. If Q_n has been defined, then define Q_{n+1} as the measure satisfying $Q_{n+1} = Q_n$ on $\mathcal{F}_{\frac{n}{m}}$, and the rcpd Q_ω^{n+1} of Q_{n+1} given $\mathcal{F}_{\frac{n}{m}}$ coincides with $\mu_{\left(\frac{n}{m}, X\left(\frac{n}{m},\omega\right), a\left(\frac{n}{m},\omega\right)\right)}$ on the σ-field $\mathcal{F}_{\frac{n}{m},\infty}$. From Lemma 7.2.6, we know the existence and uniqueness of Q_{n+1}.

Define the measure $P_n = Q_n$ on $\mathcal{F}_{\frac{n}{m}}$. Then $P_{n+1} = P_n$ on $\mathcal{F}_{\frac{n}{m}}$ for all n. Since Ω is a Polish space, and $\cup_n \mathcal{F}_{\frac{n}{m}}$ generates $\mathcal{F}$, there exists a probability measure P on $(\Omega, \mathcal{F})$ such that $P = P_n$ on $\mathcal{F}_{\frac{n}{m}}$ for all n. We now show that P is the solution of the martingale problem stated in the lemma.

Clearly, $P\{X_0 = x\} = 1$. To show that X_θ is a P-martingale, one can take successive conditional expectations to reduce the problem to one of proving that for any t_1, t_2 such that

$$\frac{n}{m} \leq t_1 \leq t_2 \leq \frac{n+1}{m},$$

we have

$$E^{Q_{n+1}}\{X_\theta(t_2) | \mathcal{F}_{t_1}\} = X_\theta(t_1) \quad Q_{n+1}\text{a.s.}$$

Toward this, let $A = B \cap C$ where $B \in \mathcal{F}_{\frac{n}{m}}$ and $C \in \mathcal{F}_{\frac{n}{m},t_1}$.

$$\begin{aligned}
\int_A X_\theta(t_2)\, dQ_{n+1} &= \int_B X_\theta\left(\frac{n}{m}, \omega\right) \int_C \frac{X_\theta(t_2)}{X_\theta\left(\frac{n}{m}, \omega\right)}\, dQ_\omega^{n+1} Q_n(d\omega) \\
&= \int_B X_\theta\left(\frac{n}{m}, \omega\right) \int_C \frac{X_\theta(t_1)}{X_\theta\left(\frac{n}{m}, \omega\right)}\, dQ_\omega^{n+1} Q_n(d\omega) \\
&= \int_A X_\theta(t_1)\, dQ_{n+1}
\end{aligned}$$

since for all $\frac{n}{m} \leq t \leq \frac{n+1}{m}$,

$$\frac{X_\theta(t)}{X_\theta\left(\frac{n}{m}, \omega\right)} = \exp\left\{\left\langle \theta, X_t - X\left(\frac{n}{m}, \omega\right)\right\rangle - \frac{1}{2}\left(t - \frac{n}{m}\right)\left\langle \theta, a\left(\frac{n}{m}, \omega\right)\theta\right\rangle\right\}$$

is a Q_ω^{n+1}-martingale for almost every ω with respect to Q_n. This is so since $Q_\omega^{n+1} = \mu_{\left(\frac{n}{m}, X\left(\frac{n}{m},\omega\right), a\left(\frac{n}{m},\omega\right)\right)}$.

The class of such sets A generates $\mathcal{F}_{t_1}$. Therefore,

$$E^{Q_{n+1}}\{X_\theta(t_2) | \mathcal{F}_{t_1}\} = X_\theta(t_1) \text{ a.s. } Q_{n+1}. \qquad \blacksquare$$

The assumption that a is m-simple is removed in the following theorem.

Theorem 7.2.8 *If $a : [0, \infty) \times \Omega \to \mathbb{R}^d \times \mathbb{R}^d$ is an adapted function which is symmetric and satisfies (7.2.6), then for each x, there exists a probability measure P on $(\Omega, \mathcal{F})$ such* *that* $P\{X_o = x\} = 1$, *and for* $t \geq 0$,

$$X_\theta(t) = \exp\left\{\langle \theta, X_t - x\rangle - \frac{1}{2}\int_0^t \langle \theta, a(s)\theta\rangle\, ds\right\}$$

is a P-martingale for all $\theta \in \mathbb{R}^d$.

Proof

Step 1 Given a, define $a_m(t) = a\left(\frac{[mt]}{m}\right)$ for all $t \geq 0$. Let P_m be the solution of the martingale problem for a_m started at x. Then,

$$X_\theta^{(m)}(t) := \exp\left\{\langle \theta, X_t - x\rangle - \frac{1}{2}\int_0^t \langle \theta, a_m(s)\theta\rangle\, ds\right\}$$

is a P_m-martingale for all $\theta \in \mathbb{R}^d$. By Theorem 7.1.5, $\langle \theta, X\rangle$ is a P_m-martingale, with a bounded quadratic variation process given by $\int_0^t \langle \theta, a_m(s)\theta\rangle\, ds$ for each θ. In particular, each coordinate of X is a martingale with its quadratic variation on $[0, t]$ bounded by Kt for all $t \geq 0$. Therefore, by the Burkhölder-Davis-Gundy inequality, there exists a constant C that depends on K such that

$$E^{P_m}\{|X_{t_2} - X_{t_1}|^4\} \leq C(t_2 - t_1)^2$$

for all $t_2 \geq t_1$. Hence, the sequence of measures $\{P_m\}$ is tight by the Kolmogorov criterion for tightness, and there exists a subsequence $\{P_{m_j}\}$ which converges weakly to a probability measure P on $(\Omega, \mathcal{F})$. For simplicity of notation, we will denote the subsequence m_j as m.

Step 2 First, we observe that $\{\omega : X_0(\omega) = x\}$ is a closed set, and hence,

$$P\{X_0 - x\} \geq \limsup_{m\to\infty} P_m\{X_0 - x\} = 1.$$

Thus, the measure P satisfies the initial condition that $X_0 = x$ P-a.s. It remains to show that $\{X_\theta(t)\}$ is a P-martingale. We know that $\{X_\theta^m(t)\}$ is a P_m-martingale for each m. Equivalently, for all $0 \leq s \leq t$, and any bounded, continuous, $\mathcal{F}_s$-measurable function, Y, defined on Ω,

$$E_{P_m}\{YX_\theta^m(t)\} = E_{P_m}\{YX_\theta^m(s)\}.$$

Therefore, it suffices to show that for all bounded, continuous, $\mathcal{F}_t$-measurable functions Y,

$$E_P\{YX_\theta(t)\} = \lim_{m\to\infty} E_{P_m}\{YX_\theta^m(t)\}. \tag{7.2.7}$$

For each fixed $N \in \mathbb{N}$, truncate $X_\theta^m(t)$ at N, and consider the sequence of functions $\{Y(X_\theta^m(t) \wedge N)\}$ indexed by m. This is a sequence of functions on Ω which are uniformly bounded and $\mathcal{F}_t$-measurable. Given any $\epsilon > 0$, choose a compact subset K in Ω such that $\sup_m P_m(K^c) + P(K^c) < \epsilon$. Then, for each fixed $\omega \in K$, there exists

$\delta > 0$ such that

$$
\begin{aligned}
&|\langle \theta, X_t(\omega) - X_t(\omega') \rangle| + \int_0^t \langle \theta, (a_m(s,\omega) - a_m(s,\omega'))\theta \rangle|\, ds \\
&\leq |\theta||\omega(t) - \omega'(t)| + |\theta|^2 \int_0^t \left| a\left(\frac{[ms]}{m}, \omega\right) - a\left(\frac{[ms]}{m}, \omega'\right)\right| ds \\
&< \epsilon \text{ for all } m \text{ if } \|\omega - \omega'\| < \delta.
\end{aligned}
$$

In the above, we have made use of uniform continuity of a on $[0, t] \times K$. It follows that

$$\sup_m |X_\theta^m(t,\omega) - X_\theta^m(t,\omega')| < \epsilon \text{ if } \|\omega - \omega'\| < \delta.$$

Hence, one obtains

$$\lim_{m\to\infty} E_{P_m}\left[Y(X_\theta^m(t) \wedge N)\right] = \lim_{m\to\infty} E_P\left[Y(X_\theta^m(t) \wedge N)\right].$$

We can therefore conclude that

$$\lim_{m\to\infty} E_{P_m}\left[Y(X_\theta^m(t) \wedge N)\right] = E_P\left[Y(X_\theta(t) \wedge N)\right].$$

Now, in order to prove (7.2.7), it suffices to show that

$$E_P\left[|X_\theta(t) - (X_\theta(t) \wedge N)|\right] \text{ and } \sup_m E_{P_m}\left[|X_\theta^m(t) - (X_\theta^m(t) \wedge N)|\right] \tag{7.2.8}$$

tend to 0 as $N \to \infty$.

Step 3 From Lemma 7.2.1, for any $m \in \mathbb{N}$,

$$P_m\left\{|X_t - x| > \lambda\right\} \leq 2d \exp\left(-\frac{\lambda^2}{2Kt}\right).$$

Therefore,

$$
\begin{aligned}
E_{P_m}\left[|X_\theta^m(t) - (X_\theta^m(t) \wedge N)\right] &= E\left[X_\theta^m(t) 1_{\{X_\theta^m(t) > N\}}\right] \\
&= \int_N^\infty P_m\left\{X_\theta^m(t) > u\right\} du \\
&\leq \int_N^\infty P_m\{|\langle \theta, X_t - x\rangle| > \ln u\}\, du \\
&\leq 2d \int_N^\infty \exp\left\{-\frac{(\ln u)^2}{2Kt}\right\} du \\
&\leq 2d \int_{\ln N}^\infty \exp\left\{y - \frac{y^2}{2Kt}\right\} dy
\end{aligned}
$$

by setting $y = \ln u$. Thus, $\sup_m E_{P_m}[|X_\theta^m(t) - (X_\theta^m(t) \wedge N)] \to 0$ as $N \to \infty$.

A similar argument doesn't apply to $E_P[|X_\theta(t) - (X_\theta(t) \wedge N)|]$ since we do not, at this stage, know that $X_\theta(t)$ is a P-martingale. Hence, we use the fact that the set $\{|X_t - x| > \lambda\}$ is an open set so that

$$
\begin{aligned}
P\{|X_t - x| > \lambda\} &\leq \liminf_{m\to\infty} P_m\{|X_t - x| > \lambda\} \\
&\leq 2d \exp\left(-\frac{\lambda^2}{2Kt}\right).
\end{aligned}
$$

Therefore, $X_\theta(t)$ is integrable, and we can conclude that $E_P[|X_\theta(t) - (X_\theta(t) \wedge N)|] \to 0$ as $N \to \infty$. ■

7.3 Analytical Tools

For any $x, y \in \mathbb{R}^d$, $\lambda > 0$, and $0 \leq s \leq t$, define

$$
p_\lambda(s, x; t, y) = \frac{1}{[2\pi(t-s)]^{d/2}} \exp\left\{-\lambda(t-s) - \frac{|y-x|^2}{2(t-s)}\right\}. \tag{7.3.1}
$$

For each $\lambda > 0$, define the operator G_λ by

$$
G_\lambda h(s,x) = \int_s^\infty \int_{\mathbb{R}^d} p_\lambda(s, x; t, y) h(t, y)\, dy\, dt
$$

for a suitable class of functions h defined on $[0, \infty) \times \mathbb{R}^d$, say, $h \in C_0([0,\infty) \times \mathbb{R}^d)$. Since s, x are the backward variables, it is easy to check that the function $f(s,x) = G_\lambda h(s,x)$ formally solves the equation:

$$
-\frac{\partial f}{\partial s} = \frac{1}{2}\sum \frac{\partial^2 f}{\partial x_i^2} - \lambda f + h. \tag{7.3.2}
$$

Lemma 7.3.1 *For each $p \in [1, \infty]$, the operator G_λ maps $L^p([0,\infty) \times \mathbb{R}^d)$ into itself, and $\|G_\lambda\|_p \leq \lambda^{-1}$.*

Proof The proof follows easily from the Jensen inequality for $1 \leq p < \infty$. The case $p = \infty$ is even simpler. ■

The next result is known as Young's inequality. The notation $h * g$ stands for the convolution of the functions h and g.

Lemma 7.3.2 *Let $1 \leq p_1, p_2, r \leq \infty$ such that*

$$
\frac{1}{p_1} - \frac{1}{p_2} = 1 - \frac{1}{r}.
$$

If $h \in L^{p_1}$ and $g \in L^r$, then $h * g \in L^{p_2}$, and

$$\|h * g\|_{p_2} \leq \|h\|_{p_1} \|g\|_r.$$

If p_1 and r are conjugate exponents, then $h * g(x)$ exists for every x, and $h * g$ is a bounded, uniformly continuous function.

Proof

Step 1 Suppose $p_1 = 1$, and p_2 is finite. Note that $p_2 = r$, so that

$$\begin{aligned}
\|h * g\|_{p_2} &= \left\| \int_{\mathbb{R}^d} h(y) g(\cdot - y)\, dy \right\|_{p_2} \\
&= \left(\int_{\mathbb{R}^d} \left| \int_{\mathbb{R}^d} h(y) g(x-y)\, dy \right|^{p_2} dx \right)^{1/p_2} \\
&\leq \int_{\mathbb{R}^d} \left(\int_{\mathbb{R}^d} |h(y) g(x-y)|^{p_2}\, dx \right)^{1/p_2} dy
\end{aligned}$$

by using the Minkowski inequality for integrals

$$\begin{aligned}
&= \int_{\mathbb{R}^d} |h(y)| \left(\int_{\mathbb{R}^d} |g(x-y)|^{p_2}\, dx \right)^{1/p_2} dy \\
&= \int_{\mathbb{R}^d} |h(y)|\, \|g\|_r\, dy = \|h\|_1 \|g\|_r.
\end{aligned}$$

Step 2 Let $p_1 > 1$ and p_1, p_2 be both finite. Then, r is finite and $p_2 > r > 1$. Let us pick any α and β in the interval $[0, 1]$, and positive number k_j such that $\sum_{j=1}^{3} \frac{1}{k_j} = 1$. By the Hölder inequality,

$$\begin{aligned}
|(h * g)(x)| &= \left| \int_{\mathbb{R}^d} h^{\alpha}(y) g^{\beta}(x-y) h^{1-\alpha}(y) g^{1-\beta}(x-y)\, dy \right| \\
&\leq \left\{ \int_{\mathbb{R}^d} |h^{\alpha}(y) g^{\beta}(x-y)|^{k_1}\, dy \right\}^{\frac{1}{k_1}} \left\{ \int_{\mathbb{R}^d} \left| h^{(1-\alpha)}(y) \right|^{k_2} dy \right\}^{\frac{1}{k_2}} \\
&\quad \left\{ \int_{\mathbb{R}^d} \left| g^{(1-\beta)}(x-y) \right|^{k_3} dy \right\}^{\frac{1}{k_3}}.
\end{aligned}$$

We will fix $\alpha = \frac{p_1}{k_1}$, and $\beta := \frac{r}{k_1}$ with $k_1 = p_2$, and $(1-\alpha)k_2 = p_1$. Then, it follows that $k_3 = \frac{r}{1-\beta}$. For this choice, the above inequality yields

$$|h * g(x)|^{p_2} \leq \left\{ \int_{\mathbb{R}^d} |h^{p_1}(y) g^{r}(x-y)|\, dy \right\} \|h\|_{p_1}^{p_2 - p_1} \|g\|_r^{p_2 - r}.$$

Integrating both sides with respect to x, it is seen that

$$\|h * g\|_{p_2} \leq \|h\|_{p_1} \|g\|_r.$$

Step 3 The remaining cases reduce to p_1 and r being a pair of conjugate exponents. Therefore, $p_2 = \infty$. By the Hölder inequality, it follows that for any x,

$$|h * g(x)| \leq \|h\|_{p_1} \|g\|_r.$$

Thus $h * g(x)$ exists and is bounded for each x, and $\|h * g\|_\infty \leq \|h\|_{p_1} \|g\|_r$. In a similar manner, one obtains, for each x and y,

$$\begin{aligned} |h * g(x - y) - h * g(x)| &= \left| \int_{\mathbb{R}^d} [h(z - y) - h(z)] g(x - z)\, dz \right| \\ &\leq \left\{ \int_{\mathbb{R}^d} |h(z - y) - h(z)|^{p_1}\, dz \right\}^{\frac{1}{p_1}} \|g\|_r. \end{aligned}$$

Therefore,

$$\|h * g(\cdot - y) - h * g\|_\infty \leq \|h(\cdot - y) - h\|_{p_1} \|g\|_r.$$

Suppose that $p_1 < \infty$. Then $h(\cdot - y) - h$ is continuous in L^{p_1}-norm. Hence $\|h(\cdot - y) - h\|_{p_1}$ converges to 0 as $y \to 0$. This proves that $h * g$ is uniformly continuous. If $p_1 = \infty$, then we can interchange the roles of h and g in the above argument. ■

In the present context, we will use Lemma 7.3.2 as follows. First, note that for the function p defined by (7.3.1),

$$\begin{aligned} &\int_s^\infty \int_{\mathbb{R}^d} |p_\lambda(s, x; t, y)|^r \, dt\, dy \\ &= \int_s^\infty \frac{1}{[2\pi(t - s)]^{(r-1)d/2}} e^{-r\lambda(t-s)} \int_{\mathbb{R}^d} \frac{1}{[2\pi(t - s)]^{d/2}} \exp\left\{ -r \frac{|y - x|^2}{2(t - s)} \right\} dy\, dt \\ &= r^{-d/2} \int_s^\infty \frac{1}{[2\pi(t - s)]^{(r-1)d/2}} e^{-r\lambda(t-s)}\, dt \end{aligned}$$

so that

$$\int_s^\infty \int_{\mathbb{R}^d} |p_\lambda(s, x; t, y)|^r dt\, dy < \infty \tag{7.3.3}$$

if $\frac{(r-1)d}{2} < 1$; that is, if $r < \frac{d+2}{d}$.

To apply Lemma 7.3.2, define $g_\lambda(t, y)$ by

$$g_\lambda(t, y) = \begin{cases} \frac{1}{(2\pi t)^{d/2}} e^{-\lambda t - \frac{|y|^2}{2t}} & \text{if } t > 0; \\ 0 & \text{if } t \leq 0. \end{cases}$$

We have shown that $g \in L^r(\mathbb{R}^1 \times \mathbb{R}^d)$ if $r < \frac{d+2}{d}$. Therefore, by Lemma 7.3.2, we obtain

Lemma 7.3.3

1. *For $p_2 > p_1 \geq 1$, the operator G_λ is a bounded map from L^{p_1} to L^{p_2} where $\frac{1}{p_1} - \frac{1}{p_2} = 1 - \frac{1}{r}$ for some r such that $1 < r < \frac{d+2}{d}$.*
2. *There exists a $1 < p_B < \infty$ such that G_λ maps L^{p_B} into $C_b(\mathbb{R}^d)$, and*

$$\sup_{s,x} |(G_\lambda h)(s,x)| \leq C_{\lambda,p_B} \|h\|_{p_B}.$$

Here, p_B and r are conjugate exponents for some $1 < r < \frac{d+2}{d}$.

3. *If $h \in C_0^\infty$, then $G_\lambda h$ is in C^∞ and vanishes as $|x| \to \infty$.*

The third statement in the above lemma follows directly from the bounded convergence theorem. Since $p_\lambda(s,x;t,y)$ is explicitly given by (7.3.1), one can find its gradient with respect to x, and verify that

$$\int_s^\infty \int_{\mathbb{R}^d} |\text{grad } p_\lambda(s,x;t,y)|^r \, dy\, dt < \infty$$

if $r < \frac{d+2}{d+1}$. Indeed,

$$\int_s^\infty \int_{\mathbb{R}^d} |\nabla p_\lambda(s,x;t,y)|^r \, dy\, dt = \int_s^\infty \int_{\mathbb{R}^d} p_\lambda(s,x;t,y)^r \frac{|x-y|^r}{(t-s)^r} \, dy\, dt$$

The proof is analogous to that of (7.3.3). As a consequence, one obtains

Lemma 7.3.4 *There exists a $1 < p_L < \infty$ such that G_λ maps L^{p_L} to $C(\mathbb{R}^d)$, and there exists a constant C_{λ,p_L} such that*

$$|(G_\lambda h)(s,x_1) - (G_\lambda h)(s,x_2)| \leq C_{\lambda,p_L} \|h\|_{p_L} |x_1 - x_2|.$$

The index p_L must satisfy

$$\frac{1}{p_L} + \frac{1}{r} = 1 \quad \text{for some } r < \frac{d+2}{d+1}.$$

Besides,

$$\begin{aligned} &|(G_\lambda h)(s,x_1+\delta) - (G_\lambda h)(s,x_1) - (G_\lambda h)(s,x_2+\delta) + (G_\lambda h)(s,x_2)| \\ &\leq w(|\delta|)\, C_{\lambda,p_L} \|h\|_{p_L} |x_1 - x_2| \end{aligned}$$

where $w(|\delta|)$ tends to 0 as $|\delta|$ decreases to 0.

Proof The proof of the first statement follows directly from the integral form of the mean value theorem. Indeed, given x_1, x_2, define the function $F(\tau) = G_\lambda h(s, x_1 +$

$\tau(x_2 - x_1))$ for $\tau \in [0, 1]$. Then $F(0) = G_\lambda h(s, x_1)$ and $F(1) = G_\lambda h(s, x_2)$.

$$F(1) - F(0) = \int_0^1 F'(\tau)\, d\tau$$
$$\int_0^1 \langle \nabla G_\lambda h(s, x_1 + \tau(x_2 - x_1)), x_2 - x_1 \rangle \, d\tau$$
$$\leq |x_2 - x_1| \int |\nabla G_\lambda h(s, x)\, dx$$
$$\leq |x_2 - x_1| \|h\|_{p_L} \|\nabla p_\lambda\|_r$$

where $1 < r < \frac{d+2}{d+1}$, and p_L, r are conjugate exponents.

The second statement follows upon observing that the ratio of

$$|(G_\lambda h)(s, x_1 + h) - (G_\lambda h)(s, x_1) - (G_\lambda h)(s, x_2 + h) + (G_\lambda h)(s, x_2)|$$

and $C_{\lambda,p_L} \|h\|_{p_L} |x_1 - x_2|$ is bounded and goes to 0 as $|h| \downarrow 0$. For,

$$|(G_\lambda h)(s, x_1 + \delta) - (G_\lambda h)(s, x_1) - (G_\lambda h)(s, x_2 + \delta) + (G_\lambda h)(s, x_2)|$$
$$= \int_0^1 \langle \nabla G_\lambda h(s, x_1 + \delta + \tau(x_2 - x_1)) - \nabla G_\lambda h(s, x_1 + \tau(x_2 - x_1)), x_2 - x_1 \rangle \, d\tau$$
$$\leq |x_1 - x_2| \, \|h\|_{p_L} \left\{ \int |\nabla p_\lambda(x + \delta) - \nabla p_\lambda(x)|^r dx \right\}^{1/r},$$

which tends to zero as $|\delta| \to 0$ by the L^r-continuity of ∇p_λ. ■

The next result is an extension to parabolic equations of the Calderon-Zygmund estimate for elliptic equations. The extension is due to Jones.

Lemma 7.3.5 *Suppose that $h \in L^p([0, \infty) \times \mathbb{R}^d)$ for some $1 \leq p \leq \infty$. Let us denote $G_\lambda h$ by f. Then*

$$\left\| \frac{\partial^2 f}{\partial x_i \partial x_j} \right\|_p \leq 2C_{d,p} \|h\|_p. \tag{7.3.4}$$

Let $a_{ij}(s, x)$ be a symmetric, positive-definite matrix. With δ_{ij} denoting the Kronecker delta, define $e_{ij} = a_{ij} - \delta_{ij}$. We will assume that $|e_{ij}| \leq \epsilon$ everywhere, and that ϵ is so small that a_{ij} is uniformly elliptic. Define the operator

$$T_\lambda h = \frac{1}{2} \sum e_{ij} \frac{\partial^2 (G_\lambda h)}{\partial x_i \partial x_j}.$$

By the estimate (7.3.4), it follows that T_λ is a bounded operator from L^p to L^p for all $1 < p < \infty$. In fact,

$$\|T_\lambda\|_p \leq d^2 \epsilon C_{d,p}.$$

Let $1 < p_1 < p_2 < \cdots < p_N < \infty$ be any set of p-values. Then choose $\epsilon > 0$ small enough such that

$$d^2 \epsilon C_{d,p} \leq \frac{1}{2} \quad \text{for } p = p_1, \ldots, p_N.$$

Thus,

$$\|T_\lambda\|_p \leq \frac{1}{2} \quad \text{for } p = p_1, \ldots, p_N.$$

Define the operator R_λ by

$$R_\lambda = G_\lambda (I - T_\lambda)^{-1}.$$

By the choice of ϵ, it is clear that $(I - T_\lambda)^{-1}$ is well defined as a bounded operator from L^{p_i} to L^{p_i} and

$$\|(I - T_\lambda)^{-1}\| \leq 2$$

for all $i = 1, 2, \ldots, N$. Let $p_1 > 1$ be any given index. Then, we can construct a finite sequence $p_2, \ldots, p_N$ such that G_λ is a bounded operator from L^{p_i} to $L^{p_{i+1}}$ for $i = 1, 2, \ldots, N - 1$, and P_N satisfies

$$\frac{1}{p_N} + \frac{1}{r} = 1 \quad \text{for some } 1 < r < \frac{d+2}{d+1}.$$

We can conclude that the above properties of G_λ also hold for R_λ with the above choice of ϵ. A similar conclusion can be reached if instead of δ_{ij}, there is a symmetric, positive-definite matrix c_{ij} such that $|a_{ij} - c_{ij}| \leq \epsilon$. In this generality, the choice of ϵ will depend on the matrix c_{ij}. Thus, we have

Theorem 7.3.6 *There exists an $\epsilon > 0$ such that if $|a_{ij} - c_{ij}| \leq \epsilon$, then a is uniformly elliptic. Define*

$$T_\lambda h = \frac{1}{2} \sum e_{ij} \frac{\partial^2 (G_\lambda h)}{\partial x_i \partial x_j} \quad \textit{where } e_{ij} = a_{ij} - c_{ij}$$

and $R_\lambda = G_\lambda (I - T_\lambda)^{-1}$. Then R_λ is a bounded operator from L^{p_i} to $L^{p_{i+1}}$ for $i = 1, 2, \ldots, N - 1$. Moreover,

$$\sup_{s,x} |R_\lambda h(s, x)| \leq C_\lambda \|h\|_{p_N}$$

and

$$|(R_\lambda h)(s, x_1) - (R_\lambda h)(s, x_2)| \leq C_\lambda \|h\|_{p_N} |x_1 - x_2|.$$

7.4 Uniqueness of Solutions

In this section, we assume that $a : [0, \infty) \times \Omega \to \mathbb{R}^d \times \mathbb{R}^d$ is such that for all t and ω, $a_{ij}(t, \omega)$ is a function of t and $\omega(t)$. Therefore, we can redefine $a : [0, \infty) \times \mathbb{R}^d \to \mathbb{R}^d \times \mathbb{R}^d$. Suppose that a is continuous and uniformly elliptic, that is, there exists positive constants k and K such that

$$k|\theta|^2 \leq \sum_{i,j=1}^{d} |a_{ij}|\theta_i\theta_j \leq K|\theta|^2 \text{ for all } \theta \in \mathbb{R}^d. \tag{7.4.1}$$

Besides, we assume that

$$\|a_{ij} - \delta_{ij}\| \leq \epsilon \tag{7.4.2}$$

where ϵ is such that

$$2C_{d,p_N}\epsilon d^2 \leq 1. \tag{7.4.3}$$

From Theorem 7.2.8, we know that there exists a solution of the martingale problem for a started at x for any $x \in \mathbb{R}^d$. We will now prove the uniqueness of this solution by explicitly identifying the Laplace transform of any solution of the martingale problem.

Lemma 7.4.1 *Let P denote any solution of the martingale problem for a started at x. Define the measures μ_λ on $[0, \infty) \times \mathbb{R}^d$ by*

$$\mu_\lambda(f) = E\left[\int_0^\infty e^{-\lambda t} f(t, X_t)\, dt\right] \tag{7.4.4}$$

for all $f \in C_b^{1,2}([0, \infty) \times \mathbb{R}^d)$. Then, there exists a a constant C_λ such that

$$|\mu_\lambda(f)| \leq 2C_\lambda \, \|f\|_{p_N}.$$

Proof

Step 1 By Corollary 7.2.5, the canonical process X satisfies the equation

$$X(t) = x + \int_0^t \sigma(s, X(s))\, dB_s \;\; P\text{-a.s.},$$

where σ is the symmetric, positive-definite square root of a. Define

$$\pi_n(t) = \begin{cases} \frac{[nt]}{n} & \text{if } t < n; \\ n & \text{if } t \geq n. \end{cases}$$

Define

$$X_n(t) = x + \int_0^t \sigma(\pi_n(s), X(\pi_n(s))\, dB_s. \tag{7.4.5}$$

Since X has continuous trajectories, and σ is a bounded, continuous function, one gets $\lim_{n\to\infty} E^P[|X_n(t) - X(t)|^2] = 0$ for all $t \geq 0$. Therefore, $X_n(t)$ converges in distribution to $X(t)$ so that $Ef(t, X_n(t)) \to Ef(t, X(t))$ for all $f \in C_b(\mathbb{R}^+ \times \mathbb{R}^d)$, and $t \geq 0$. By the bounded convergence theorem,

$$\lim_{n\to\infty} E\left[\int_0^\infty e^{-\lambda t} f(t, X_n(t))\, dt\right] = E\left[\int_0^\infty e^{-\lambda t} f(t, X(t))\, dt\right].$$

If $\mu_\lambda^n(f)$ denotes $E\left[\int_0^\infty e^{-\lambda t} f(t, X_n(t))\, dt\right]$, then we have shown that μ_λ^n converges weakly to μ_λ.

Step 2 Denote $\sigma(\pi_n(s), X(\pi_n(s)))$ and $a_{ij}(\pi_n(s), X(\pi_n(s)))$ by $\sigma^{(n)}(s)$ and $a_{ij}^{(n)}(s)$ respectively. Let $f \in C_b^2(\mathbb{R}^+ \times \mathbb{R}^d)$. Let the subscripts beneath f denote the corresponding partial derivatives of f. Using the Itô formula,

$$\begin{aligned} f(t, X_n(t)) &= f(0, x) + \int_0^t \left\langle \nabla f(s, X_n(s)), \sigma^{(n)}(s)\, dB_s \right\rangle \\ &\quad + \int_0^t \left\{ f_s + \frac{1}{2} \sum_{i,j} a_{ij}^{(n)}(s) f_{ij} \right\} (s, X_n(s))\, ds. \end{aligned}$$

Multiply both sides by $\lambda e^{-\lambda t}$, integrate from 0 to ∞, and then take expectation to obtain

$$\begin{aligned} \lambda \mu_\lambda^n(f) &= f(0, x) + E^P \int_0^\infty \lambda e^{-\lambda t} \int_0^t \left\{ f_s + \frac{1}{2} \sum_{i,j} a_{ij}^{(n)}(s) f_{ij} \right\} (s, X_n(s))\, ds\, dt \\ &= f(0, x) + E^P \int_0^\infty e^{-\lambda s} \left\{ f_s + \frac{1}{2} \sum_{i,j} a_{ij}^{(n)}(s) f_{ij} \right\} (s, X_n(s))\, ds \end{aligned}$$

by changing the order of integration. Using the notation $e_{ij}^n = (a_{ij}^n - \delta_{ij})$, we can write the above equation as

$$\mu_\lambda^{(n)} \left(\lambda f - f_s - \frac{1}{2} \sum f_{jj} \right) = f(0, x) + \frac{1}{2} E^P \int_0^\infty e^{-\lambda s} \left\{ \sum_{i,j} e_{ij}^{(n)}(s) f_{ij} \right\} (s, X_n(s))\, ds. \tag{7.4.6}$$

Take $f = G_\lambda h$ where $h \in C_0^\infty(\mathbb{R}^d)$. Then f is in $C_b^2(\mathbb{R}^d)$ so that Equation (7.4.6) is valid. Recall that f satisfies the equation:

$$\lambda f - f_s - \frac{1}{2} \sum f_{jj} = h.$$

Therefore, Equation (7.4.6) yields, upon recalling the definition of p_N,

$$|\mu_\lambda^{(n)}(h)| \leq C_\lambda \|h\|_{p_N} + \frac{1}{2}\epsilon E^P \int_0^\infty e^{-\lambda s} \sum_{i,j} |f_{ij}(s, X_n(s))| \, ds$$

$$= C_\lambda \|h\|_{p_N} + \frac{1}{2}\epsilon \mu_\lambda^{(n)}(g_n) \tag{7.4.7}$$

where

$$g_n(s,x) = \sum_{i,j} |(G_\lambda h)_{ij}(s, X_n(s))|. \tag{7.4.8}$$

From Lemma 7.3.5, we know that

$$\|g_n\|_{p_N} \leq 2d^2 C_{d,p_N} \|h\|_{p_N}. \tag{7.4.9}$$

Step 3 From the definition of X_n given by Equation (7.4.5), it is clear that X_n is the stochastic integral of a simple function. Note that the distribution of $X_n(s)$ has a probability density function for each $s \geq 0$, since a and hence σ are positive definite. When $0 \leq s \leq 1/n$, the density function is the d-dimensional, normal density $\eta(s)$ with mean x and variance-covariance matrix given by $s\sigma(0,x)\sigma(0,x)^*$. Therefore, by the Hölder inequality,

$$E^P\left[\int_0^{1/n} e^{-\lambda s} h(s, X_n(s)) \, ds\right] = \int_0^{1/n} e^{-\lambda s} E^P[h(s, X_n(s))] \, ds$$

$$\leq \int_0^\infty e^{-\lambda s} \|h\|_{p_N} \|\eta(s)\|_{q_N} \, ds$$

$$\leq k_\lambda^{(n)} \|h\|_{p_N}$$

by the choice of p_N. Here q_N is the conjugate of p_N, and $k_\lambda^{(n)} > 0$ is a constant that depends on n and λ. It is simpler to obtain that there exists another constant $r_\lambda^{(n)}$ such that

$$E^P\left[\int_{1/n}^\infty e^{-\lambda s} h(s, X_n(s)) \, ds\right] \leq r_\lambda^{(n)} \|h\|_{p_N},$$

since the magnitude of the determinant of the variance-covariance matrix of $X_n(s)$ is uniformly bounded away from 0 for all $s \geq 1/n$. Putting the above two estimates together, we get

$$|\mu_\lambda^{(n)}(h)| \leq K_\lambda^{(n)} \|h\|_{p_N}.$$

Let $K_\lambda^{(n)}$ denote the best possible constant.

In particular, $|\mu_\lambda^{(n)}(g_n)| \leq K_\lambda^{(n)} \|g_n\|_{p_N}$, where g_n is defined by (7.4.8). This estimate together with (7.4.9) yields

$$|\mu_\lambda^{(n)}(g_n)| \leq K_\lambda^{(n)} 2d^2 C_{d,p_N} \|h\|_{p_N}.$$

Therefore, the estimate in (7.4.7) becomes

$$|\mu_\lambda^{(n)}(h)| \leq \left[C_\lambda + \epsilon K_\lambda^{(n)} d^2 C_{d,p_N}\right] \|h\|_{p_N} \leq \left[C_\lambda + \frac{1}{2} K_\lambda^{(n)}\right] \|h\|_{p_N},$$

since $\epsilon d^2 C_{d,p_N} \leq \frac{1}{2}$. By the requirement that $K_\lambda^{(n)}$ is the best such constant, we get

$$K_\lambda^{(n)} \leq C_\lambda + \frac{1}{2} K_\lambda^{(n)}.$$

Since $K_\lambda^{(n)}$ is finite,

$$K_\lambda^{(n)} \leq 2C_\lambda < \infty$$

for all n. By Step 1, $\mu_\lambda^{(n)} \to \mu_\lambda$ weakly and hence,

$$|\mu_\lambda(h)| \leq 2C_\lambda \|h\|_{p_N}$$

for all $h \in C_0^\infty$, and hence for all $h \in L^{p_N}(\mathbb{R}^d)$. ■

Lemma 7.4.2 *Let P be any solution of the martingale problem for a started at x. Define*

$$\mu_\lambda(f) = E^P \int_0^\infty e^{-\lambda s} f(s, X(s))\, ds \quad \text{for } f \in L^{p_N}(\mathbb{R}^d).$$

Then, $\mu_\lambda(f) = R_\lambda(f)(0, x)$, where R_λ is the operator constructed in the previous section.

Proof By Corollary 7.2.5, X can be viewed as a solution of a stochastic differential equation. Therefore, using the Itô formula, we obtain

$$\mu_\lambda \left[\lambda f - f_s - \frac{1}{2} \sum f_{jj}\right] = f(s, x) + \mu_\lambda \left(\frac{1}{2} \sum e_{ij} f_{ij}\right).$$

Choosing $f = G_\lambda h$ for $h \in C_0^\infty$, we have

$$\mu_\lambda(h) = (G_\lambda h)(s, x) + \mu_\lambda(T_\lambda h);$$

that is,

$$\mu_\lambda(h - T_\lambda h) = (G_\lambda h)(s, x).$$

By definition of R_λ, $R_\lambda(I - T_\lambda) = G_\lambda$, so that

$$[R_\lambda(h - T_\lambda h)](s, x) = (G_\lambda h)(s, x).$$

Thus, we can conclude that $\mu_\lambda(f) = (R_\lambda f)(s, x)$ with $f = (I - T_\lambda)h$ for $h \in C_0^\infty$. By the density of such functions in L^{p_N} and the boundedness of the operators in L^{p_N} (shown in Section 7.3), the proof is completed. ■

Theorem 7.4.3 *Let P be any solution of the martingale problem for a started at x. Then, P is uniquely determined.*

Proof From the above lemma, for each fixed closed set F, the Laplace transform $\int_0^\infty e^{-\lambda t} P(0, x; t, F)\, dt$ is uniquely determined. By the inversion of Laplace transforms, $P(0, x; t, F)$ is therefore uniquely identified for almost every t. By continuity of paths, $P(0, x; t, F)$ is continuous in t, so that $P(0, x; t, F)$ is uniquely determined for all t. Thus, we obtain unique one-dimensional marginals of P. This is equivalent to determining the finite-dimensional projections of P. The proof is thus complete. ■

So far, the matrix $a_{ij}(t, x)$ was assumed to be near the identity matrix I. Instead of I, suppose we are given a positive-definite, symmetric matrix C. We can define the corresponding operator G_λ by replacing $p_\lambda(s, x; t, y)$ by $e^{-\lambda(t-s)} q_C(s, x; t, y)$ where q_C denotes the multivariate normal density function with mean x and variance-covariance matrix $(t-s)C$. One can write $a_{ij}(t, x) = C_{ij} + e_{ij}(t, x)$, and define the operators

$$T\lambda^C h = \frac{1}{2} \sum_{i,j} e_{ij} \frac{\partial^2 (G_\lambda h)}{\partial x_i \partial x_j}$$

and $K_\lambda^C = G_\lambda (I - T_\lambda^C)^{-1}$. All of the properties proved for the operator K_λ in the previous section continue to hold for K_λ^C. The only change is that the constants in the bounds will depend on the lowest and highest eigenvalues of C.

Thus, there exists $\epsilon > 0$ such that if a_{ij} satisfies

$$\|a_{ij} - C_{ij}\| \leq \epsilon \quad \forall \ i, j,$$

then the solution to the martingale problem for a started at x has a unique solution. The choice ϵ is given in Theorem 7.3.6. We will later return to further results on the uniqueness of solutions. In the next section, we establish the Markov property of solutions of martingale problems.

7.5 Markov Property of Solutions

Let P be a probability measure on the canonical space $(\Omega, \mathcal{F})$, and Y_t, any $\mathbb{R}^d$-valued, adapted process. Let $a : [0, \infty) \times \Omega \to \mathbb{R}^d \times \mathbb{R}^d$ be jointly measurable and adapted to $\mathcal{F}_t$. Let us assume the following:

1. There exist a positive constant K such that

$$0 \leq \sum_{i,j=1}^{d} a_{ij}(s) \theta_i \theta_j \leq K|\theta|^2 \quad \text{for all } \theta \in \mathbb{R}^d.$$

2. $X_\theta(t) = \exp\{\langle\theta, Y_t - Y_0\rangle - \frac{1}{2}\int_0^t \langle\theta, a(s)\theta\rangle\, ds\}$ is a P-martingale for all $\theta \in \mathbb{R}^d$.

Theorem 7.5.1

1. *If τ_1 and τ_2 are two stopping times such that*

$$\tau_1 \leq \tau_2 \leq C$$

where C is a finite constant, then

$$E(X_\theta(\tau_2)|\mathcal{F}_{\tau_1}) = X_\theta(\tau_1) \ \ P\text{-a.s.} \tag{7.5.1}$$

2. *Let τ be a bounded stopping time, and Q_ω the regular conditional probability distribution of P given $\mathcal{F}_\tau$. Then there exists an $N \in \mathcal{F}_\tau$ such that $P(N) = 0$, and for any fixed $\omega \notin N$, the process*

$$X_\theta^{\tau(\omega)}(t) = \exp\left\{\langle\theta, Y_t - Y_{\tau(\omega)}\rangle - \frac{1}{2}\int_{\tau(\omega)}^t \langle\theta, a(s)\theta\rangle\, ds\right\} \tag{7.5.2}$$

is a Q_ω-martingale for all $t \geq \tau(\omega)$.

Proof The first assertion follows from the hypotheses and Doob's optional sampling theorem. To prove (7.5.2), consider any two time points $0 \leq t_1 \leq t_2$ and two events $A \in \mathcal{F}_\tau$ and $B \in \mathcal{F}_{\tau+t_1}$. Then,

$$\begin{aligned}\int_\Omega \left\{1_A(\omega)E_{Q_\omega}\left(1_B X_\theta^{\tau(\omega)}(\tau(\omega)+t_2)\right)\right\} dP &= E_P\left[1_{A\cap B}\frac{X_\theta(\tau+t_2)}{X_\theta(\tau)}\right] \\ &= E_P\left[1_{A\cap B}\frac{1}{X_\theta(\tau)}E\left(X_\theta(\tau+t_2)|\mathcal{F}_{\tau+t_1}\right)\right] \\ &= E_P\left[1_{A\cap B}\frac{1}{X_\theta(\tau)}X_\theta(\tau+t_1)\right] \\ &= \int_\Omega 1_A(\omega)E_{Q_\omega}\left(1_B X_\theta^{\tau(\omega)}(\tau(\omega)+t_1)\right)\Big\} \, dP.\end{aligned}$$

Hence, there exists a P-null set $N_1 \in \mathcal{F}_\tau$ that depends on t_1, t_2, θ, and the set B such that for all $\omega \notin N_1$,

$$E_{Q_\omega}\left[1_B X_\theta^{\tau(\omega)}(\tau(\omega)+t_2)\right] = E_{Q_\omega}\left[1_B X_\theta^{\tau(\omega)}(\tau(\omega)+t_1)\right] \tag{7.5.3}$$

The union of the sets N_1 as we vary $B \in \mathcal{F}_{\tau+t_1}$ is a P-null set, since $\mathcal{F}_{\tau+t_1}$ is generated by a countable number of finite-dimensional cylinder sets. We will call this P-null set N_2. Therefore, Equation (7.5.3) holds for all B if $\omega \notin N_2$. Thus,

$$E_{Q_\omega}\left[X_\theta^{\tau_\omega}(\tau(\omega)+t_2)|\mathcal{F}_{\tau+t_1}\right] = X_\theta^{\tau_\omega}(\tau(\omega)+t_1)$$

for all $\omega \notin N_2$. Still, N_1 depends on t_1, t_2 (where $t_1 \leq t_2$) and on θ. However, upon taking the union of N_2 over t_1, t_2 (with $t_1 \leq t_2$) and θ varying over a countable dense set, we can obtain a single P-null set N outside of which $X_\theta^{\tau(\omega)}$ is a Q_ω-martingale. ■

In the existence and uniqueness of solutions to martingale problems, the initial time was taken as 0 for convenience. We can take any time point $t_0 \geq 0$ as the initial time.

Theorem 7.5.2 *Let the hypotheses of Section 7.4 hold. Suppose that P is any solution of the martingale problem starting at time 0 and initial state x. Then P is the law of a Markov process and is unique.*

Proof Consider any time $t_0 > 0$. Let Q_ω denote the regular conditional probability of P given $\mathcal{F}_{t_0}$. By Theorem 7.5.1, for P-a.a. ω, Q_ω is a solution of the martingale problem started at t_0 with initial state given by $X(t_0, \omega)$. By Theorem 7.4.3, we obtain for any $t > t_0$ and Borel set B

$$Q_\omega\{X(t) \in B\} = P(t_0, X(t_0, \omega); t, B) \quad P\text{-a.s..}$$

Here $P(t_0, x_0; t, dy)$ denotes the marginal distribution at time t of the solution of the martingale problem started at time t_0 and state x_0. Thus, $P\{X(t) \in B | \mathcal{F}_{t_0}\} = P(t_0, X(t_0); t, B)$ P-a.s. Hence P is the law of a Markov process with $P(t_0, x_0; t, dy)$ as the transition probabilities. As a consequence, P is unique. ■

Theorem 7.5.3 *Let the hypotheses of Section 7.4 hold. Then, the martingale problem for a has a unique solution, $P_{t,x}$ for any initial time t and state x. The family of measures $P_{t,x}$ is jointly continuous in t, x.*

Proof First, note that $P_{t,x}$ is a probability measure on $(\Omega, \mathcal{F}_{t,\infty})$. We can treat it as a measure on $(\Omega, \mathcal{F})$ by requiring

$$P_{t,x}\{X(s) = x \;\; \forall \;\; s \leq t\} = 1.$$

Suppose (t_n, x_n) converges to (t, x). Consider the sequence of probability measures $P_n = P_{t_n, x_n}$. Since

$$E_{P_n}|X(u) - X(v)|^4 \leq C|u - v|^2$$

where C depends only on K, we obtain the relative compactness of $\{P_n\}$. Therefore, there exists a subsequence which converges weakly. Let us denote its limit by Q. The arguments in the proof of Theorem 7.2.8 can be used here to show that

$$Q\{X(s) = x \;\; \forall \;\; s \leq t\} = 1,$$

and Q is a solution of the martingale problem for a with initial time t and initial state x. By uniqueness of solutions, any weak limit is equal to $P_{t,x}$. The proof is over upon noting that

$$\lim_{n\to\infty} P_{t_n,x_n} = P_{t,x}$$

in the sense of weak convergence of probability measures. ■

7.6 Further Results on Uniqueness

So far, we have assumed that the matrix a_{ij} is close to a symmetric, positive-definite matrix C. In this section, we will first prove uniqueness of solutions to the martingale problem for a started at x by assuming that the matrix (a_{ij}) is bounded, continuous, and uniformly elliptic. Later, we will drop uniform ellipticity, and replace it with strict ellipticity. We start with a result which uses the idea of patching measures.

Theorem 7.6.1 *Let P be a probability measure on the canonical space $(\Omega, \mathcal{F})$, and τ, a bounded stopping time. Suppose that $a^{(1)}$ and $a^{(2)}$ are two measurable functions from $[0,\infty) \times \mathbb{R}^d$ to $\mathbb{R}^d \times \mathbb{R}^d$, and $b^{(1)}$ and $b^{(2)}$ are measurable functions from $[0,\infty) \times \mathbb{R}^d$ to $\mathbb{R}^d$ which satisfy the following conditions:*

(i) $0 \le \langle \theta, a^{(j)}\theta\rangle \le K|\theta|^2$.

(ii) $a^{(1)}(s, X(s)) = a^{(2)}(s, X(s))$ *for* $0 \le s \le \tau$ *P-a.s.*

(iii) $b^{(1)}(s, X(s)) = b^{(2)}(s, X(s))$ *for* $0 \le s \le \tau$ *P-a.s.*

(iv) $X_\theta(t) = \exp\{\langle \theta, X(t) - X(0)\rangle - \int_0^t \langle \theta, b^{(1)}(s, X(s))\theta\rangle\, ds$
$-\frac{1}{2}\int_0^t \langle \theta, a^{(1)}(s, X(s))\theta\rangle\, ds\}$ *is a P-martingale for each* $\theta \in \mathbb{R}^d$.

For each ω, let P_ω be a probability measure on $(\Omega, \mathcal{F}_{\tau(\omega),\infty})$ such that at time $\tau(\omega)$, the measure P_ω is concentrated at the single value $X(\tau(\omega), \omega)$, and the map $\omega \to P_\omega(A)$ is $\mathcal{F}_t$-measurable on the set $\{\tau \le t\}$ for all $A \in \mathcal{F}_{[t,\infty)}$. Besides, the process $X_\theta^{\tau(\omega)}(t)$ defined as

$$\exp\left\{\langle \theta, X(t) - X(\tau(\omega))\rangle - \int_{\tau(\omega)}^t \left\langle \theta, b^{(2)}(s, X(s))\theta\right\rangle ds - \frac{1}{2}\int_{\tau(\omega)}^t \left\langle \theta, a^{(2)}(s, X(s))\theta\right\rangle ds\right\}$$

is a P_ω-martingale for all $\theta \in \mathbb{R}^d$ and $\omega \in \Omega$.

Then there is a probability measure Q on $(\Omega, \mathcal{F})$ such that $Q = P$ on $\mathcal{F}_\tau$, and the process $Y_\theta(t)$ defined by

$$\exp\left\{\left\langle \theta, X(t) - X((0))\right\rangle - \int_0^t \left\langle \theta, b^{(2)}(s, X(s))\theta\right\rangle ds - \frac{1}{2}\int_0^t \left\langle \theta, a^{(2)}(s, X(s))\theta\right\rangle ds\right\}$$

is a Q-martingale for all $\theta \in \mathbb{R}^d$.

Proof

Step 1 Let Q be the probability measure on $(\Omega, \mathcal{F})$ such that $Q = P$ on $\mathcal{F}_\tau$, and the rcpd of Q given $\mathcal{F}_\tau$, denoted Q_ω coincides with P_ω on $\mathcal{F}_{\tau(\omega),\infty}$ for each ω. Such a measure Q exists by Lemma 7.2.6. We need to show that

$$\int_A Y_\theta(t)\, dQ = \int_A Y_\theta(s)\, dQ$$

for all $0 \le s \le t$, and $A \in \mathcal{F}_s$. It suffices to take sets of the form

$$A = \{X(s_0) \in B_0, \ldots, X(s_n) \in B_n\} \text{ where } 0 = s_0 < s_1 < \cdots < s_n = s$$

and $B_0, \ldots B_n \in \mathcal{B}(\mathbb{R}^d)$. For each $k < n$, define the sets

$$B_k = \{X(s_0) \in B_0, \ldots, X(s_k) \in B_k\}$$

and

$$C_k = \{X(s_{k+1}) \in B_{k+1}, \ldots, X(s_n) \in B_n\}.$$

Let $D_k = \{s_k \le \tau < s_{k+1}\}$ for all $0 \le k \le n-1$. By the definition of Q_ω,

$$\begin{aligned}
\int_{D_k} E_{Q_\omega}\left[1_A Y_\theta(t)\right] P(d\omega) &= \int_{D_k} 1_{B_k} X_\theta(\tau(\omega),\omega) E_{P_\omega}\left[1_{C_k} X_\theta^{\tau(\omega)}(t)\right] P(d\omega) \\
&= \int_{D_k \cap B_k} X_\theta(\tau(\omega),\omega) E_{P_\omega}\left[1_{C_k} X_\theta^{\tau(\omega)}(s)\right] P(d\omega) \\
&= \int_{D_k} E_{Q_\omega} E_{Q_\omega}\left[1_A Y_\theta(s)\right] P(d\omega).
\end{aligned}$$

Adding over k,

$$\int_{\{\tau < s\}} E_{Q_\omega}\left[1_A Y_\theta(t)\right] P(d\omega) = \int_{\{\tau < s\}} E_{Q_\omega}\left[1_A Y_\theta(s)\right] P(d\omega). \tag{7.6.1}$$

Step 2 Consider

$$\begin{aligned}
\int_{\{s \le \tau < t\}} E_{Q_\omega}\left[1_A Y_\theta(t)\right] P(d\omega) &= \int_{\{s \le \tau < t\}} 1_A X_\theta(\tau(\omega),\omega) E_{P_\omega}\left[X_\theta^{\tau(\omega)}(t)\right] P(d\omega) \\
&= \int_{\{s \le \tau < t\}} 1_A X_\theta(\tau(\omega),\omega)\, P(d\omega)
\end{aligned}$$

since $E_{P_\omega}[X_\theta^{\tau(\omega)}(t)] = 1$. Hence,

$$\int_{\{s \le \tau < t\}} E_{Q_\omega}\left[1_A Y_\theta(t)\right] P(d\omega) = \int_{\{s \le \tau < t\}} E_{Q_\omega}\left[1_A Y_\theta(s)\right] P(d\omega) \tag{7.6.2}$$

By the definition of Q_ω,

$$\begin{aligned}\int_{\{\tau\geq t\}} E_{Q_\omega}[1_A Y_\theta(t)]\,P(d\omega) &= \int_{\{\tau\geq t\}\cap A} X_\theta(t)\,dP\\ &= \int_{\{\tau\geq t\}\cap A} X_\theta(t\wedge\tau)\,dP\\ &= \int_{\{\tau\geq t\}\cap A} X_\theta(\tau)\,dP\end{aligned}$$

since $\{\tau \geq t\} \cap A \in \mathcal{F}_{\tau\wedge t}$, and hence Doob's optional sampling theorem can be applied. Thus,

$$\int_{\{\tau\geq t\}} E_{Q_\omega}\left[1_A Y_\theta(t)\right]\,P(d\omega) = \int_{\{\tau\geq t\}} E_{Q_\omega}\left[1_A Y_\theta(s)\right]\,P(d\omega). \tag{7.6.3}$$

Combining Equations (7.6.1), (7.6.2), and (7.6.3),

$$\begin{aligned}\int_A Y_\theta(t)\,dQ &= \int E_{Q_\omega}\left[1_A Y_\theta(t)\right]\,P(d\omega)\\ &= \int E_{Q_\omega}\left[1_A Y_\theta(s)\right]\,P(d\omega)\\ &= \int_A Y_\theta(s)\,dQ.\end{aligned}$$

■

Let the matrix (a_{ij}) be bounded, symmetric, continuous, and uniformly elliptic. Let the initial time be t_0 and the initial state be x_0. Define the stopping time τ by

$$\tau = \inf\{t : t_0 \leq t \leq t_0 + 1 \ \text{ and } \ \sup_{i,j} |a_{ij}(t, X(t)) - a_{ij}(t_0, x_0)| \geq \epsilon\}.$$

In the context of Theorem 7.3.6, the matrix $(a_{ij}(t_0, x_0))$ should be viewed as playing the role of (C_{ij}).

Lemma 7.6.2 *Suppose P_1 and P_2 are two solutions of the martingale problem for (a_{ij}) starting at t_0, x_0. Then P_1 and P_2 coincide on the σ-field $\mathcal{F}_{t_0,\tau}$.*

Proof Let C_{ij} denote $a_{ij}(t_0, x_0)$. Let F denote the closure of the open set

$$G = \{t, x : |a_{ij}(t, x) - C_{ij}| < \epsilon \ \text{ for all } \ i, j\}.$$

Then a_{ij} satisfies $|a_{ij} - C_{ij}| \leq \epsilon$ on F and can therefore be extended to a continuous function b_{ij} on $[t_0, \infty) \times \mathbb{R}^d$ satisfying $\|b_{ij} - C_{ij}\| \leq \epsilon$. Let us choose b_{ij} to be symmetric. For the martingale problem for (b_{ij}) started at t_0, x_0, there exists a unique solution Q_{t_0,x_0} by Theorem 7.4.3. The choice of t_0, x_0 is arbitrary, and, therefore, the above procedure can be be used for any t, x.

Define two probability measures Q_1 and Q_2 as follows: Q_j is P_j on $\mathcal{F}_{t_0,\tau}$, and the conditional distribution of Q_j given $\mathcal{F}_{t_0,\tau}$ is defined for each ω by $Q_{\tau(\omega),X(\tau(\omega))}$ on $\mathcal{F}_{\tau(\omega),\infty}$. The construction of such measures is given in Lemma 7.2.6. The measures Q_1 and Q_2 are solutions of the martingale problem for (b_{ij}) started at t_0, x_0 by Theorem 7.6.1. Since such solutions are unique by the results in Section 7.4, $Q_1 = Q_2$. Therefore, their restrictions to $\mathcal{F}_{t_0,\tau}$ coincide; that is, $P_1 = P_2$ on $\mathcal{F}_{t_0,\tau}$. ■

Remark 7.6.1 Suppose that P_1 and P_2 are two probability measures on $\mathcal{F}$ such that $\{\omega : X(s,\omega) = f(s) \ \forall \ s \leq t_0\}$ has measure 1 under both P_1 and P_2. Here, f is a deterministic function with values in $\mathbb{R}^d$. If P_1 and P_2 are solutions of the martingale problem for (a_{ij}) starting at $t_0, f(t_0)$, then $P_1 = P_2$ on $\mathcal{F}_\tau$ by the above lemma.

Let P_1 and P_2 be two solutions to the martingale problem (a_{ij}) starting at $0, x_0$. Let (a_{ij}) satisfy the conditions of Lemma 7.6.2. Define

$$\tau_1 = \inf\left\{t : 0 \leq t \leq 1 \text{ and } \sup_{i,j} |a_{ij}(t, X(t)) - a_{ij}(t_0, x_0)| \geq \epsilon\right\}$$

and

$$\tau_2 = \inf\left\{t : \tau_1 \leq t \leq \tau_1 + 1 \text{ and } \sup_{i,j} |a_{ij}(t, X(t)) - a_{ij}(\tau_1, x_1)| \geq \epsilon\right\},$$

where $x_1 = X(\tau_1)$. Successively, one can define for any $n \in \mathbb{N}$,

$$\tau_{n+1} = \inf\left\{t : \tau_n \leq t \leq \tau_n + 1 \text{ and } \sup_{i,j} |a_{ij}(t, X(t)) - a_{ij}(\tau_n, x_n)| \geq \epsilon\right\},$$

where $x_n = X(\tau_n)$.

Lemma 7.6.3 *If P_1 and P_2 are two solutions to the martingale problem (a_{ij}) starting at $0, x_0$, then $P_1 = P_2$ on $\mathcal{F}_{\tau_n}$ for every n.*

Proof The proof is by induction. For $n = 1$, the assertion follows from Lemma 7.6.2. Suppose $P_1 = P_2$ on $\mathcal{F}_{\tau_n}$. Let $Q_\omega^{(j)}$ denote the rcpd of P_j given $\mathcal{F}_{\tau_n}$. Then

$$P_j(A) = \int_\Omega Q_\omega^{(j)}(A)\, dP_j.$$

Since $Q_\omega^{(j)}$ is $\mathcal{F}_{\tau_n}$-measurable, and $P_1 = P_2$ on $\mathcal{F}_{\tau_n}$, it is enough to show that, for any $A \in \mathcal{F}_{\tau_{n+1}}$, we have $Q_\omega^{(1)}(A) = Q_\omega^{(2)}(A)$ P-a.s., where P is the restriction of P_i to $\mathcal{F}_{\tau_n}$.

Fix ω_0 to be any fixed member of Ω. By Theorem 7.5.1, if ω_0 is not in a set of measure zero, then $Q_{\omega_0}^{(j)}$ is a solution of the martingale problem for (a_{ij}) started at

$\tau_n(\omega_0), x_n(\omega_0)$. Besides,

$$Q^{(j)}_{\omega_0}\left[\omega : X(t,\omega) = X(t,\omega_0) \text{ for } 0 \le t \le \tau_n(\omega_0)\right] = 1.$$

Therefore, by Remark 7.6.1, $Q^{(1)}_{\omega_0} = Q^{(2)}_{\omega_0}$ on $\mathcal{F}_{\tau_{n+1}(\omega_0)}$ where

$$\tau_{n+1}(\omega_0) = \inf\left\{t : t \in [\tau_n(\omega_0), \tau_n(\omega_0)+1] \text{ and } \sup_{i,j} |a_{ij}(t, X(t)) - a_{ij}(\tau_n(\omega_0), x_n(\omega_0))| \ge \epsilon\right\}.$$

If $A \in \mathcal{F}_{\tau_{n+1}}$, then define

$$A_{\omega_0} = A \cap \{\omega : X(t,\omega) = X(t,\omega_0) \text{ for all } 0 \le t \le \tau_n(\omega_0)\}.$$

Then $A_{\omega_0} \in \mathcal{F}_\tau$, and $Q^{(j)}_{\omega_0}(A) = Q^{(j)}_{\omega_0}(A_{\omega_0})$.

Thus $Q^{(1)}_{\omega_0}(A) = Q^{(2)}_{\omega_0}(A)$ for all $A \in \mathcal{F}_{\tau_{n+1}}$. ∎

Theorem 7.6.4 *Let the matrix (a_{ij}) be bounded, symmetric, continuous, and uniformly elliptic. Suppose that P_1 and P_2 are two solutions of the martingale problem for (a_{ij}) started at time 0 and state x_0. Then $P_1 = P_2$.*

Proof We have already shown that $P_1 = P_2$ on $\mathcal{F}_{\tau_n}$ for every n. Since the functions a_{ij} are jointly continuous, and $X(t,\omega) = \omega(t)$ is a continuous function of t, these functions are uniformly continuous on any given compact set $[0, T]$. Therefore, $\lim_{n\to\infty} \tau_n(\omega) \ge T$ for each ω, so that $\lim_{n\to\infty} \tau_n = \infty$ for all ω.

Take any $A \in \mathcal{F}_t$. Then,

$$A = [A \cap \{\tau_n < t\}] \cup [A \cap \{\tau_n \ge t\}].$$

Since $[A \cap \{\tau_n \ge t\}] \in \mathcal{F}_{\tau_n}$, and $P_1 = P_2$ on $\mathcal{F}_{\tau_n}$,

$$|P_1(A) - P_2(A)| \le P_1[A \cap \{\tau_n < t\}] + P_2[A \cap \{\tau_n < t\}].$$

Allowing $n \to \infty$, we obtain that $P_1 = P_2$ on $\mathcal{F}_t$. The proof is over since t is arbitrary. ∎

Now, we can drop the uniform ellipticity condition and require that the matrix of functions (a_{ij}) is strictly elliptic at each point t, x. Note that such a hypothesis implies uniform ellipticity on compact subsets.

Theorem 7.6.5 *Let (a_{ij}) be a symmetric matrix of functions which is bounded, continuous, and strictly elliptic at each t, x. Then the solution P_{t_0,x_0} to the martingale problem for (a_{ij}) started at t_0, x_0 is unique.*

Proof The existence of $P_{t,x}$ is already known from Theorem 7.2.8. Let P_1 and P_2 be two solutions. Then, take a large compact set K given by $|t - t_0| \leq R$, $|x - x_0| \leq M$ and change the matrix (a_{ij}) outside K to obtain a matrix (b_{ij}) that is symmetric, uniformly elliptic, and continuous. There is a unique solution Q for the (b_{ij}) problem for all initial conditions. Therefore, arguing as in Lemma 7.6.2, $P_1 = P_2$ on $\mathcal{F}_{\tau_{R,M}}$ where

$$\tau_{R,M} = \inf_{t:|t-t_0|\leq R} \left\{t : |X(t,\omega) - x_0| \geq M\right\}.$$

Allowing $R, M \to \infty$, we get $P_1 = P_2$. ■

Theorem 7.6.6 *Under $P_{t,x}$, the solution of the martingale problem started at t, x, $X(t)$ is a Markov process:*

$$P_{t,x}\{X(t_2) \in B|\mathcal{F}_{t_1}\} = P_{t_1,X(t_1)}\{X(t_2) \in B\} \quad a.s.\ P_{t,x}.$$

In fact, it is a strong Markov process.

Proof From Theorem 7.5.1, we know that the conditional distribution of $P_{t,x}$ given $\mathcal{F}_{t_1}$ is a solution of the martingale problem starting at $t_1, X(t_1)$ a.s. By uniqueness, this must be $P_{t_1,X(t_1)}$. The same conclusion holds if t_1 is replaced by a stopping time. ■

Thanks to the Girsanov theorem, one can easily add a drift term whose coefficient is given by a bounded, Borel-measurable function $b(s,x)$ and study the resulting martingale problem. Such an extension is left as an exercise.

8 Probability Theory and Partial Differential Equations

The interplay between probability theory and partial differential equations (pdes) has a rich history dating back to 1931, when A. N. [45] established the connection between diffusion processes and a set of parabolic equations known as the forward and backward equations. In 1944, [37] studied harmonic functions via the two-dimensional Brownian motion, which led to a probabilistic approach to the Doob problem. Deep and elaborate study of the natural bridge between probability and potential theory can be found in the works of [14, 15, 16] as well as [57]. For finer aspects of the Dirichlet problem, the avid reader can consult [20], and [35].

Probabilistic representations of solutions of elliptic and parabolic pdes have been quite useful in forming approximate solutions of such equations by the Monte Carlo method. In this method, one uses a simulation of a path of the associated diffusion process to first evaluate a suitable functional of the path, and then, to form the arithmetic mean of the functional over several simulations. The Kolmogorov backward equation has found an unexpected albeit an important application in the pricing of derivatives, such as call and put options in finance theory.

In this chapter, we will discuss the basic connections between stochastic analysis and pdes. In Section 8.1, the Dirichlet problem is studied with several examples to illustrate the theory. Next, we briefly discuss the Feynman-Kac formula and a class of parabolic pdes. The Kolmogorov equations are derived in Section 8.3. After introducing the Feynman-Kac formula, an Application to finance theory is discussed Section 8.4. In the final section, a rigorous derivation of the Kolmogorov equation is presented.

8.1 The Dirichlet Problem

Let G be a bounded, open, connected set in $\mathbb{R}^d$. We will denote the closure of G by D. Let $(\Omega, \mathcal{F}, (\mathcal{F}_t), P)$ be a probability space equipped with a filtration that satisfies the usual conditions. Consider the d-dimensional $\mathcal{F}_t$-adapted process X given by

$$dX_t = b(X_t)\,dt + \sigma(X_t)\,dW_t$$

where W is a d-dimensional Wiener process, and the initial condition $X_0 = x_0 \in D$ is specified. The matrix σ and the vector-valued function b satisfy the conditions for the existence and uniqueness of X. For instance, we can assume that σ and b are Lipschitz continuous functions. We will denote the distribution of X as P_{x_0}.

With $\sigma\sigma^* = (a_{ij})$, and the ith-coordinate of b denoted by b_i, let L denote the differential operator for X. That is,

$$L = \frac{1}{2}\sum_{i,j=1}^{d} a_{ij}(x)\frac{\partial^2}{\partial x_i \partial x_j} + \sum_{i=1}^{d} b_i(x)\frac{\partial}{\partial x_i} \tag{8.1.1}$$

defined on $C_b^2(\mathbb{R}^d)$.

Definition 8.1.1 *Let G and L be as described above. Suppose we are given a function f on G, and a function g on ∂G. Then a function u is called the solution of the* **Dirichlet problem** *for (L, f, g) on the domain G if u satisfies*

$$Lu(x) = -f(x) \quad on \quad G, \tag{8.1.2}$$

$$u(x) = g(x) \quad on \quad \partial G, \tag{8.1.3}$$

and u is continuous up to the boundary.

To solve the Dirichlet problem as stated above, we obviously need the functions f and g to be smooth enough. If we take L as Δ and $f \equiv 0$, the problem reads as

$$\Delta u(x) = 0 \text{ on } G$$

and

$$u(x) = g(x) \text{ on } \partial G,$$

which is the original form of the Dirichlet problem. Such a problem corresponds to the steady-state heat equation in G with the boundary temperature specified by g. In this form, the problem reduces to finding a harmonic function u in G with boundary values specified by the function g. There is a natural connection of this problem to probability theory. To understand this without much ado, define τ as the first time that the d-dimensional Wiener process W started at x escapes from G. Thus, if $x \notin G$, then $\tau = 0$. If $x \in G$, by continuity of paths of W, it follows that W_τ is a point on the boundary of G. Since the Wiener paths are unbounded a.s., we have $\tau < \infty$. Define the function u on D by

$$u(x) = E_x[g(W_\tau)].$$

It is clear that the boundary condition of the Dirichlet problem is satisfied by u. If $x \in G$, let r be small enough such that the open ball $B_r(x)$ of radius r centered at x is contained in G. Let σ_r denote the first exit time of the process from B_r. By rotational invariance of W, the distribution of W_{σ_r} is the uniform distribution on S_r, the surface of the ball. We will denote the uniform distribution on S_r by μ. Then

$$\begin{aligned} u(x) &= E_x\left[E_x\left\{g(W_\tau)|\mathcal{F}_{\sigma_r}\right\}\right] \\ &= E_x\left[E_x\left\{g(W_\tau)|W_{\sigma_r}\right\}\right] \quad \text{by the strong Markov property} \\ &= \int_{S_r} E_y\{g(W_\tau)\}\mu\,(dy) \\ &= \int_{S_r} u(y)\,\mu\,(dy). \end{aligned}$$

Thus, u has the mean-value property and is therefore harmonic in G. In other words, $u(x) = E_x[g(W_\tau)]$ is a candidate for the solution of the Dirichlet problem for $(\Delta, 0, g)$. A more delicate point is to establish that u can be extended continuously to the closure of G. We will show later that such an extension is possible only when the domain G is "regular".

In solving the Dirichlet problem for (L, f, g), there are two hurdles to overcome: finiteness of τ and continuous extension of u up to the boundary of G. To give a probabilistic representation for the solution u, we first obtain sufficient conditions that guarantee the finiteness of the first exit time of the process X from G. Define

$$\tau = \inf\{t : X_t \notin G\}.$$

Lemma 8.1.1 *Let h be a $C^2(\mathbb{R}^d)$ function which is nonnegative and satisfies the condition*

$$Lh(x) < -c < 0 \text{ for all } x \in G$$

for a positive constant c. Then for any $x \in D$,

$$E_x\tau \leq \frac{h(x)}{c}.$$

Proof Applying the Itô formula for the function h,

$$h(X_{t\wedge\tau}) = h(x_0) + \sum_{i,j} \int_0^{t\wedge\tau} \sigma_{ij}(X_s)\frac{\partial h}{\partial x_i}(X_s)\,dW_j(s) + \int_0^{t\wedge\tau} Lh(X_s)\,ds.$$

Rearranging the terms and taking expectation,

$$\begin{aligned} h(x_0) &= E_x\left[h(X_{t\wedge\tau}) - \int_0^{t\wedge\tau} Lh(X_s)\,ds\right] \\ &\geq E_x[h(X_{t\wedge\tau})] + cE_x(t\wedge\tau) \\ &\geq cE_x(t\wedge\tau). \end{aligned}$$

The proof is over upon letting $t \to \infty$. ■

In the above lemma, the set G is bounded. The assumption that G is bounded can be removed easily.

Corollary 8.1.2 *If G is an open set in $\mathbb{R}^d$, and h, a function satisfying the conditions of Lemma 8.1.1, then for any $x \in D$,*

$$E_x\tau \leq \frac{h(x)}{c}.$$

Proof Consider the sets $G_n = G \cap \{x : |x| \leq n\}$, and the exit times $\tau_n := \inf\{t : X_t \notin G_n\}$. By Lemma 8.1.1,

$$E_x\tau_n \leq \frac{h(x)}{c} \quad \text{for all } \; x \in D_n.$$

Given any $x \in G$, there exists an m such that $x \in G_m$ and,

$$E_x\tau_n \leq \frac{h(x)}{c} \quad \text{for all } \; n \geq m.$$

Allowing $n \to \infty$, by the monotone convergence theorem, the proof is completed, since $\tau_n \uparrow \tau$. ■

We will illustrate the results by an example from Wentzell that is quite useful for us later on.

Example 8.1.3 *Let $G \subset A_R := \{x : |x_1| < R\}$ in $\mathbb{R}^d$. Suppose that there exist constants a_0 and B such that, for all $x \in G$,*

(i) $a_{11}(x) \geq a_0 > 0$ *and*
(ii) $|b_1(x)| \leq B$.

Let τ denote the first exit time from G. We will show that $E_x\tau$, as a function of $x \in D$ is a bounded function.

Define $h(x) = e^{CR} - e^{Cx_1}$ where C is a positive constant. Then for any $x \in G$, we have $h(x) \geq 0$ and

$$\begin{aligned} Lh(x) &= -\frac{a_{11}(x)}{2}C^2e^{Cx_1} - Cb_1(x)e^{Cx_1} \\ &\leq \left(-\frac{a_0}{2}C^2 + CB\right)e^{Cx_1}. \end{aligned}$$

If we choose $C = \frac{4B}{a_0}$, then

$$Lh(x) \leq -\frac{a_0}{4}C^2e^{Cx_1} \leq -\frac{a_0}{4}C^2e^{-CR}.$$

By the corollary 8.1.2,

$$E_x S \leq \frac{4e^{CR}}{a_0 C^2}\left(e^{CR} - e^{Cx_1}\right) \leq \frac{4}{a_0 C^2}\left(e^{2CR} - 1\right)$$
$$\leq \frac{a_0}{4B^2} \exp\left\{\frac{8BR}{a_0}\right\}.$$

Hence, $E_x\tau$ is a bounded function of x.

Theorem 8.1.4 *Let L be the differential operator given by Equation (8.1.1) with Lipschitz continuous coefficients a_{ij} and b, and let $(a_{ij}(x))$ be positive-definite uniformly on the closed domain $D = \bar{G}$. Let X be the associated process for the operator L. Suppose that the Dirichlet problem for (L,f,g) on G admits a unique solution u. Then u can be represented as*

$$u(x) = E_x\left[g(X_\tau) + \int_0^\tau f(X_s)\,ds\right]. \tag{8.1.4}$$

where τ is the first exit time from G.

Proof By the Itô formula, for each $x \in G$,

$$u(X_{t\wedge\tau}) - u(x) - \int_0^{t\wedge\tau} Lu(X_s)\,ds$$

is a martingale. Therefore,

$$u(x) = E_x\left[u(X_{t\wedge\tau}) - \int_0^{t\wedge\tau} Lu(X_s)\,ds\right]. \tag{8.1.5}$$

Clearly, $\left|u(X_{t\wedge\tau}) - \int_0^{t\wedge\tau} Lu(X_s)\,ds\right| \leq \sup_{x\in D}|u(x)| + \tau\sup_{x\in D}|Lu(x)|$.

Since D is compact, u and Lu are both bounded. Choose an R such that G is contained in A_R. Therefore, $E_x\tau < \infty$ for all $x \in D$, by the preceding example. Therefore, we can let $t \to \infty$ and use the dominated convergence theorem in Equation (8.1.5) to complete the proof. ■

Remark 8.1.1

1. In the above theorem, we have not explicitly stated the conditions satisfied by the functions f, g, (a_{ij}), and b that would guarantee the existence of a unique solution to the Dirichlet problem. From the theory of partial differential equations, if f, (a_{ij}), and b are continuously differentiable, and g is three times continuously differentiable, then the Dirichlet problem for (L,f,g) on G admits a unique solution.
2. If we set the boundary function $g \equiv 0$, then $u(x) = E_x \int_0^\tau f(X_s)\,ds$. In particular, if $f \equiv 1$, then $u(x) = E_x\tau$.

3. If we take $f \equiv 0$ in G, then, as a particular case of the theorem, we obtain $u(x) = E_x[g(X_\tau)]$. This is useful in understanding the location of exit, namely X_τ.

We will illustrate the usefulness of the probabilistic representation of solutions of the Dirichlet problem in the following examples.

Example 8.1.5 *Consider the one-dimensional Wiener process started at $x \in [a, b]$. We will find $E_x\tau$ and $P_x\{W_\tau = a\}$.*

To find $E_x\tau$, consider

$$\frac{1}{2}\frac{d^2u}{dx^2} = -1 \quad in \quad (a, b) \quad with \quad u = 0 \quad at \quad a, b.$$

The general solution of the pde is of the form $u(x) = -x^2 + \alpha x + \beta$, where α and β are constants. The boundary conditions are $u(a) = u(b) = 0$, so that the solution is $u(x) = (x - a)(b - x)$. Thus $E_x\tau = (x - a)(b - x)$.

Consider the problem

$$\frac{d^2u}{dx^2} = 0 \quad in \quad (a, b) \quad with \quad u = \begin{cases} 1 & at\ x = a; \\ 0 & at\ x = b. \end{cases}$$

The general solution of the pde is $u(x) = \alpha x + \beta$. The boundary conditions yield

$$u(x) = \frac{b - x}{b - a}.$$

That is, $P_x\{W_\tau = a\} = \dfrac{b - x}{b - a}$.

Example 8.1.6 *Let W_t denote the planar Brownian motion started at x with $0 < r < |x| < R$. Let τ denote the first exit time from the annular region. We will find the probability that the process exits through the inner boundary C_r. We need to find a harmonic function u in $r < |x| < R$ subject to the boundary condition*

$$u = \begin{cases} 1 & if\ |x| = r; \\ 0 & if\ |x| = R. \end{cases}$$

The function $\log|x|$ is harmonic everywhere except at the origin. Using it, we can write the solution

$$u(x) = \frac{\log R - \log|x|}{\log R - \log r}.$$

Thus

$$P_x\{W_\tau \in C_r\} = \frac{\log R - \log|x|}{\log R - \log r}. \tag{8.1.6}$$

Keeping r fixed and allowing $R \to \infty$, *we get* $P_x\{W_\tau \in C_r\} = 1$. *In other words,*

$$P_x\left\{|W_t| = r \ \ \textit{for some}\ t\right\} = 1. \tag{8.1.7}$$

Thus, the planar Brownian motion comes arbitrarily close to the origin with probability one. By the same argument, almost surely, the Brownian paths come arbitrarily close to any rational on the plane. In other words, a typical Brownian path is everywhere dense in the plane.

Since almost every Brownian path is unbounded, $P\left\{\limsup_{t\to\infty} |W_t| = \infty\right\} = 1$. *Using this along with* (8.1.7), *we have*

$$P\left\{|W_t| = r \ \ \textit{infinitely often}\right\} = 1.$$

Thus a typical path comes arbitrarily close to any fixed point infinite number of times. This is known as recurrence of the two-dimensional Brownian motion.

On the other hand, if we keep R *fixed and let* $r \to 0$ *in* (8.1.6), *then* $P_x\left\{|W_\tau| = (0,0)\right\} = 0$, *which implies that the Brownian paths do not hit the origin. Even if we start at the origin, there are no further visits to the origin.*

Example 8.1.7 *In three-dimensions, let us solve* $\Delta u = 0$ *in* $0 < r < |x| < R$ *subject to*

$$u = \begin{cases} 1 & \textit{if}\ |x| = r; \\ 0 & \textit{if}\ |x| = R. \end{cases}$$

Using the harmonic function $1/|x|$, *we arrive at*

$$u(x) = \frac{\dfrac{1}{|x|} - \dfrac{1}{R}}{\dfrac{1}{r} - \dfrac{1}{R}}.$$

We have thus obtained the probability of exit (from the annular region $G = \{x : r < |x| < R\}$) *via* S_r, *the sphere of radius* r, *for the three-dimensional Brownian motion started at* $x \in G$. *If* τ *denotes the first time that the process exits* G, *then*

$$P_x\left\{|W_\tau| = r\right\} = \frac{\dfrac{1}{|x|} - \dfrac{1}{R}}{\dfrac{1}{r} - \dfrac{1}{R}}. \tag{8.1.8}$$

If we let $R \to \infty$ *in Equation* (8.1.8), *then*

$$P_x\left\{|W_\tau = r\right\} = \frac{r}{|x|}.$$

Thus, the paths of W_t *come arbitrarily close to the origin with positive probability, but not with probability* 1.

Note that by the unboundedness of Brownian paths, if we start at x with $|x| \leq r$ *then*

$$P\left\{|W_t| = r \text{ for some } t\right\} = 1.$$

If $|x| \geq r$, $P\left\{|W_t| = r \text{ for some } t\right\} = \frac{r}{|x|}$. *Define a function* ϕ *by*

$$\phi_r(x) = \begin{cases} 1 & \text{if } |x| < r; \\ \frac{r}{|x|} & \text{if } |x| \geq r. \end{cases}$$

We can now write

$$P_x\{W \text{ hits } S_r \text{ after time } t\} = \int \phi_r(y) \frac{1}{(2\pi t)^{3/2}} \exp\left\{-\frac{|y-x|^2}{2t}\right\} dy,$$

which tends to zero as $t \to \infty$. *Thus,*

$$P_x\{W \text{ hits } S_r \text{ for arbitrarily large } t\} = 0,$$

which shows the transient behavior of W in three dimensions. Therefore, $P_x\left\{\liminf_{t\to\infty} |W_t| \geq r\right\} = 1$. *Since this is true for all* $r > 0$, *we can conclude that the Brownian motion wanders away to infinity a.s.; that is,*

$$P_x\left\{\liminf_{t\to\infty} |W_t| = \infty\right\} = 1.$$

If we let $r \to 0$ *in Equation (8.1.8), then* $\lim_{r\to 0} P_x\left\{|W_\tau| = r\right\} = 0$. *Thus, probability that the Brownian path hits the origin is zero.*

Remark 8.1.2 The conclusions of Example 8.1.7 hold in all dimensions $d \geq 3$. Indeed, the expression given by Equation (8.1.8) must be replaced by

$$P_x\left\{|W_\tau| = r\right\} = \frac{\frac{1}{|x|^{d-2}} - \frac{1}{R^{d-2}}}{\frac{1}{r^{d-2}} - \frac{1}{R^{d-2}}}$$

when $d > 3$.

Now, we will give simple instances where the probabilistic representations of solutions for Dirichlet problems help in the theory of differential equations.

Example 8.1.8 **Exit from angular regions:** *Consider a two-dimensional Wiener process W started at x. If x is on the bisecting line of an angle AOB, then by symmetry of the process,*

$$P\{W_t \text{ exits } AOB \text{ through } OB\} = \frac{1}{2}.$$

Using this, one can infer in general that if $\angle AOB = \alpha$, *and* $\angle AOx = \theta$, *then*

$$P\{W_t \text{ exits } AOB \text{ through } OB\} = \frac{\theta}{\alpha}. \tag{8.1.9}$$

Let $x = (x_1, x_2)$ with $x_2 > 0$, and let τ denote the exit from the upper half-plane. Take any point $C = (z, 0)$ on the x-axis. Let A and B be points far out on the positive and negative side of the x-axis respectively. Let $\angle ACx = \theta$. Then by Equation (8.1.9),

$$P\{W_t \text{ exits } ACB \text{ through } CB\} = \frac{\theta}{\pi};$$

that is,

$$P\{W_\tau < z\} = \frac{\theta}{\pi}.$$

Since $\theta = \tan^{-1}\left(\frac{x_2}{x_1 - z}\right)$, we obtain the density of the distribution for W_τ as

$$p((x_1, x_2); (z, 0)) = \frac{1}{\pi} \frac{x_2}{x_2^2 + (x_1 - z)^2}.$$

Hence, the solution of the Dirichlet problem $\Delta u = 0$ on the upper half-plane subject to $u = g$ on the x-axis can be written explicitly as follows:

$$u(x) = \frac{1}{\pi} \int_{-\infty}^{\infty} \frac{g(z) x_2}{x_2^2 + (x_1 - z)^2} \, dz.$$

This is a particular case of an important formula known as the Poisson integral formula for a half-space (see [41]).

The following illustration of the usefulness of probabilistic solutions is due to Wentzell. First, we quote a theorem from A. Friedman. Here, the domain G is the open set

$$G = \{(x, y) : 0 < y < T, \ h_1(y) < x < h_2(y)\}$$

where ϕ_1 and ϕ_2 are functions that are three times continuously differentiable. Let $F_1 = \bar{G} \cap \{(x, y) : x = h_1(y), y < T\}$ and $F_2 = \bar{G} \cap \{(x, y) : x = h_2(y), y < T\}$ denote the two lateral boundaries of G. Let D denote the closure of G.

Theorem 8.1.9 *Let $f \in C^1(G)$, and let g_1, g_2 be three times continuously differentiable on F_1 and F_2 respectively. Let g be a three times continuously differentiable function defined on $D \cap \{(x, y) : y = T\}$, namely the upper boundary of G. Suppose that*

$$g_i'(T) + 1/2 g''(h_i(T)) + f(h_i(T), T) = 0 \ \text{ for } \ i = 1, 2.$$

Then there exists a unique solution u for the problem

$$\frac{\partial u}{\partial y}(x, y) + \frac{1}{2} \frac{\partial^2 u}{\partial x^2}(x, y) = -f(x, y), \ \ (x, y) \in G \tag{8.1.10}$$

subject to the following conditions on the two lateral sides and the upper boundary of G:

$$u(h_i(y), y) = g_i(y) \qquad i = 1, 2 \tag{8.1.11}$$

and

$$u(x, T) = g(x) \tag{8.1.12}$$

where the solution is twice continuously differentiable up to the boundary.

The uniqueness of solutions in the above result should come as a surprise since there is no condition on the lower boundary $D \cap \{(x, y) : y = 0\}$. However, by probability theory, it would be obvious that there is no need for a condition on the lower boundary.

Define $X_t = x + W_t$ and $Y_t = y + t$ where W is a standard one-dimensional Wiener process. Then the process $Z_t = (X_t, Y_t)$ is started at (x, y) and corresponds to the differential operator $L = \frac{1}{2}\frac{\partial^2}{\partial x^2} + \frac{\partial}{\partial y}$.

Theorem 8.1.10 *Let τ be the first time of exit for Z_t from the domain G given in Theorem 8.1.9. Then the unique solution u for the Equation 8.1.10 subject to 8.1.11 and 8.1.12 is given by*

$$u(x, y) = E_{(x,y)}\left[\int_0^{\tau} f(Z_s)\, ds + g(X_\tau)1_{\{Y_\tau = T\}} + g_1(Y_\tau)1_{\{Z_\tau \in F_1\}} + g_2(Y_\tau)1_{\{Z_\tau \in F_2\}}\right].$$

Proof Let $G_n = \{z : d(z, G^c) > 1/n\}$. Then, there exists a $u_n \in C_b^2(\mathbb{R}^2)$ such that on G_n, we have $u_n = u$, where u is given by Theorem 8.1.9. If τ_n denotes the first exit time of Z from G_n, then $\tau_n \uparrow \tau$ as $n \to \infty$. Applying the Itô formula to u_n and then taking expectation,

$$u(x, y) = E_{x,y}[u(Z_{t\wedge\tau_n})] + E_{x,y}\int_0^{t\wedge\tau_n} f(Z_s)\, ds.$$

Allowing t and then $n \to \infty$, the proof is completed. ■

Note that the process Z_t doesn't exit through the lower boundary of G, and hence a boundary condition for the lower face is not needed. In fact, such a specification would have made the problem over determined.

Next, we show a probabilistic version of Fatou's theorem on radial limits of harmonic functions. We first state Fatou's theorem:

Theorem 8.1.11 *Let u be harmonic and bounded on the open unit ball in R^2. Then $\lim_{r\uparrow 1} u(r, \theta)$ exists for almost all θ.*

The analogous result based on stochastic analysis works for any bounded open domain. Though Fatou's theorem follows from it, we will omit the proof of this fact.

Theorem 8.1.12 *Suppose that u is a bounded harmonic function on a bounded open domain G. Let W be a two-dimensional Wiener process started at any $z = (x, y) \in G$. Then $\lim_{t\to\tau} u(W_t)$ exists a.s., where τ is the first exit time of the process from G.*

Proof Let $G_n = \{z : d(z, G^c) > 1/n\}$. If $z \in G_n$, let τ_n denote the first exit time of W from G_n. We know that $u(W_{t\wedge\tau_n})$ is a real-valued martingale. It is L^1-bounded, since

u is a bounded function. Let us denote this bound by C. Consider any interval (a, b). Then

$$E[\text{number of upcrossings of } (a,b) \text{ by } u(W_{t\wedge\tau_n})] \le \frac{C}{b-a}.$$

Since $\tau_n \uparrow \tau$, it follows that

$$E[\text{number of upcrossings of } (a,b) \text{ by } u(W_{t\wedge\tau})] \le \frac{C}{b-a}.$$

Hence, the number of upcrossings of (a, b) by $u(W_{t\wedge\tau})$ is finite a.s. In fact, the number of upcrossings of any interval is finite a.s. Therefore, $\lim_{t\to\tau} u(W_t)$ exists a.s. ■

8.2 Boundary Regularity

Any solution of the Dirichlet problem is required to be a continuous function on the closure of the domain G. Such a continuous extension from G to $\bar{G}$ is not always possible even when the boundary function is continuous, as illustrated by the following example.

Example 8.2.1 *Consider the domain in $\mathbb{R}^3$ given by*

$$G = \{x : |x| < 1\} \cap \{x : x = (x_1, 0, 0) \text{ and } x_1 \ge 1/2\}^c.$$

On G, consider the problem $\Delta u = 0$ subject to

$$u(x) = \begin{cases} 1 & \text{if } x \in \partial G \text{ and } |x| = 1; \\ x_1 & \text{if } x \in \partial G \text{ and } |x| < 1. \end{cases}$$

Since the three-dimensional Wiener process doesn't hit a given line,

$$P\{|W_\tau| = 1\} = 1$$

where τ is the first exit time from G. Thus, a.s., τ coincides with the first exit time from the unit ball punctured at the origin. The function $v(x) = \frac{1}{|x|}$ is harmonic in $\{x : 0 < |x| < 1\}$ with $v \equiv 1$ on the boundary. However, v does not satisfy the boundary conditions required of u. In other words, there is no harmonic function u on G with a continuous extension to $\bar{G}$ that satisfies the stated boundary conditions.

In view of the above example, we have the following definition:

Definition 8.2.1 *Suppose that $\{X_t\}$ is a diffusion process with L as its differential operator. Let G be a bounded, open, connected set. If X is started at $x \in \partial G$, define $\sigma = \inf\{t > 0 : X_t \in G^c\}$. Then x is called a* **regular point** *of ∂G if $P_x\{\sigma = 0\} = 1$. If $P_x\{\sigma = 0\} < 1$, the point x is called a* **singular point** *of the boundary.*

If all boundary points of a domain are regular, the domain is said to be regular. Since the event $\{\sigma = 0\}$ belongs to $\mathcal{F}_{0+}$, by the Blumenthal $0 - 1$ law, $P_x\{\sigma = 0\}$ is either 0 or 1. Hence we could have defined singularity of x by the requirement $P_x\{\sigma = 0\} = 0$.

It is worthwhile to note the difference between the stopping times τ and σ. When the process is started at $x \in \partial G$, then $\tau \equiv 0$, whereas σ need not be zero. In Example 8.2.1, the line segment $\left\{(x_1, 0, 0) : 1 > x_1 > 1/2\right\}$ consists of singular points. For the Dirichlet problem in the unit disc in $\mathbb{R}^2$ with the origin removed, the point $(0, 0)$ is a singular point.

Lemma 8.2.2 *Let z be a regular point of the boundary of G. Then, for any $\epsilon > 0$, $P_x\{\tau \geq \epsilon\} \to 0$ as $x \to z$.*

Proof Let Ω denote the space of all continuous functions defined on $[0, \infty)$ with $\mathcal{F}$ as its Borel σ-field. With $X_t(\omega) = \omega(t)$, let P_x denote the probability measure under which X_t is the diffusion started at x with L as its differential operator. Then, we know from Chapter 7 that P_x converges weakly to P_z as $x \to z$.

On Ω, define the function

$$\psi(\omega) = 1_{\{\sigma \geq \epsilon\}}(\omega) = 1_{\{X_t \in G \,\forall\, 0 < t < \epsilon\}}(\omega).$$

With z being a regular point, $\psi \equiv 0$ a.s. with respect to P_z and is therefore continuous P_z-a.s. Hence,

$$E_x\psi \to E_z\psi = 0 \quad \text{as} \quad x \to z;$$

that is, $P_x\{\sigma \geq \epsilon\} \to 0$ as $x \to z$, and hence $P_x\{\tau \geq \epsilon\} \to 0$. ■

Theorem 8.2.3 *Let $\{X_t\}$ be a diffusion process with L as its differential operator. Suppose that z is a regular point of the boundary of G, and g, a real-valued, bounded, Borel-measurable function defined on ∂G. Then*

$$\lim_{\substack{x \to z \\ x \in G}} E_x g(X_\tau) = g(z)$$

if g is continuous at z.

Proof For any $\eta > 0$, consider the event $A = \left\{|X_\tau - z| > \eta\right\}$. Then

$$P_x(A) \leq P_x[A \cap \{\tau < \epsilon\}] + P_x[tau \geq \epsilon]. \tag{8.2.1}$$

As $x \to z$, we know that $P_x(tau \geq \epsilon) \to 0$. To bound the first term on the right side of (8.2.1), observe that

$$X_{t\wedge\tau} - x = \int_0^{t\wedge\tau} b(X_s)\, ds + \int_0^{t\wedge\tau} \sigma(X_s)\, dW_s,$$

and b and σ are bounded and continuous on $\bar{G}$. Therefore, the Burkhölder-Davis-Gundy inequality yields

$$E \sup_{t \leq t \leq \epsilon} |X_{t\wedge\tau} - x| \leq C_1\epsilon + C_2\sqrt{\epsilon} < C\epsilon$$

for all $\epsilon \leq 1$ and $C = \max\{C_1, C_2\}$. Taking $x \in G$ such that $|x - z| < \eta/2$, we have

$$\begin{aligned} P_x(A \cap \{\tau < \epsilon\}) &\leq P_x\left[\sup_{0 \leq t \leq \epsilon} |X_{t\wedge\tau} - z| > \eta\right] \\ &\leq P_x\left[\sup_{0 \leq t \leq \epsilon} |X_{t\wedge\tau} - x| > \eta/2\right] \\ &< C\epsilon. \end{aligned}$$

Thus $P_x(A \cap \{\tau < \epsilon\}) \to 0$ uniformly in x as $\epsilon \to 0$. Letting $x \to z$, and then $\epsilon \to 0$ on the right side of Equation (8.2.1), we obtain

$$\lim_{\substack{x \to z \\ x \in G}} P_x\left\{|X_\tau - z| > \eta\right\} = 0$$

for any $\eta > 0$.

By continuity of g at z, given any $\delta > 0$, there exists an $\eta > 0$ such that

$$|g(z*) - g(z)| < \delta \quad \text{for all} \quad |z * -z| < \eta, \; z* \in \partial G.$$

Let M denote $\|g\|_\infty$. Then

$$|E_x g(X_\tau) - g(z)| \leq 2MP_x(A) + \delta P_x(A^c).$$

Allowing $x \to z$, the proof is over, since δ is arbitrary. ■

Combining the results obtained above, we can state the following theorem on the solvability of the (L, f, g) Dirichlet problem:

Theorem 8.2.4 *Let G be a regular domain, and $\{X_t\}$, a diffusion process with L as its differential operator. Let h be a $C^2(\mathbb{R}^d)$ function which is nonnegative and satisfies the condition*

$$Lh(x) < -c \quad \textit{for all} \quad x \in G$$

for a positive constant c. Then the solution of the (L, f, g) Dirichlet problem is given by $u(x) = E_x g(X_\tau)$.

The only remaining question before us is to find sufficient conditions that help us to identify regular points of a domain. We will give a criterion due to Poincaré which is geometric and easy to state. First, we define a d-dimensional **cone** to be a translation and rotation of an open set of the form $\{x : x_1^2 + x_2^2 + \cdots + x_{d-1}^2 < cx_d^2\}$ for some constant c.

Definition 8.2.2 *Let G be a bounded, open, connected set in $\mathbb{R}^d$. A point $z \in \partial G$ is said to satisfy the* **cone condition of Zaremba** *if there exists a cone V with vertex at z which is contained in V^c.*

Theorem 8.2.5 *Let W be a d-dimensional Wiener process. If $z \in \partial G$ satisfies the Zaremba's condition for a cone V, then it is regular.*

Proof Suppose that z is a singular point. Consider W started at z. Define $S_V = \inf\{t > 0 : W_t \in V\}$. Then $S_V \geq S$ where $S = \inf\{t > 0 : W_t \in G^c\}$. Hence, $P_z[S_V > 0] = 1$. We can construct a sufficient number of cones $V_1, \ldots V_n$, each congruent to V with vertex z, such that the entire neighborhood of z is covered. For each $j = 1, 2, \ldots, n$, define the stopping time $S_j = \inf\left\{t > 0 : W_t \in V_j\right\}$. Since the Wiener process is invariant under rotation,

$$P_z\left\{S_j > 0\right\} = 1 \; \forall\, j = 1, 2, \ldots, n.$$

Therefore, $P\left\{\min_j S_j > 0\right\} = 1$. That is,

$$P\left\{W_t \equiv z \;\text{ on the non-empty interval }\; [0, \min_j S_j)\right\} = 1.$$

A contradiction ensues, since almost every Wiener path has no interval of constancy. Therefore, z must be a regular point. ■

In the context of a diffusion process X with differential operator L, the above result holds. In the following, we lay the groundwork to prove the result.

Lemma 8.2.6 *Let $X_t = M_t + A_t$ be a continuous one-dimensional semimartingale satisfying*

$$\left|\frac{dA_t}{dt}\right| \leq C_1 \;\; \text{and} \;\; \frac{d\langle M\rangle_t}{dt} \geq C_2 > 0$$

where C_1 and C_2 are finite constants. Then, for any given $\epsilon > 0$ and u, there exists a constant $C_3 > 0$ such that

$$P\left\{\sup_{0\leq s\leq u} |X_s| < \epsilon\right\} \geq C_3$$

and C_3 depends only on C_1, C_2, ϵ, and u.

Proof Define $T_t = \inf\{s : \langle M\rangle_s > t\}$. Then T_t is continuous and strictly increasing in t. Besides, $W_t = M_{T_t}$ is a continuous martingale with t as its quadratic variation at time t. Therefore, by the Lévy characterization theorem, W is a standard Wiener process.

Define $Y_t := X_{T_t} = W_t + A_{T_t}$. Since $T_t \leq \frac{t}{C_2}$ for all t, we have the set inclusion

$$\left\{\sup_{\{0\leq s\leq C_2 u\}} |Y_s| < \epsilon\right\} \subset \left\{\sup_{\{0<s<u\}} |X_s| < \epsilon\right\}.$$

Therefore, it suffices to prove that

$$P\left\{\sup_{0\leq s\leq C_2 u} |Y_s| < \epsilon\right\} \geq C_3.$$

Since T_t and $\langle M\rangle_t$ are inverses of each other,

$$\frac{dA_{T_t}}{dt} = \frac{dA_{T_t}}{dT_t}\frac{dT_t}{dt} = \frac{dA_{T_t}}{dT_t}\left[\frac{d\langle M\rangle_{T_t}}{dT_t}\right]^{-1},$$

which exists and is bounded by $\frac{C_1}{C_2}$. Therefore, we can write

$$Y_t = W_t + \int_0^t H_s\,ds$$

where H is an adapted process bounded by $\frac{C_1}{C_2}$. We can now use the Girsanov theorem. Define a probability measure Q by

$$\frac{dQ}{dP} = \exp\left\{-\int_0^u H_s\,dW_s - \frac{1}{2}\int_0^u H_s^2\,ds\right\}$$

on $\mathcal{F}_u$. Under Q, $W_t + \int_0^t H_s\,ds$ is a standard Wiener process. As a consequence of the reflection principle,

$$Q[\sup_{0\le s\le C_2 u} |Y_s| < \epsilon] \ge C = e^{-\frac{\pi^2 C_2 u}{8\epsilon^2}}.$$

Let G denote the event $\left\{\sup_{0\le s\le C_2 u}|Y_s| < \epsilon\right\}$. Then

$$C \le Q(G) \le \int_G \frac{dQ}{dP}\,dP \le \left\{E\left[\frac{dQ}{dP}\right]^2\right\}^{1/2} P(G)^{1/2}$$

by the Cauchy-Schwarz inequality. By the boundedness of H, $E[(\frac{dQ}{dP})^2]$ is finite so that

$$P(G) \ge C^2\left[E\left(\frac{dQ}{dP}\right)^2)\right]^{-1} > 0. \qquad \blacksquare$$

Lemma 8.2.7 *Let X be a d-dimensional diffusion process with generator L, whose coefficients σ and b are bounded. Assume that σ^{-1} is bounded as well. Suppose their common bound is denoted by k. Then, for any given ϵ and $t \ge 0$, there exists a constant $c > 0$ such that*

$$P\left\{\sup_{0\le s\le t}|X_s - X_0| < \epsilon\right\} \ge c.$$

The constant c depends only on ϵ, t, and k.

Proof We will take $X_0 = 0$, and $\epsilon < 1$ without loss of generality. Let x denote $(2\epsilon, 0, \cdots, 0) \in \mathbb{R}^d$. Apply the Itô formula to the function $f(y) = |y - x|^2$ to obtain

$$|X_t - x|^2 = 4\epsilon^2 + 2\sum_{j=1}^{d}\int_0^t (X_j(s) - x_j)\, dX_j(s) + \sum_{j=1}^{d}\langle X_j\rangle_t.$$

Let Y_t denote $|X_t - x|^2$, and $\tau = \inf\{s : |Y_t - Y_0| > \epsilon^2\}$. Define a semimartingale Z by

$$Z_t = \begin{cases} Y_t & \text{if } t \leq \tau, \\ Y_\tau + W_{t+\tau} - W_\tau & \text{if } t > \tau \end{cases}$$

where W is a Wiener process independent of Y. Then Z is a semimartingale that satisfies the hypotheses of Lemma 8.2.6. Therefore, there exists a $c > 0$ such that

$$P\left\{\sup_{0\leq s\leq t} |Z_s - Z_0| < \epsilon^2\right\} \geq c > 0.$$

The proof is over upon observing that

$$P\left\{\sup_{0\leq s\leq t} |Y_s - Y_0| < \epsilon^2\right\} = P\left\{\sup_{0\leq s\leq t} |Z_s - Z_0| < \epsilon^2\right\}. \qquad \blacksquare$$

We are now ready to prove a theorem which states that with positive probability, the trajectories of X lie inside an ϵ-tube of the graph of ϕ till time t. This is a *support theorem* for solutions of stochastic differential equations.

Theorem 8.2.8 *Let X satisfy the hypotheses of Lemma 8.2.7. Suppose that $\phi : [0, t] \to \mathbb{R}^d$ is a differentiable function with a bounded derivative and with $\phi(0) = x$. Then, for any given ϵ, there exists a constant c which depends on ϵ, t, ϕ', and the bounds for σ, σ^{-1}, and b such that*

$$P\left\{\sup_{0\leq s\leq t} |X_s - \phi(s)| < \epsilon\right\} \geq c.$$

Proof Define a probability measure Q by

$$\frac{dQ}{dP} = \exp\left\{\int_0^t \phi'(0)\sigma^{-1}(X_s)\, dW_s - \frac{1}{2}\int_0^t |\phi'(s)\sigma^{-1}(X_s)|^2\, ds\right\}.$$

By the Girsanov theorem, $B_t = W_t - \int_0^t \phi'(0)\sigma^{-1}(X_s)\, ds$ is a Q-Wiener process so that

$$X_t - \phi(t) = X_0 + \int_0^t \sigma(X_s)\, dB_s + \int_0^t b(X_s)\, ds \quad Q\text{-a.s.}$$

By Lemma 8.2.7, we obtain

$$Q\left\{\sup_{0\leq s\leq t} |X_s - \phi(s)| < \epsilon\right\} \geq c.$$

By using the Cauchy-Schwarz inequality as in Lemma 8.2.6, we can conclude that

$$P\left\{\sup_{0\le s\le t} |X_s - \phi(s)| < \epsilon\right\} \ge c.$$

■

Theorem 8.2.9 *Let X_t be a diffusion process with differential operator L. Let σ and b be bounded and Lipschitz continuous such that σ^{-1} exists and is bounded. Suppose that G is an open-bounded, connected set in $\mathbb{R}^d$, and x is a point on ∂G satisfying the Zaremba cone condition. Then $P\{\tau = 0\} = 1$ where $\tau = \inf\{t > 0 : X_t \in G^c\}$.*

Proof Let V denote the open cone with vertex x that lies outside G. Take a function $\phi : [0,\infty) \to \mathbb{R}^d$ such that $\phi(0) = x$, $\phi(1) \in V$, and ϕ' is bounded. We can take ϵ small enough so that the ϵ-neighborhood of $\phi(1)$, namely $B_\epsilon(\phi(1)) \subseteq V$. By Theorem 8.2.8,

$$P\{X_1 \in V\} \ge P\left\{X_1 \in B_\epsilon(\phi(1))\right\} \ge c > 0.$$

We will now use scaling. Fix any $a > 1$ and define $Y_t = aX_{\frac{t}{a^2}}$. Then

$$\begin{aligned} Y_t &= aX_0 + a\int_0^{\frac{t}{a^2}} \sigma(X_s)\, dW_s + a\int_0^{\frac{t}{a^2}} b(X_s)\, ds \\ &= Y_0 + \int_0^t \sigma(a^{-1}Y_u)\, dB_u + a^{-1}\int_0^t b(a^{-1}Y_u)\, du \end{aligned}$$

by setting $u = a^2 s$ and $B_u = aW_{u/a^2}$. Thus Y solves a stochastic differential equation whose coefficients have the same bound as σ and b.

Hence $P\{Y_1 \in V\} \ge c$; that is, $P\{aX_{\frac{1}{a^2}} \in V\} \ge c$. The cone is invariant under scaling so that $P\{X_{\frac{1}{a^2}} \in V\} \ge c$. Thus,

$$P\left\{\tau \le \frac{1}{a^2}\right\} \ge c > 0.$$

Allowing $a \to \infty$, we obtain $P\{\tau = 0\} \ge c$. Invoking the Blumenthal $0-1$ law (Ex. 11 in Chapter 3), the proof is over. ■

8.3 Kolmogorov Equations: The Heuristics

In this section, we will formally derive a pair of parabolic pdes known as the Kolmogorov backward and forward equations. The forward equation is also called the Fokker-Planck equation. Consider a d-dimensional stochastic differential equation

$$dX_t = b(t, X_t)\, dt + \sigma(t, X_t)\, dW_t$$

with X_0 as the initial random variable. Here W is a k-dimensional Wiener process, and X_0 is independent of W. Let $b(t, x)$ and $\sigma(t, x)$ be such that a solution X exists. We know that X is a Markov process. Let the transition probabilities $p(s, x; t, dy)$ admit a density so that

$$p(s, x; t, dy) = p(s, x; t, y)\, dy$$

for all $0 \leq s < t$ and $x \in \mathbb{R}^d$. The variables s, x are known as backward variables, and t, y, the forward variables. We will assume that $p(s, x; t, y)$ is smooth in s, x. If $s > 0$, then for a small $h > 0$, we have by the Chapman-Kolmogorov equation,

$$p(s-h, x; t, y) = \int p(s-h, x; s, z)\, p(s, z; t, y)\, dz. \tag{8.3.1}$$

Let us expand $p(s, z; t, y)$ around x as

$$\begin{aligned} p(s, z; t, y) &= p(s, x; t, y) + \sum_{i=1}^{d} (z_i - x_i) \frac{\partial}{\partial x_i} p(s, x; t, y) \\ &\quad + \frac{1}{2} \sum_{i,j=1}^{d} (z_i - x_i)(z_j - x_j) \frac{\partial^2}{\partial x_i \partial x_j} p(s, x; t, y) + o\left(|z - x|^3\right). \end{aligned}$$

Plugging in this expansion on the right side of (8.3.1), we have

$$\begin{aligned} p(s-h, x; t, y) &= \int p(s-h, x; s, z)\, dz\, p(s, x; t, y) \\ &\quad + \sum_{i=1}^{d} \int (z_i - x_i)\, p(s-h, x; s, z)\, dz \frac{\partial}{\partial x_i} p(s, x; t, y) \\ &\quad + \frac{1}{2} \sum_{i,j=1}^{d} \int (z_i - x_i)(z_j - x_j)\, p(s-h, x; s, z)\, dz \frac{\partial^2}{\partial x_i \partial x_j} p(s, x; t, y) \\ &\quad + \int o(|z - x|^3)\, p(s-h, x; s, z)\, dz. \end{aligned}$$

The first term on the right side is simply $p(s, x; t, y)$. Transfer it to the left side, then divide the resulting equation by h. Letting $h \to 0$ yields

$$-\frac{\partial p(s, x; t, y)}{\partial s} = b(s, x) \cdot \nabla_x p(s, x; t, y) + \frac{1}{2} (\sigma(s, x), D^2 p(s, x; t, y) \sigma(s, x))$$

where D^2 denotes the matrix consisting of second-order partial derivatives of p with respect to x. We have thus formally derived the Kolmogorov backward equation:

$$-\frac{\partial p(s,x;t,y)}{\partial s} = b(s,x)\cdot\nabla_x p(s,x;t,y) + \frac{1}{2}\sum_{i,j=1}^{d} a_{ij}(s,x)\frac{\partial p(s,x;t,y)}{\partial x_i \partial x_j} \tag{8.3.2}$$

for $0 < s < t$, and with $p(s,x;t,y) = \delta_y(x)$ when $s = t$. Here, the matrix (a_{ij}) stands for $\sigma\sigma^*$.

To obtain the Kolmogorov forward equation, we write

$$p(s,x;t+h,y) = \int p(s,x;t,z)\,p(t,z;t+h,y)\,dz$$

using the Chapman-Kolmogorov equation. Then, by using $p(t+h,z;t+h,y) = \delta_y(z)$,

$$\begin{aligned}&\frac{1}{h}\left\{p(s,x;t+h,y) - p(s,x;t,y)\right\}\\ &\quad= \int p(s,x;t,z)\frac{1}{h}\left[p(t,z;t+h,y) - p(t+h,z;t+h,y)\right]\,dz.\end{aligned} \tag{8.3.3}$$

For small h, the right side of (8.3.3) is formally equal to

$$\int p(s,x;t,z)\left\{-\frac{\partial}{\partial s}p(s,z;t+h,y)|_{s=t}\right\}dz + o(1).$$

Using the backward Kolmogorov equation in the above expression, the above expression

$$= \int p(s,x;t,z)\{b(t,z)\cdot\nabla_z p(t,z;t+h,y) + \frac{1}{2}\sum_{i,j=1}^{d} a_{ij}(t,z)\frac{\partial p(t,z;t+h,y)}{\partial z_i\partial z_j}\,dz\Bigg\} + o(1),$$

which, using integration by parts,

$$\begin{aligned}= -\int\Bigg[&\nabla_z\cdot p(s,x;t,z)\,b(t,z)\\ &+\frac{1}{2}\sum_{i,j=1}^{d}\frac{\partial^2}{\partial z_i\partial z_j}\left\{p(s,x;t,z)\,a_{ij}(t,z)\right\}\Bigg]\,p(t,z;t+h,y)\,dz + o(1).\end{aligned}$$

Plugging in the above expression on the right side of Equation (8.3.3) and letting $h \to 0$,

$$\begin{aligned}\frac{\partial p(s,x;t,y)}{\partial t} = -\int\Bigg[&\nabla_z\cdot p(s,x;t,z)\,b(t,z)\\ &+\frac{1}{2}\sum_{i,j=1}^{d}\frac{\partial^2}{\partial z_i\partial z_j}\left\{p(s,x;t,z)\,a_{ij}(t,z)\right\}\Bigg]\delta_y(z)\,dz.\end{aligned}$$

Thus we have the Kolmogorov forward equation:

$$\frac{\partial p(s,x;t,y)}{\partial t} = -\nabla_z \cdot \{p(s,x;t,y)\, b(t,y)\} + \frac{1}{2}\sum_{i,j=1}^{d} \frac{\partial^2}{\partial y_i \partial y_j} \{p(s,x;t,y)\, a_{ij}(t,y)\} \tag{8.3.4}$$

with the initial condition $p(s,x;t,y) = \delta_x(y)$ when $t = s$. The Equation (8.3.4) is the Fokker-Planck equation.

8.4 Feynman-Kac Formula

Fix any time point $s \in [0,T]$, and $x \in \mathbb{R}^d$. Let W be a k-dimensional Wiener process. Consider the d-dimensional stochastic differential equation given by

$$dX_t = b(t,X_t)\, dt + \sigma(t,X_t)\, dW_t \quad \text{for } s \le t \le T$$

with $X_s = x$. Here, b is $\mathbb{R}^d$-valued, and σ is a $\mathbb{R}^d \times \mathbb{R}^k$-valued function. The associated partial differential operator on $C^2([0,T] \times \mathbb{R}^d)$ given by

$$Lu(t,x) = \frac{1}{2}\sum_{i,j=1}^{d} a_{ij}(t,x) \frac{\partial^2 u}{\partial x_i \partial x_j}(t,x) + \sum_{i=1}^{d} b_i(t,x) \frac{\partial u}{\partial x_i}(t,x).$$

Suppose that u solves the terminal value problem

$$\frac{\partial u}{\partial t}(t,x) + Lu(t,x) = 0 \tag{8.4.1}$$

and

$$u(T,x) = \phi(x) \tag{8.4.2}$$

where ϕ is a given function. By the Itô formula applied to this function u, we obtain

$$u(T,X_T) = u(s,x) + \int_s^T \left\{\frac{\partial u}{\partial t}(t,X_t) + Lu(t,X_t)\right\} dt + \int_s^T \nabla u(t,X_t) \cdot \sigma(t,X_t)\, dW_t. \tag{8.4.3}$$

If $\sum_{i=1}^{d} \sum_{j=1}^{k} E[\int_0^T |\frac{\partial u(t,X_t)}{\partial x_i} \sigma_{ij}(t,X_t)|^2\, dt] < \infty$, then the stochastic integral in the above equality has expectation zero. Hence, upon taking expectation in (8.4.3), we get

$$u(s,x) = E_{s,x} f(T,X_T),$$

which by u solves (8.4.1). Recalling the terminal value of u at time T, we obtain

$$u(s,x) = E_{s,x}\phi(X_T).$$

This is a probabilistic representation of the solution of (8.4.1). We have thus shown

Theorem 8.4.1 *Let u be a solution of the terminal value problem*

$$\frac{\partial u}{\partial t}(t,x) + Lu(t,x) = 0$$

and

$$u(T,x) = \phi(x).$$

Then u can be represented as

$$u(s,x) = E_{s,x}\phi(X_T)$$

where X solves

$$dX_t = b(t,X_t)\,dt + \sigma(t,X_t)\,dW_t \quad \textit{for} \quad s \le t \le T \tag{8.4.4}$$

with $X_s = x$, and the condition

$$\sum_{i=1}^{d}\sum_{j=1}^{k} E\left[\int_0^T \left|\frac{\partial u(t,X_t)}{\partial x_i}\sigma_{ij}(t,X_t)\right|^2 dt\right] < \infty$$

holds.

We can also obtain the following theorem along similar lines of reasoning.

Theorem 8.4.2 (Feynman-Kac) *Let u be a solution of the terminal value problem*

$$\frac{\partial u}{\partial t}(t,x) + Lu(t,x) + v(x)u(t,x) = 0$$

and

$$u(T,x) = \phi(x)$$

where v is a given function of x. Then, u can be represented as

$$u(s,x) = E_{s,x}\left(e^{\int_s^T v(X_t)dt}\,\phi(X_T)\right)$$

where X solves

$$dX_t = b(t,X_t)\,dt + \sigma(t,X_t)\,dW_t \quad \textit{for} \quad s \le t \le T \tag{8.4.5}$$

with $X_s = x$, provided that the condition

$$\sum_{i=1}^{d}\sum_{j=1}^{k} E\left[\int_0^T e^{\int_0^t v(X_s)\,ds}\left|\frac{\partial u(t,X_t)}{\partial x_i}\sigma_{ij}(t,X_t)\right|^2 dt\right] < \infty$$

holds.

8.5 An Application to Finance Theory

Let us consider a market consisting of a riskless asset known as a bond, and a risky asset known as a stock. Their prices at time t are denoted by B_t and S_t. Suppose our portfolio at time t consists of $\pi_1(t)$ units of bonds and $\pi_2(t)$ shares of the stock. Then the value of our portfolio at time t is $V(t) = \pi_1(t)B_t + \pi_2(t)S_t$. The vector $\pi(t) = (\pi_1(t), \pi_2(t))$ is called the portfolio vector at time t.

The portfolio is called a **self-financing** portfolio if $\pi(t)$ is formed out of the wealth $V(t-)$; that is, we will assume that there is no external infusion of wealth into the portfolio, or consumption from the wealth $V(t)$. We will assume that our portfolio is self-financing.

The term **arbitrage** refers to the possibility of creating a positive wealth at any given time $t > 0$, though the initial wealth was zero; that is, $V(0) = 0$, and $V(t) > 0$ with positive probability for any $t > 0$. We assume that there are no arbitrage opportunities in the market.

Then, one can infer that

$$dV(t) = \pi_1(t)\,dB_t + \pi_2(t)\,dS_t.$$

Let the time at present be denoted by t. Suppose that a trader sells today a financial instrument that pays an amount $\Psi(S_T)$ at a future time T. Such a financial asset is called a **simple T-claim**. It derives its value from the stock price S_T, and hence, is called a **financial derivative**. There is a price at time t for this T-claim. The problem is to determine a fair price for the T-claim, where fairness refers to a price that is acceptable to both the buyer and the seller. This is called derivative pricing.

The natural estimate at time t of $\Psi(S_T)$ is $E(\Psi(S_T) \mid \mathcal{F}_t)$. If the stock price is a Markov process and $\mathcal{F}_t$ is the σ-field generated by the stock price up to time t, then

$$E(\Psi(S_T) \mid \mathcal{F}_t) = E_{t,s}[\Psi(S_T)]$$

where $S_t = s$, where the subscripts t, s denote the initial time and the value of S_t respectively. Suppose the interest rate on the bond is r. Then, we can write the present value of $E_{t,s}[\Psi(S_T)]$ as $e^{-r(T-t)}E_{t,s}[\Psi(S_T)]$ since $T-t$ is the elapsed time.

By the assumption of absence of arbitrage, one would expect the local rate of return on the stock to be r as well. With this ansatz, let us model the market (that is, B and S) by

$$dB_u = rB_u\,du \tag{8.5.1}$$

$$dS_u = rS_u\,du + \sigma S_u\,dW_u. \tag{8.5.2}$$

where $r > 0$, and $\sigma \neq 0$ are constants. Here W is a one-dimensional Wiener process. Then the above model is called the Black-Scholes model, and one can find the fair price of the T-claim at time t by solving for S_T from

$$\begin{aligned} dS_u &= rS_u\, du + \sigma S_u\, dW_u \quad \text{for} \quad u > t, \\ S_t &= s, \end{aligned}$$

and computing $e^{-r(T-t)} E_{t,s}[\Psi(S_T)]$. Note that the linearity of the stochastic differential equation yields

$$S_T = s \exp\left\{\sigma(W_T - W_t) + (r - \frac{\sigma^2}{2})(T - t)\right\}.$$

Alternately, one can find the price p from the Feynman-Kac formula:

$$p_t + rsp_s + \frac{1}{2}\sigma^2 s^2 p_{ss} - rp = 0$$

and

$$p(T, s) = \Psi(s)$$

where p_t denotes the partial derivative of p with respect to t, etc. Then $p(t, x)$ gives us the required fair price. We have thus obtained the solution of a terminal value problem as the fair price of a stochastic pay-off at a future time. In the particular case when $\Psi(S_T) = (S_T - K)^+$, where $K > 0$ is a constant, the resulting financial derivative is known as a European call option, with maturity time T and strike price K.

8.6 Kolmogorov Equations

In this section, we will first study the solvability of stochastic differential equations that depend on a parameter θ. Our main interest in such systems is on the differentiability properties of solutions with respect to the parameter θ. Here, differentiability would be taken in the $L^2(P)$ sense. The goal is to obtain sufficient conditions on the coefficients of the stochastic differential equations that ensure differentiability of the solution with respect to θ. We will then proceed to the particular case where θ is taken as the initial condition of the stochastic system. This would set the stage to solve the backward Kolmogorov equation. Its adjoint equation is known as the Kolmogorov forward equation.

Let $(\Omega, \mathcal{F}, P)$ be a complete probability space with a filtration $(\mathcal{F}_t)$ that satisfies the usual conditions. On it, consider the d-dimensional stochastic differential equations on the time interval $[0, T]$:

$$X(t) = \xi(t) + \int_0^t b(s, X(s))\, ds + \int_0^t \sigma(s, X(s))\, dW_s \tag{8.6.1}$$

where W is a k-dimensional Wiener process, and the following conditions hold for all $t \in [0, T]$, $x, y \in \mathbb{R}^d$, and $\omega \in \Omega$.

Hypotheses 8.6.1

(i) The process $\xi(t)$ is an adapted process with continuous paths, and there exists a finite constant K such that

$$\sup_{0 \le t \le T} E\left[|\xi(t)|^2\right] \le K.$$

(ii) The coefficient $b(t, x, \omega)$ is $\mathbb{R}^d$-valued, and $\sigma(t, x, \omega)$ is $\mathbb{R}^{d \times k}$ valued. They are measurable in (t, x, ω). For each fixed (t, x), the coefficients are $\mathcal{F}_t$-measurable.

(iii) For each fixed ω, $b(t, x, \omega)$ and $\sigma(t, x, \omega)$ are continuous in (x, t).

(iv) For each fixed (t, x, ω),

$$|b(t, x, \omega)|^2 + |\sigma(t, x, \omega)|^2 \le K(1 + |x|^2) \tag{8.6.2}$$

and

$$|b(t, x, \omega) - b(t, y, \omega)|^2 + |\sigma(t, x, \omega) - \sigma(t, y, \omega)|^2 \le K|x - y|^2. \tag{8.6.3}$$

The argument ω in the coefficients b and σ will be suppressed as usual. The *novelty* in this stochastic differential equation consists in the addition of the process $\xi(t)$ to the right side of the equation.

Proposition 8.6.1 *Under Hypotheses 8.6.1, there exists a unique strong solution of the stochastic differential equation* (8.6.1), *and it has continuous paths.*

Proof The existence of a solution is proved by the Picard iteration scheme. Indeed, define

$$X_0(t) = \xi(t)$$

and

$$X_n(t) = \xi(t) + \int_0^t b(s, X_{n-1}(s))\, ds + \int_0^t \sigma(s, X_{n-1}(s))\, dW_s$$

for all $n \ge 2$. Using the condition (8.6.2), one obtains

$$\begin{aligned} E|X_1(t) - X_0(t)|^2 &\le 2(t+1)K \int_0^t E(1 + |\xi(s)|^2)\, ds \\ &\le 2T(T+1)K(1 + \sup_{0 \le s \le T} E[|\xi(s)|^2]. \end{aligned} \tag{8.6.4}$$

Let us call the above bound L, which is finite by the hypothesis on $\xi(t)$. For any $n \geq 2$, one can use the condition (8.6.3) to obtain

$$\begin{aligned} E\left[|X_n(t) - X_{n-1}(t)|^2\right] &\leq 2(T+1)K^2 \int_0^t E\left[|X_{n-1}(s) - X_{n-2}(s)|^2\right] ds \\ &\leq L \int_0^t E\left[|X_{n-1}(s) - X_{n-2}(s)|^2\right] ds. \end{aligned} \tag{8.6.5}$$

Iterating the inequality (8.6.7), and using (8.6.4), we conclude that

$$E\left[|X_n(t) - X_{n-1}(t)|^2\right] \leq \frac{(LT)^n}{n!}.$$

Using this estimate, the proof of existence of a solution can be completed. Uniqueness of solutions is a consequence of the Lipschitz continuity of the coefficients in the variable x, and the Gronwall inequality.

Since $\sup_{0 \leq t \leq T} E[|\xi(t)|^2] < \infty$, the solution $X(t)$ satisfies

$$\sup_{0 \leq t \leq T} E[|X(t)|^2] < \infty.$$

Therefore, the stochastic integral on the right side of (8.6.1) is a martingale with continuous paths. We are given that $\xi(t)$ has continuous paths, and hence, one obtains the continuity in t of $X(t)$ a.s. ∎

The next result is on the stability of solutions of equations of the form (8.6.1).

Proposition 8.6.2 *For any $\alpha \in [0, 1]$, let $\xi_\alpha(t)$, $b_\alpha(t, x)$, and $\sigma_\alpha(t, x)$ satisfy Hypotheses 8.6.1 for a constant K that is independent of α. Suppose that for each $N > 0$, $t \in [0, T]$, and $\epsilon > 0$, we have*

$$\lim_{\alpha \to 0} P\left\{\sup_{|x| \leq N} \{|b_\alpha(t, x) - b_0(t, x)| + |\sigma_\alpha(t, x) - \sigma_0(t, x)|\} > \epsilon\right\} = 0, \tag{8.6.6}$$

and

$$\lim_{\alpha \to 0} \sup_{0 \leq t \leq T} E[|\xi_\alpha(t) - \xi_0(t)|^2] = 0. \tag{8.6.7}$$

Then,

$$\lim_{\alpha \to 0} \sup_{0 \leq t \leq T} E\left[|X_\alpha(t) - X_0(t)|^2\right] = 0 \tag{8.6.8}$$

where $X_\alpha(t)$ is the solution of

$$X_\alpha(t) = \xi_\alpha(t) + \int_0^t b_\alpha(s, X_\alpha(s))\, ds + \int_0^t \sigma_\alpha(s, X_\alpha(s))\, dW_s. \tag{8.6.9}$$

Proof It is clear that

$$X_\alpha(t) - X_0(t) = H_\alpha(t) + \int_0^t [b_\alpha(s, X_\alpha(s)) - b_\alpha(s, X_0(s))]\, ds$$
$$+ \int_0^t [\sigma_\alpha(s, X_\alpha(s)) - \sigma_\alpha(s, X_0(s))]\, dW_s$$

where

$$H_\alpha(t) = \xi_\alpha(t) - \xi_0(t) + \int_0^t \left[b_\alpha(s, X_0(s)) - b_0(s, X_0(s))\right] ds$$
$$+ \int_0^t [\sigma_\alpha(s, X_0(s)) - \sigma_0(s, X_0(s))]\, dW_s.$$

Using Hypotheses 8.6.1, there exists a constant L such that

$$E|X_\alpha(t) - X_0(t)|^2 \leq 3E|H_\alpha(t)|^2 + L \int_0^t E|X_\alpha(s) - X_0(s)|^2\, ds.$$

Therefore, if we prove that $\lim_{\alpha \to 0} \sup_{0 \leq t \leq T} E[|H_\alpha(t)|^2] = 0$, then the Gronwall lemma would allow us to conclude (8.6.8). By assumption (8.6.7), it suffices to show that both

$$E\left[\left|\int_0^t \left[b_\alpha(s, X_0(s)) - b_0(s, X_0(s))\right] ds\right|^2\right] \quad \text{and}$$

$$E\left[\left|\int_0^t \left[\sigma_\alpha(s, X_0(s)) - \sigma_0(s, X_0(s))\right] dW_s\right|^2\right]$$

tend to 0 uniformly in t as $\alpha \to 0$. The proof of each of these statements is similar, and hence we will prove the latter. By the Itô isometry, it is enough to show that

$$\int_0^T E\left[\left|\sigma_\alpha(s, X_0(s)) - \sigma_0(s, X_0(s))\right|^2\right] ds.$$

converges to 0 as $\alpha \to 0$.

First, we write the integrand as $E[1_{\{|X_0(s)| \leq N\}} |\sigma_\alpha(s, X_0(s)) - \sigma_0(s, X_0(s))|^2]$ $+E[1_{\{|X_0(s)| > N\}} ||\sigma_\alpha(s, X_0(s)) - \sigma_0(s, X_0(s))|^2]$. As $N \to \infty$, the second term converges to 0 since

$$|\sigma_\alpha(s, X_0(s)) - \sigma_0(s, X_0(s))|^2 \leq K(1 + |X_0(s)|^2)$$

which is an integrable random variable. Besides,

$$E\left[|\sigma_\alpha(s, X_0(s)) - \sigma_0(s, X_0(s))|^2\right] \leq K\left(1 + \sup_{0 \leq s \leq T} E\left[|X_0(s)|^2\right]\right) < \infty \tag{8.6.10}$$

so that

$$\lim_{N\to\infty}\int_0^T E\left[1_{\{|X_0(s)|>N\}}||\sigma_\alpha(s,X_0(s))-\sigma_0(s,X_0(s))|^2\right]ds=0.$$

Therefore, given any $\delta > 0$, there exists an N_0 such that for all $N \geq N_0$,

$$\int_0^T E\left[1_{\{|X_0(s)|>N\}}||\sigma_\alpha(s,X_0(s))-\sigma_0(s,X_0(s))|^2\right]ds<\delta. \tag{8.6.11}$$

For any such N, consider $1_{\{|X_0(s)|\leq N\}}|\sigma_\alpha(s,X_0(s))-\sigma_0(s,X_0(s))|^2$, which is bounded by an integrable random variable as shown earlier, and converges to 0 in probability as $\alpha \to 0$ by condition (8.6.6). By the dominated convergence theorem, it follows that

$$\lim_{\alpha\to 0}E\left[1_{\{|X_0(s)|\leq N\}}|\sigma_\alpha(s,X_0(s))-\sigma_0(s,X_0(s))|^2\right]=0,$$

which along with (8.6.10) yields

$$\lim_{\alpha\to 0}\int_0^T E\left[1_{\{|X_0(s)|\leq N\}}|\sigma_\alpha(s,X_0(s))-\sigma_0(s,X_0(s))|^2\right]ds=0. \tag{8.6.12}$$

We can conclude from (8.6.11) and (8.6.12) that

$$\lim_{\alpha\to 0}\int_0^T E\left[|\sigma_\alpha(s,X_0(s))-\sigma_0(s,X_0(s))|^2\right]ds<\delta.$$

The proof is over by the arbitrariness of δ. ■

Definition 8.6.1 *If $\{Y(a_1,\ldots,a_d)\}$ is a family of random variables indexed by a d-dimensional parameter $a=(a_1,\ldots,a_d)$, then Y is called* **differentiable** *(in the $L^2(P)$ sense) with respect to a if there exist random variables Z_i for each $i = 1,\ldots,d$ such that*

$$E\left[\left|\frac{1}{h}[Y((a_1,\ldots,a_{i-1},a_i+h,a_{i+1},\ldots,a_d)-Y((a_1,\ldots,a_d)]-Z_i\right|^2\right]\to 0$$

as $h \to 0$. We will also denote Z_i as $\frac{\partial}{\partial a_i}Y(a)$, and call it the derivative of $Y(a)$ with respect to a_i.

Consider the d-dimensional stochastic differential equation

$$X_a(t)=a+\int_0^t b(s,X_a(s))\,ds+\int_0^t\sigma(s,X_a(s))\,dW_s$$

where W is a k-dimensional Wiener process. Here the initial state a is the d-dimensional parameter. In what follows, we employ the following notation:

If $z \in \mathbb{R}^d$ and $b(t,x)$, an $\mathbb{R}^d$-valued function, then $z \cdot \nabla_x b$ denotes a d-dimensional vector whose r^{th} element is given by $z \cdot \nabla b_r$.

If $\sigma(t,x)$ is a $d \times k$ matrix-valued function, then $z \cdot \nabla_x \sigma \, dW$ denotes a d-dimensional vector whose r^{th} element is given by $z \cdot \nabla_x \sigma_r \, dW$ where σ_r is the rth row of σ.

Theorem 8.6.3 *Let b and σ satisfy the conditions (ii), (iii), and (iv) in Hypotheses 8.6.1. In addition, let $b(t,x)$ and $\sigma(t,x)$ be differentiable with respect to x_i for all $i = 1, \ldots, d$, and let the derivatives be bounded and continuous in both t and x. Then $Z_i(t) := \frac{\partial}{\partial a_i} X_a(t)$ exists in the $L^2(P)$ sense and satisfies*

$$Z_i(a,t) = e_i + \int_0^t Z_i(a,s) \cdot \nabla_x b(s, X_a(s))\, ds + \int_0^t Z_i(a,s) \cdot \nabla_x \sigma(s, X_a(s))\, dW_s \quad (8.6.13)$$

where e_i is the ith unit vector in the standard basis for $\mathbb{R}^d$.

Proof We will consider $i = 1$ and take $h = (h_1, 0, \ldots, 0)$ where $h_1 \neq 0$. Then,

$$\begin{aligned}[X_{a+h}(t) - X_a(t)] = h &+ \int_0^t [b(s, X_{a+h}(s)) - b(s, X_a(s))]\, ds \\ &+ \int_0^t [\sigma(s, X_{a+h}(s)) - \sigma(s, X_a(s))]\, dW_s. \end{aligned} \quad (8.6.14)$$

Divide both sides by h_1, and consider each term on the right. For instance, the stochastic integral term is

$$\begin{aligned} &\frac{1}{h_1}\int_0^t [\sigma(s, X_{a+h}(s)) - \sigma(s, X_a(s))]\, dW_s \\ &= \frac{1}{h_1}\int_0^t \int_0^1 \frac{d}{dy}\sigma(s, X_a(s) + y(X_{a+h}(s) - X_a(s)))\, dy\, dW_s \\ &= \int_0^t \frac{X_{a+h}(s) - X_a(s)}{h_1} \cdot \left[\int_0^1 \nabla_x \sigma(s, X_a(s) + y(X_{a+h}(s) - X_a(s)))\, dy\right] dW_s. \end{aligned}$$

A similar expression can be written for the drift term in (8.6.14). Let us denote $\frac{X_{a+h}(s) - X_a(s)}{h_1}$ as $Y_h(s)$. Then Y_h satisfies the stochastic differential equations

$$\begin{aligned} Y_h(t) = e_1 &+ \int_0^t Y_h(s) \cdot \left[\int_0^1 \nabla_x b(s, X_a(s) + y(X_{a+h}(s) - X_a(s)))\, dy\right] ds \\ &+ \int_0^t Y_h(s) \cdot \left[\int_0^1 \nabla_x \sigma(s, X_a(s) + y(X_{a+h}(s) - X_a(s)))\, dy\right] dW_s. \end{aligned} \quad (8.6.15)$$

Let $Y_0(t)$ be another symbol for $Z_1(t)$, the solution of (8.6.13) when $i = 1$. Note that Equation (8.6.13) has a unique strong solution by Proposition 8.6.1, which applies by the linearity of the stochastic differential equations and the condition that b and σ have bounded, continuous, first-order derivatives. Thus,

$$Y_0(t) = e_1 + \int_0^t Y_0(s) \cdot \nabla_x b(s, X_a(s))\, ds + \int_0^t Y_0(s) \cdot \nabla_x \sigma(s, X_a(s))\, dW_s. \tag{8.6.16}$$

We would like to invoke Proposition 8.6.2 to complete the proof. The role of α is played by $|h_1|$. By continuity of solutions with respect to initial data, $E|X_{a+h}(t) - X_a(t)|^2 \to 0$ as $h \to 0$, and hence

$$X_{a+h}(t) - X_a(t) \to 0$$

in probability as $h \to 0$. Therefore,

$$\left[\int_0^1 \nabla_x b(s, X_a(s) + y(X_{a+h}(s) - X_a(s)))\, dy\right] \to \nabla_x b(s, X_a(s))\, ds$$

in probability, since b has continuous first-order derivatives. A similar statement holds for the function σ. Besides, the coefficients in Equations (8.6.15) and (8.6.16) are linear in the solutions Y_h and Y_0, respectively. This allows one to verify condition (8.6.6) of Proposition 8.6.2, and the proof is over upon noting that $Y_0(t)$ is the same as $Z_1(a, t)$ that appears in the statement of the theorem. ■

The next theorem is a statement about the existence of second-order derivatives (in the mean-square sense) of $X_a(t)$ with respect to a. We will prove the result when $d = 1$ and $k = 1$ to avoid notational complexity. From the above theorem, we know that the first derivative $Z(a, t)$ satisfies

$$Z(a, t) = 1 + \int_0^t Z(a, s) b_x(s, X_a(s))\, ds + \int_0^t Z(a, s)\sigma_x(s, X_a(s))\, dW_s.$$

All moments $Z(a, t)$ exists. In fact, $\sup_{0\le t\le T} E|Z(a, t)|^r < \infty$ for any positive integer r. Define

$$Z_h(t) = \frac{Z(a + h, t) - Z(a, t)}{h}$$

so that

$$Z_h(t) = \xi_h(t) + \int_0^t Z_h(s) b_x(s, X_{a+h}(s))\, ds + \int_0^t Z_h(s)\sigma_x(s, X_{a+h}(s))\, dW_s \tag{8.6.17}$$

where

$$\begin{aligned}\xi_h(t) &= \int_0^t Z(a,s)\frac{1}{h}\left[b_x(s,X_{a+h}(s)) - b_x(s,X_a(s))\right] ds \\ &+ \int_0^t Z(a,s)\frac{1}{h}\left[\sigma_x(s,X_{a+h}(s)) - \sigma_x(s,X_a(s))\right] dW_s \\ &= \int_0^t Z(a,s)Y_h(s)\left[\int_0^1 b_{xx}(s,X_a(s) + y(X_{a+h}(s) - X_a(s)))\,dy\right] ds \\ &+ \int_0^t Z(a,s)Y_h(s)\left[\int_0^1 \sigma_{xx}(s,X_a(s) + y(X_{a+h}(s) - X_a(s)))\,dy\right] dW_s \end{aligned} \tag{8.6.18}$$

where Y_h is as in Equation (8.6.15). As $h \to 0$, we would expect $Z_h(t)$ to converge in some sense to $Z_0(t)$ where

$$Z_0(t) = \xi(t) + \int_0^t Z_0(s)b_x(s,X_a(s))\,ds + \int_0^t Z_0(s)\sigma_x(s,X_a(s))\,dW_s \tag{8.6.19}$$

where

$$\xi(t) = \int_0^t Z^2(a,s)b_{xx}(s,X_a(s))\,ds + \int_0^t Z^2(a,s)\sigma_{xx}(s,X_a(s))\,dW_s. \tag{8.6.20}$$

We will use the full force of Proposition 8.6.2 to prove that

$$\lim_{h\to 0}\sup_{0\le t\le T} E\left[|Z_h(t) - Z_0(t)|^2\right] = 0.$$

In other words, $Z_0(t)$ is the mean-square derivative of $Z(a,t)$ so that $X_a(t)$ is twice differentiable with respect to a.

Theorem 8.6.4 *Let b and σ satisfy the conditions of Theorem 8.6.3. In addition, suppose that $b(t,x)$ and $\sigma(t,x)$ have continuous second derivatives in x, with polynomial growth given by*

$$|b_{xx}(t,x)|^2 + |\sigma_{xx}(t,x)|^2 \le K(1 + |x|^r) \tag{8.6.21}$$

where K and r are positive constants. Then, the second derivative (in the mean-square sense) of $X_a(t)$ with respect to a exists and is given by Z_0, the solution of Equation (8.6.19). It has moments of any order.

Proof

Step 1 First, we will show that Equation (8.6.19) has a unique strong solution. Since b_x and σ_x are bounded and continuous, the coefficients $zb_x(t,x)$ and $z\sigma_x(t,x)$ are

Lipschitz continuous in z with linear growth and are jointly continuous in both t and z. Also, by the condition 8.6.21, we have

$$\begin{aligned} E|\xi(t)|^2 &\leq 2(t+1)KE\left\{\int_0^t Z^4(a,s)(1+||X_a(s)|^r)\right\} ds \\ &\leq 2(t+1)K\sqrt{2}\int_0^t \left\{\left[EZ^8(a,s)\right]\left[1+E|X_a(s)|^{2r}\right]\right\}^{1/2} ds \\ &< \infty \end{aligned}$$

since all moments of $Z(a,s)$ and $X_a(s)$ exist and are uniformly bounded in s. Therefore,

$$\sup_{0\leq t\leq T} E|\xi(t)|^2 < \infty.$$

Therefore, one can invoke Proposition 8.6.1 to conclude that there exists a unique strong solution of the stochastic differential equation (8.6.19).

Step 2 We need to show that $\lim_{h\to 0}\sup_{0\leq t\leq T} E[|Z_h(t)-Z_0(t)|^2] = 0$ to complete the proof of the theorem. For this, we will use Proposition 8.6.2. By Equation (8.6.18) and the condition 8.6.21, it is easy to see that there exists a constant K such that

$$\sup_{0\leq t\leq T} E(|\xi_h(t)|^r) \leq K_r$$

for all positive r, and K_r doesn't depend on the parameter h. Likewise, as $h \to 0$, we have

$$\sup_{0\leq t\leq T} E|\xi_h(t)-\xi(t)|^2 \to 0.$$

Besides, condition (8.6.6) in Proposition 8.6.2 follows easily as in the proof of Theorem 8.6.3 by using the continuity of b_{xx} and σ_{xx} and the fact that $X_{a+h}(t) \to X_a(t)$ in probability as $h \to 0$. ■

In a similar manner, one can prove the following multi dimensional version of the above theorem:

Theorem 8.6.5 *Suppose that hypotheses H hold. Let the coefficients b and σ be differentiable with respect to x_i for all $i = 1, 2, \ldots, d$ with the derivatives being bounded and continuous in (t, x). In addition, let b and σ have continuous second-order derivatives satisfying*

$$|b_{x_ix_j}(t,x)| + |\sigma_{x_ix_j}(t,x)| \leq K(1+|x|^\rho) \tag{8.6.22}$$

for some positive constants K and ρ. Then

$$\eta_{ij}(a,t) = \frac{\partial^2 X_a(t)}{\partial a_i \partial a_j}$$

exists in the $L^2(P)$ sense and satisfies the stochastic differential equation:

$$\eta_{ij}(a,t) = A_{ij}(t) + \int_0^t \eta_{ij}(a,t)\, b_x(s, X_a(s))\, ds + \int_0^t \eta_{ij}(s)\, \sigma(s, X_a(s))\, dW_s \tag{8.6.23}$$

where

$$A_{ij}(t) = \int_0^t Z_i Z_j\, b_{x_i x_j}(s, X_a(s))\, ds + \int_0^t Z_i Z_j\, \sigma_{x_i x_j}(s, X_a(s))\, dW_s.$$

Remark 8.6.1 If $H(a,t)$ denotes any of the processes $X_a(t)$, $Z_i(a,t)$, and $\eta_{ij}(a,t)$ given by Equations (8.6.12), (8.6.13), and (8.6.23), then for any finite $p \geq 1$, there exists a constant $C_{p,T,a}$ such that

$$\sup_{0 \leq t \leq T} |H(a,t)|^p \leq C_{p,T,a}. \tag{8.6.24}$$

There exists a finite constant $C_{p,T}$ such that $C_{p,T,a+h} \leq C_{p,T}$ for all h with $|h| \leq 1$. The bound (8.6.24) is quite simple for $X_a(t)$. Since Equation (8.6.13) is linear in Z_i, the bound (8.6.24) for $Z_i(a,t)$ follows quite easily from the boundedness of the derivatives of b and σ. To prove the bound for the process $\eta_{ij}(a,t)$, one needs to use the condition (8.6.22).

Remark 8.6.2 The results proved thus far in this section hold if the initial time is $s > 0$ instead of 0. We will denote the solution of Equation (8.6.12) with initial time s as $X_{s,a}(t)$ where $t \geq s$.

Theorem 8.6.6 *Suppose that the conditions of Theorem 8.6.5 hold. Let f be a twice continuously differentiable function defined on $\mathbb{R}^d$ that satisfies the bound*

$$|D_x^\alpha f(x)| \leq C(1 + |x|^r) \ \text{ for all multi-indices } \alpha \ \text{ such that } |\alpha| \leq 2 \tag{8.6.25}$$

for some positive constants C and r. Define, for any fixed s, t such that $s \leq t$, the function

$$g(x) = Ef(X_{s,x}(t)). \tag{8.6.26}$$

Then g is twice continuously differentiable, and satisfies

$$|D_x^\alpha g(x)| \leq C_1(1 + |x|^p) \ \text{ for all multi-indices } \alpha \ \text{ such that } |\alpha| \leq 2 \tag{8.6.27}$$

for suitable positive constants C_1 and p.

Proof

Step 1 The proof is similar to that of Theorems 8.6.3 and 8.6.4. Indeed, let us take $h = (h_1, 0, \ldots, 0)$. Then,

$$g(x+h) - g(x) = E\left[\int_0^1 \frac{d}{dy} f(X_{s,x}(t) + y(X_{s,x+h}(t) - X_{s,x}(t))\, dy\right]$$

$$= E\left[\int_0^1 \nabla f(X_{s,x}(t) + y(X_{s,x+h}(t) - X_{s,x}(t))\, dy \cdot (X_{s,x+h}(t) - X_{s,x}(t))\right].$$

Divide both sides by h_1 and let $h_1 \to 0$. Observe that

$$\frac{X_{s,x+h}(t) - X_{s,x}(t)}{h_1} \to \frac{\partial}{\partial x_1} X_{s,x}(t)$$

in $L^2(P)$. Next, we claim that

$$\int_0^1 f_x(X_{s,x}(t) + y(X_{s,x+h}(t) - X_{s,x}(t))\, dy \to \nabla f(X_{s,x}(t))$$

in $L^2(P)$. Indeed, for any $0 \le y \le 1$, we have

$$f_x(X_{s,x}(t) + y(X_{s,x+h}(t) - X_{s,x}(t)) \to f_x(X_{s,x}(t))$$

in probability as $h \to 0$ since, for instance,

$$\lim_{h\to 0} E|X_{s,x+h}(t) - X_{s,x}(t)|^2 = 0.$$

Therefore, the boundedness of f_x allows us to use the bounded convergence theorem to prove the claim made above.

Thus, as $h \to 0$, $\frac{g(x+h)-g(x)}{h}$ converges to $E[\nabla f(X_{s,x}(t)) \cdot \frac{\partial}{\partial x_1} X_{s,x}(t)]$; that is,

$$\frac{\partial}{\partial x_1} g(x) = E\left[\nabla f(X_{s,x}(t)) \cdot \frac{\partial}{\partial x_1} X_{s,x}(t)\right] \tag{8.6.28}$$

The derivative of g is thus found by formally differentiating with respect to x inside the expectation. Since $\frac{\partial}{\partial x_1} X_{s,x}(t)$ has finite moments of all orders, and ∇f is bounded, the bound (8.6.27) for first-order derivatives of g with respect to x is straightforward.

Let us call $\frac{\partial}{\partial x_1} X_{s,x}(t)$ as $Z_1(s,x,t)$. Since Z_1 is the solution of a linear stochastic differential equation started at x at time s, it is continuous with respect to initial data x. This along with boundedness of ∇f proves the continuity of first order derivatives of g.

Step 2 We need to show that g is twice differentiable and that

$$D_{ij} g(x) = E\left[\nabla f(X_{s,x}(t)) \cdot \frac{\partial^2}{\partial x_i \partial x_j} X_{s,x}(t) + (Z_i(s,x,t), \nabla^2 f(X_{s,x}(t)) Z_j(s,x,t))\right].$$

where, in the second term on the right side, we have used the inner product in $\mathbb{R}^d$. We can proceed as in step 1 with the necessary changes. First, using Equation (8.6.28), we can write an expression for $\frac{\partial}{\partial x_1} g(x+h) - \frac{\partial}{\partial x_1} g(x)$. Let us take $h = (0, h_2, 0, \ldots, 0)$.The expression can be rewritten as

$$E\left[\nabla f(X_{s,x+h}(t))\cdot\left(\frac{\partial}{\partial x_1}X_{s,x+h}(t)-\frac{\partial}{\partial x_1}X_{s,x}(t)\right)\right]$$
$$+E\left[(\nabla f(X_{s,x+h}(t))-\nabla f(X_{s,x}(t)))\cdot\frac{\partial}{\partial x_1}X_{s,x}(t)\right]. \tag{8.6.29}$$

Divide by h and take the limit as $h\to 0$. We need to take the limit inside the expectation. In the first term in (8.6.29), this is quite easy since $|\nabla f|$ is bounded, and the $L^2(P)$-limit of

$$\frac{1}{h}\left(\frac{\partial}{\partial x_1}X_{s,x+h}(t)-\frac{\partial}{\partial x_1}X_{s,x}(t)\right)$$

exists as $h\to 0$. For the second term, we can use condition (8.6.25) to obtain uniform integrability in h. For,

$$\begin{aligned}&(\nabla f(X_{s,x+h}(t))-\nabla f(X_{s,x}(t)))\\&=\int_0^1\frac{\partial}{\partial y}\nabla f(X_{s,x}(t)+y(X_{s,x+h}(t)-X_{s,x}(t))\,dy\\&=\int_0^1\nabla^2 f(X_{s,x}(t)+y(X_{s,x+h}(t)-X_{s,x}(t))\cdot(X_{s,x+h}(t)-X_{s,x}(t))\end{aligned}$$

so that the second term in Equation (8.6.29) when divided by h becomes

$$E\left[\left(\frac{1}{h_2}(X_{s,x+h}(t)-X_{s,x}(t)),\int_0^1\nabla^2 f\left(X_{s,x}(t)+y(X_{s,x+h}(t)-X_{s,x}(t))\,dy\frac{\partial}{\partial x_1}X_{s,x}(t)\right)\right)\right]. \tag{8.6.30}$$

First, as $h\to 0$, $\int_0^1(\frac{1}{h_2}(X_{s,x+h}(t)-X_{s,x}(t))\to\frac{\partial}{\partial x_2}X_{s,x}(t))$ in $L^2(P)$. Next, by the condition (8.6.25),

$$\begin{aligned}&E\left|\int_0^1\nabla^2 f(X_{s,x}(t)+y(X_{s,x+h}(t)-X_{s,x}(t))\,dy\right|^4\\&\le 4C_1(1+|X_{s,x}(t)+X_{s,x+h}(t)-X_{s,x}(t)|^{4p})\,dy\\&\le C_2(1+2|X_{s,x}(t)|^{4p}+|X_{s,x+h}(t)|^{4p})\\&\le C_3\end{aligned}$$

where C_3 doesn't depend on h. In addition, $\frac{\partial}{\partial x_1}X_{s,x}(t))$ has moments of all orders. Hence, by Hölder's inequality, one obtains the uniform integrability of

$$\left|\int_0^1\nabla^2 f(X_{s,x}(t)+y(X_{s,x+h}(t)-X_{s,x}(t))\,dy\frac{\partial}{\partial x_1}X_{s,x}(t)\right|^2.$$

Besides, the above expression converges in $L^2(P)$ to $\nabla^2 f(X_{s,x}(t))\frac{\partial}{\partial x_1}X_{s,x}(t)$. Therefore, the expression displayed in Equation (8.6.30) converges to

$$(Z_2(s,x,t), \nabla^2 f(X_{s,x}(t))Z_1(s,x,t)).$$

The bound (8.6.27) for $|D_x^\alpha g(x)|$, and the continuity of $D_x^\alpha g(x)$ in x follow from the above arguments. ■

Let $T > 0$ be a fixed time. Define the function

$$u(t,x) = E[f(X_{t,x}(T))].$$

Recall that there exists a family of measures $\{P_{t,x} : t \geq 0;\ x \in \mathbb{R}^d\}$ (on the canonical space Ω of $\mathbb{R}^d$-valued continuous functions) with respect to which the canonical process X is a Markov process and a solution of the stochastic differential equation (8.6.1) started at time t and state x. Hence, we can write

$$u(t,x) = E[f(X_{t,x}(T))] = E_{t,x}[f(X(T)]. \tag{8.6.31}$$

Theorem 8.6.7 *Let the hypotheses of Theorem 8.6.6 hold. Then u is continuously differentiable once with respect to t and twice with respect to x in $[0,T) \times \mathbb{R}^d$. The function u solves the backward partial differential equation in $[0,T) \times \mathbb{R}^d$:*

$$u_t + \sum_{i=1}^{d} b_i D_i u + \frac{1}{2}\sum_{i,j=1}^{d} a_{ij} D_{ij} u = 0 \tag{8.6.32}$$

under the terminal condition

$$u(t,x) \to f(x) \ \ as \ \ t \uparrow T. \tag{8.6.33}$$

Equation (8.6.32) is known as the Kolmogorov backward equation.

Proof

Step 1 For any fixed $t \in (0,T)$ and $h > 0$ such that $t - h > 0$, consider $u(t-h,x) = E_{t-h,x} f(X(T))$. By the Markov property, we can write

$$\begin{aligned} E_{t-h,x} f(X(T)) = E_{t-h,x}[E_{t-h,x}\{f(X(T)|\mathcal{F}_t\}] &= E_{t-h,x}E_{t,X(t)}[f(X(T)] \\ &= E_{t-h,x}[u(t,X(t))] \\ &= E\left[u(t, X_{t-h,x}(t))\right]. \end{aligned}$$

Therefore, we can write

$$\frac{1}{h}[u(t,x) - u(t-h,x)] = -\frac{1}{h}\left[Eu(t, X_{t-h,x}(t)) - u(t,x)\right]. \tag{8.6.34}$$

On the right side of the above equation, with t fixed, let us view u as simply a function of the second argument. In fact, let us call $u(t,x)$ as $v(x)$. From the previous theorem, we know that v has continuous second order derivatives. Therefore, we can apply the Itô formula to v so that

$$E[v(X_{t-h,x}(t))] - v(x) = E\left[\int_{t-h}^{t} L_s v(X_{t-h,x}(s))\, ds\right]$$

where L_s is the differential operator corresponding to the process $X_{t-h,x}$. Dividing both sides by h, and letting $h \to 0$, we claim that

$$\frac{1}{h}E\left[\int_{t-h}^{t} L_s v(X_{t-h,x}(s))\, ds\right] \to L_t v(x).$$

Indeed, we can write

$$\frac{1}{h}E\left[\int_{t-h}^{t} L_s v(X_{t-h,x}(s))\, ds\right] = E\int_{0}^{1} L_{(t-h)+yh} v(X_{t-h,x}((t-h)+yh)\, dy. \tag{8.6.35}$$

As $h \downarrow 0$, $L_{(t-h)+yh}v(X_{t-h,x}((t-h)+yh) \to L_t v(x)$ in probability. In addition,

$$E\left[|L_{(t-h)+yh}v(X_{t-h,x}((t-h)+yh)|^2\right] \le CE(1 + |X_{t-h,x}((t-h)+yh)|^{\beta})$$

for a suitable $\beta > 0$ by the following two observations. Recall the bound given by (8.6.27):

$$|D_x^{\alpha} v(x)| \le C_1(1 + |x|^p) \text{ if } |\alpha| \le 2,$$

and that the coefficients b and σ satisfy a linear growth condition. Next, $X_{t-h,x}(t-h+yh)$ has moments of all orders, and the second moment, for instance, can be bounded uniformly in $h \in [0,t]$ and $y \in [0,1]$. Therefore, we have uniform integrability of the family

$$\int_{0}^{1} L_{(t-h)+yh} v(X_{t-h,x}((t-h)+yh)\, dy$$

indexed by h. Letting $h \to 0$ in Equation (8.6.35), the claim is proved. We have thus shown that

$$\lim_{h\downarrow 0}\frac{1}{h}[u(t-h,x) - u(t,x)] = L_t u(t,x). \tag{8.6.36}$$

Step 2 Existence of $\frac{\partial u}{\partial t}$: From the condition (8.6.27) on f, and the linear growth condition on b and σ, one obtains using the Itô formula that there exists a constant C such that

$$\frac{1}{h}|E[v(X_{t-h,x}(t))] - v(x)| \le C$$

where C is independent of h; that is,

$$|u(t,x) - u(t-h,x)| \le\le Ch.$$

Thus, u is absolutely continuous with respect to t so that $\frac{\partial u}{\partial t}$ exists for almost every t. For such t,

$$\begin{aligned} u(t,x) &= u(0,x) + \int_0^t \frac{\partial u}{\partial t}(s,x)\,ds \\ &= u(0,x) - \int_0^t L_s u(x)\,ds. \end{aligned}$$

Since $L_s u$ is continuous in s, $\frac{\partial u}{\partial t}$ exists everywhere. Equation (8.6.36) yields for all $0 < t < T$ that

$$\frac{\partial u}{\partial t} = -L_t u.$$

Finally, observe that we can write

$$u(t,x) - f(x) = Ef(X_{t,x}(T)) - Ef(X_{T,x}(T)).$$

By uniform integrability of $\{X_{t,x}(T) : 0 \le t \le T\}$ and the condition (8.6.25) on f,

$$\lim_{t \uparrow T} Ef(X_{t,x}(T)) = Ef(X_{T,x}(T))$$

which proves the terminal condition (8.6.33).

■

Remark 8.6.3 1. In the context of Theorem 8.6.5, if we consider the function $v(t,x) = E_{T-t,x} f(X_T)$, then v is the solution of the initial value problem:

$$v_t = \frac{1}{2}\sum_{i,j=1}^{d} a_{ij} D_{ij} v + \sum_{i=1}^{d} b_i D_i v$$

for $0 < t < T$ with $v(0,x) = f(x)$. When X is the solution of a stochastic differential equation with autonomous coefficients, then X is a time-homogeneous Markov process, and hence

$$v(t,x) = E_{T-t,x} f(X_T) = Ef(X_t^x)$$

where X_t^x is the solution started at time 0 in state x. Therefore, v solves the equation (8.6.3) for all $t > 0$ subject to the initial condition $v(0,x) = f(x)$.

2. The adjoint of equation (8.6.32) is known as the Kolmogorov forward equation or the Fokker-Planck equation.

Exercises

1. Let $G \subset \mathbb{R}^d$ be a domain as in Section 6.1. Using the Itô formula, show that a harmonic function u defined on G has the mean-value property in G.
2. Let G be as in the previous problem. If u is a bounded measurable function in G, and has the mean-value property, show that u is harmonic in G.
3. With the notation and hypotheses of Theorem 2.5, show that $P\{z + W_t \in V\}$ is a positive constant for all t. Use it along with the Blumenthal $0-1$ law to prove Theorem 8.2.5.
4. Consider the unit disc $D = \{z : |z| < 1\}$. Let W be a two-dimensional Wiener process, and $\tau_D = \inf\{t : |W_t| = 1\}$. Without using harmonic functions, show that for any $\epsilon > 0$,

$$P_z\{|W_{\tau_D} - 1| > \epsilon\} \to 0$$

as $z \in D$ tends to 1.
5. Let W be a Wiener process in $\mathbb{R}^2$. Let D denote the open unit disc with the subset $\{(x, 0) : x \geq 0\}$ removed from it. Show that $(0, 0)$ is a regular boundary point.
6. Let W be a d-dimensional Wiener process. Define $X_t = |W_t|$ for all $t \geq 0$. Show that X_t is a solution of

$$dX_t = dM_t + \frac{n-1}{2X_t}\,dt$$

with $X_0 = 0$. The process X is known as the Bessel process.
7. Let $X(t) = W(t) + t$ for all $t \geq 0$. Find a function f such that $f(X(t))$ is a martingale. Let $\tau = \inf\{t : X(t) = -a \text{ or } b\}$. Find $P\{X(\tau) = b\}$. What is $E(\tau)$?
8. Let τ be a stopping time such that the process $\{W(\tau \wedge t)\}$ is bounded by a finite constant. Show that $E(W(\tau)) = 0$ and $E\tau = E[W^2(\tau)]$.
9. Use a probabilistic representation to solve the following one-dimensional boundary value problem:

$$\frac{\partial u}{\partial t} + bx\frac{\partial u}{\partial x} + \frac{1}{2}\sigma^2 x^2 \frac{\partial^2 u}{\partial x^2} = 0 \text{ for } t < T$$

and $u(T, x) = \ln(x^2)$. Here, b and σ are constants.

9 Gaussian Solutions

The theory of stochastic differential equations whose solutions are Gaussian processes is an instructive special case of the general theory of functional stochastic differential equations. The theory of nonanticipative representations of equivalent Gaussian measures provides the framework within which Gaussian solutions of stochastic equations are studied in this chapter. The results of this chapter are related to "stochastic infinitesimal equations" proposed by Paul Lévy over fifty years ago. In his study of stochastic infinitesimal equations, Lévy discusses, by means of examples, Gaussian processes given by an equation of the form:

$$\delta x(t) \simeq dt \int_0^t F(t,u)\, dx(u) + \zeta \sqrt{d\omega(t)}$$

where ζ is a standard normal random variable, and $\omega(t)$ is a positive increasing function of t. If $\omega(t) = t$, the results of this chapter provide a complete answer to the following questions raised by him [50], page 349:

1. If $dZ_t = b(t,Z)\,dt + dW_t$, how can one pass from this equation to the covariance function of Z_t, and conversely?
2. Given the covariance, how can the process Z_t be expressed by an infinitesimal equation?

In this chapter, we begin with a discussion of linear operators on Hilbert spaces that leads us to a special factorization theorem of Gohberg and Krein [28]. Such a representation is derived explicitly for a Gaussian measure equivalent to Wiener measure. Using it, the theory of Gaussian solutions of stochastic equations is studied.

9.1 Introduction

Let H be a separable Hilbert space with inner product denoted by $(\cdot, \cdot)$ and the norm, by $|\cdot|$. We know that a linear map $T : H \to H$ is a bounded linear operator if there exists a finite $c \geq 0$ such that $|Tx| \leq c|x|$ for all $x \in H$. Let $L(H)$ denote the space of all bounded linear operators. For $T \in L(H)$, define

$$\|T\| = \sup_{0<|x|\leq 1} \frac{|Tx|}{|x|}.$$

This defines a norm on $L(H)$. Equipped with this norm, the linear space $L(H)$ is a Banach space.

Definition 9.1.1 *A sequence $\{T_n\}$ in $L(H)$ is said to converge to a bounded linear operator T*

(i) **uniformly** *if $\|T_n - T\| \to 0$ as $n \to \infty$,*

(ii) **strongly** *if $|T_n x - Tx| \to 0$ as $n \to \infty$ for all $x \in H$, and*

(iii) **weakly** *if $(T_n x, y) \to (Tx, y)$ as $n \to \infty$ for all $x, y \in H$.*

Let T be any element of $L(H)$. Fix any $z \in H$. Define a function f on H by $f(x) = (Tx, z)$. Then f is a bounded linear functional and hence admits the representation $f(x) = (x, y)$ where $y \in H$ is uniquely determined. Thus

$$(Tx, z) = (x, y) \quad \forall x \in H.$$

Naming y as $T^* z$, we get

$$(Tx, z) = (x, T^* z) \quad \forall x, z \in H.$$

The operator $T^* : H \to H$ is called the **adjoint** of T. It is well known that T^* defined above is a bounded linear operator and that $\|T^*\| = \|T\|$. An operator $T \in L(H)$ is called **self-adjoint** if $T^* = T$.

Example 9.1.1 *$H = \mathbb{C}^n$, the n-dimensional complex space. For any integer $1 \leq j \leq n$, let $e_j = (0, \ldots, 1, 0, \ldots, 0)$ where 1 appears in the jth coordinate.*

Suppose $T : H \to H$. Define $\alpha_{jk} = (Te_k, e_j)$. If $x \in H$ with $x = \sum_{k=1}^{n} x_k e_k$,

$$(Tx, e_j) = \sum_{k=1}^{n} x_k (Te_k, e_j) = \sum_{k=1}^{n} \alpha_{jk} x_k.$$

Define the $n \times n$ matrix $A = (\alpha_{jk})$. If A_j denotes the jth row of A, then the jth coordinate of Tx, namely, $(Tx, e_j) = A_j x$, where x is the column vector $(x_1, x_2, \ldots, x_n)$. Thus, T corresponds to matrix multiplication by A and conversely.

Clearly, $\|T\|^2 = \sum_{j,k} |\alpha_{jk}|^2$. *If T were self-adjoint, then*

$$(Te_k, e_j) = (e_k, Te_j) = \bar{\alpha}_{kj}.$$

where the bar denotes complex conjugation. Thus $\alpha_{jk} = \bar{\alpha}_{kj}$, *that is, A is a Hermitian matrix.*

Example 9.1.2 *Let* $H = L^2[a, b]$. *Define an operator T on H by* $(Tf)(t) = h(t)f(t)$ *where h is a bounded real-valued function on* $[a, b]$. *Then for any* $f, g \in H$,

$$(Tf, g) = \int_a^b h(t)f(t)\bar{g}(t)\,dt = (f, hg) = (f, Tg).$$

Thus T is a self-adjoint operator.

Example 9.1.3 *Let k be a complex-valued function in* $L^2([a, b] \times [a, b])$. *Let I denote the interval* $[a, b]$. *Define an operator T on* $H = L^2(I)$ *by*

$$(Tf)(s) = \int_I k(s, t)f(t)\,dt.$$

Then T is in $L(H)$, *and for any* $g \in H$,

$$\begin{aligned}(Tf, g) &= \int_I \int_I k(s, t)f(t)\overline{g(s)}\,dt\,ds \\ &= \int_I \int_I k(s, t)f(t)\overline{g(s)}\,ds\,dt \qquad \text{by Fubini's theorem} \\ &= \int_I f(t)\overline{\int_I \overline{k(s, t)}g(s)\,ds}\,dt \\ &= (f, T^*g)\end{aligned}$$

where $T^*g(t) = \int_I \bar{k}(s, t)g(s)\,ds$. *Thus T is self-adjoint if* $k(s, t) = \overline{k(s, t)}$.

Given an operator $T \in L(H)$, consider the equation $Tf = \lambda f$, where λ is a complex number. If a solution $f \neq 0$ exists, then λ is called an eigenvalue of T, and f, an eigenfunction for λ. The collection of all the eigenvalues of T is known as the **point spectrum** of T.

It should be noted that self-adjoint operators on H need not have eigenvalues. In fact, take $H = L^2(0, 1)$. Define $(Tf)(x) = xf(x)$. T is a bounded self-adjoint operator. However, there is no λ for which $xf(x) = \lambda f(x)$ for a.e. $x \in (0, 1)$ for any non-zero f. Towards obtaining fruitful results on the eigenvalue problems, known as spectral theorems, we need the idea of compact operators.

Definition 9.1.2 *An operator* $T \in L(H)$ *is called a* **compact** *operator if for any bounded set* $B \subset H$, *the closure of the image* $T(B)$ *is a compact set.*

The usefulness of a compact operator arises from the fact that it converts the weak convergence of a sequence in H to a strong convergence. In fact,

Theorem 9.1.4 *The following are equivalent:*

1. *T is a compact operator.*
2. *$f_n \to f$ weakly in $H \implies Tf_n \to Tf$ in H.*

A proof of this can be found in [61]. It is worthwhile to note that the second statement given in Theorem 9.1.4 is Hilbert's definition of compact operators. A compact operator was called a *completely continuous* operator since it has more nicety than a continuous operator. Indeed, $Tf_n \to Tf$ even if f_n converges to f just weakly.

A famous result of Banach and Alaoglu, in the context of Hilbert spaces, states that bounded subsets of H are weakly sequentially precompact, that is, given a bounded sequence $\{f_n\}$ in H, there exists a weakly convergent subsequence $\{f_{n_j}\}$. If T is a compact operator on H, then by the definition of Hilbert, Tf_{n_j} converges to an element of H. Conversely, given a bounded sequence $\{f_n\}$, if there exists a subsequence $\{f_{n_j}\}$ such that Tf_{n_j} converges, then T is a compact operator.

The next result is a well-known spectral representation theorem of Hilbert and Schmidt. The reader can find a proof of it in [61].

Theorem 9.1.5 *Let T be a compact, self-adjoint operator on H. There exists a complete orthonormal system (CONS) $\{\phi_n\}$ of eigenfunctions of T such that $T\phi_n = \lambda_n\phi_n$ with the real numbers $\lambda_n \to 0$ as $n \to \infty$. Besides, each eigenvalue is of finite multiplicity.*

Definition 9.1.3 *A bounded linear operator T is called* **positive** *or* **positive definite** *if $(Tx, x) \geq 0$ for all $x \in H$.*

If T is a positive, compact, self-adjoint operator on H, then its eigenvalues are nonnegative. Next, we prove a useful proposition which gives the algebraic properties of compact operators.

Proposition 9.1.6 *Let $T, T_1, T_2, \ldots$ be compact operators on H. Let S be a bounded linear operator on H. Then the following hold:*

(i) The set of compact operators is a linear space.

(ii) ST and TS are compact operators.

(iii) T^* is compact.

(iv) If T_n converges uniformly to an operator R, then R is compact.

Proof By using Theorem 9.1.4, it is straightforward to prove (i) and (ii).

Proof of (iii): Let $f_n \to f$ weakly in H. Then the set $\{f, f_n : n \geq 1\}$ is bounded by the Banach-Steinhaus theorem. Consider

$$\begin{aligned} \|T^*f_n - T^*f\|^2 &= (T^*f_n - T^*f, T^*f_n - T^*f) \\ &= (TT^*f_n - TT^*f, f_n - f). \end{aligned}$$

Clearly, $T^*f_n \to T^*f$ weakly in H. Therefore, $TT^*f_n \to TT^*f$ since T is compact. Hence

$$\|T^*f_n - T^*f\|^2 \leq \|TT^*f_n - TT^*f\|(\|f_n\| + \|f\|),$$

which tends to zero since f_n and f are bounded.

Proof of (iv): The definition of compact operators is used in proving this part. Given any $\epsilon > 0$, choose N such that $\|T_n - R\| < \epsilon/2$ if $n \geq N$. Since T_n is compact, for any bounded set B, $T_n(B)$ has compact closure. Take B to be the closed unit ball in H. Then, $T_N(B)$ can be covered by a finite number of balls of radius $\epsilon/2$. Therefore, $R(B)$ can be covered by the same balls provided their radius is increased to ϵ. The proof is over. ■

By Theorem 9.1.5, it follows that the subspace of all compact operators on H is a closed subspace of $L(H)$, and is therefore a Banach space.

Example 9.1.7 *Consider the operator T introduced in Example 9.1.3. We will show that T is a compact operator. For any $f \in L^2(I)$,*

$$(Tf)(s) = \int_I k(s,t)f(t)\,dt.$$

Let $\{e_j\}$ be a CONS in H. Fix any $s \in I$. We can write $k(s,t) = \sum_j k_j(s)e_j(t)$ so that

$$(Tf)(s) = \sum_j k_j(s)\,(e_j, f).$$

By the Parseval theorem,

$$\|T\|^2 = \sum_j \int_I |k_j(s)|^2\,ds < \infty. \tag{9.1.1}$$

Fix an n. Define $(T_nf)(s) = \sum_{j=1}^n k_j(s)(e_j, f)$. T_n is a compact operator. To see this, let $f_k \to f$ weakly in H. Then,

$$\begin{aligned}\|T_nf_k - T_nf\|^2 &\leq n\sum_{j=1}^{n}\int_I |k_j(s)|^2\,|(e_j, f_k - f)|^2\,ds\\ &\leq n \sup_{1\leq j\leq n} |(e_j, f_k - f)|^2 \sum_{j=1}^{n}\int_I |k_j(s)|^2\,ds,\end{aligned}$$

which goes to zero as $k \to \infty$ by the weak convergence of f_k and (9.1.1).

As $n \to \infty$, T_n tends to T. Indeed,

$$\|T_n - T\|^2 \leq \int_I\int_I |k_n(s,t) - k(s,t)|^2\,ds\,dt = \sum_{j=n+1}^{\infty}\int_I k_j(s)^2\,ds.$$

which tends to zero as $n \to \infty$ *by* (9.1.1). *By part (iv) of the above proposition, we obtain the compactness of* T.

We end this section with the polar decomposition of compact operators. Just as a complex number z can be written as $z = re^{i\theta}$ where $r \geq 0$, we can write a compact operator T as US where $U * U = I$ on the range of S, and S is a positive operator. The following theorem can be shown using the spectral representation theorem 9.1.5 for T^*T.

Theorem 9.1.8 *Let* T *be a compact operator on* H. *Then* T *can be written in the form* $T = US$ *where* S *is a compact, self-adjoint, positive operator, and* U, *an isometry defined on the range of* S.

Remark 9.1.1 The operator S is very often denoted by $(T^*T)^{1/2}$ or as $|T|$. The polar representation theorem 9.1.8 is useful in defining trace class operators in the next section.

9.2 Hilbert-Schmidt Operators

The space of Hilbert-Schmidt operators provides us with an important subspace of compact operators. To define such operators, we need the following lemma.

Lemma 9.2.1 *Let* T *be a bounded operator on* H. *If* $\{e_n\}$ *and* $\{f_n\}$ *are any two complete orthonormal systems for* H, *then*

$$\sum_{n=1}^{\infty} \|Te_n\|^2 = \sum_{n=1}^{\infty} \|Tf_n\|^2. \tag{9.2.1}$$

Proof Clearly, $\|Te_n\|^2 = \sum_{m=1}^{\infty} |(Te_n, f_m)|^2$ so that

$$\begin{aligned} \sum_n \|Te_n\|^2 &= \sum_n \sum_m |(e_n, T^*f_m)|^2 \\ &= \sum_m \sum_n |(e_n, T^*f_m)|^2 \\ &= \sum_m \|T^*f_m\|^2. \end{aligned}$$

The above calculation holds for any two orthonormal bases. In particular, if $f_m = e_m$ for all m, we obtain

$$\sum_n \|Tf_n\|^2 = \sum_m \|T^*f_m\|^2. \tag{9.2.2}$$

Thus,

$$\sum_n \|Te_n\|^2 = \sum_m \|T^*f_m\|^2 = \sum_n \|Tf_n\|^2. \qquad \blacksquare$$

Definition 9.2.1 *A bounded linear operator T on H is called a* **Hilbert-Schmidt operator** *if, for any given CONS $\{e_n\}$ for H, $\sum_n \|Te_n\|^2 < \infty$. The Hilbert-Schmidt norm of T is defined as*

$$\|T\|_2 = \left(\sum_{n=1}^{\infty} \|Te_n\|^2\right)^{1/2}.$$

By lemma 9.2.1, $\|T\|_2$ doesn't depend on the choice of $\{e_n\}$, and hence, is well defined.

Proposition 9.2.2 *If T is a Hilbert-Schmidt operator on H, then it is compact.*

Proof Let $\{e_n\}$ be a CONS for H. Define, for each m, the operator T_m by $T_m x := \sum_{n=1}^{m} (x, e_n) Te_n$. Being an operator with a finite-dimensional range, T_m is compact. For any $x \in H$,

$$\begin{aligned}\|T - T_m\| x &\leq \sum_{n>m} |(x, e_n)| \, \|Te_n\| \\ &\leq \left(\sum_{n>m} |(x, e_n)|^2\right)^{1/2} \left(\sum_{n>m} \|Te_n\|^2\right)^{1/2} \\ &\leq \|x\| \left(\sum_{n>m} \|Te_n\|^2\right)^{1/2}\end{aligned}$$

so that as $m \to \infty$, $\|T - T_m\| \leq (\sum_{n>m} \|Te_n\|^2)^{1/2} \to 0$. ■

The set of all Hilbert-Schmidt operators is a linear space which is denoted by $L_2(H)$ or simply by L_2 if H remains fixed. Note that if $T, S \in L_2$, then

$$\|T + S\|_2 \leq \|T\|_2 + \|S\|_2$$

by the Minkowski inequality. Also, $\|T^*\|_2 = \|T\|_2$ by Equation (9.2.2) so that $T^* \in L_2$.

Proposition 9.2.3 *If $T \in L_2$, and S is any bounded operator on H, then*

(i) $\|T\| \leq \|T\|_2$.

(ii) $\|ST\|_2 \leq \|S\| \, \|T\|_2$ *and* $\|TS\|_2 \leq \|T\|_2 \|S\|$. *Thus, ST and TS are in L_2.*

Proof For any $x \in H$,

$$\begin{aligned}\|Tx\|^2 &= \left\|\sum_n (x, e_n) Te_n\right\|^2 \\ &\leq \left(\sum_n |(x, e_n)|^2\right) \left(\sum_n \|Te_n\|^2\right) = \|x\|^2 \|T\|_2^2\end{aligned}$$

so that $\|Tx\| \leq \|x\| \, \|T\|_2$.

For part (ii),

$$\|ST\|_2^2 = \sum_n \|STe_n\|^2 \leq \|S\|^2 \|Te_n\|^2$$

so that $\|ST\|_2 \leq \|S\| \|T\|_2$.

Since $T^* \in L_2$, the above calculation applies to T^* and S^*. Hence,

$$\|S^* T^*\|_2 \leq \|S^*\| \|T^*\|_2;$$

that is, $\|TS\|_2 \leq \|S\| \|T\|_2$. ∎

It can be shown that $L_2(H)$ equipped with the inner product

$$\langle T, S\rangle = \sum_n \langle Te_n, Se_n\rangle$$

is a Hilbert space.

Definition 9.2.2 *A compact operator T on H is called a* **trace class operator** *if $\sum_n \lambda_n < \infty$ where λ_n's are the eigenvalues of $(T^*T)^{1/2}$.*

In contrast, a bounded operator T is Hilbert-Schmidt if $\sum_n \lambda_n^2 < \infty$ where λ_n's are the eigenvalues of $(T^*T)^{1/2}$, and, in such a case, $\|T\|_2 = (\sum_n \lambda_n^2)^{1/2}$.

The collection of all trace class operators on H is denoted by $L_1(H)$ or simply L_1. If $T \in L_1$, define the trace class norm of T by

$$\|T\|_1 = \sum_n \lambda_n.$$

Definition 9.2.3 *If $T \in L_1$, for any CONS $\{f_n\}$ of H, the sum $\sum_n (Tf_n, f_n)$ is called the* **trace** *of T.*

It is a simple exercise to show that L_1 is a linear space.

Proposition 9.2.4

(i) If $S, T \in L_2(H)$, then $ST \in L_1(H)$ and

$$\|ST\|_1 \leq \|S\|_2 \|T\|_2.$$

(ii) If $R \in L_1(H)$, then

$$\|R\| \leq \|R\|_2 \leq \|R\|_1,$$

and $\|R^*\|_1 = \|R\|_1$.

Also, one can show that if A is a bounded operator on H and B, a trace class operator, then AB is a trace class operator and

$$\|AB\|_1 \leq \|A\| \|B\|_1.$$

Likewise, BA is in $L_1(H)$ and $\|BA\|_1 \leq \|A\| \|B\|_1$.

9.3 The Gohberg-Krein Factorization

The motivation for the Gohberg-Krein factorization comes from linear algebra. Let A be an $n \times n$ nonsingular matrix. It is well known that if all the principal minors of A are nonsingular, then A^{-1} admits a unique factorization:

$$A^{-1} = S_+ D S_-$$

where D is a diagonal matrix and $S_\pm$ are right and left triangular matrices whose diagonal elements are equal to unity. An abstract formulation of this factorization for certain operators in a separable Hilbert space H is known as the Gohberg-Krein factorization and is described below.

Define an integral operator as in Example 9.1.7 where the function k is a scalar or matrix-valued kernel in $L^2(I)$. Such an operator V is an example of the Volterra integral operator. A subspace $L \subset H$ is known as an invariant subspace of V if $f \in L \implies Vf \in L$. The Volterra integral operator V has an explicit increasing family of invariant subspaces, namely

$$L_r = \{f : f \in L^2(a,b), f(t) = 0 \text{ for } r \leq t \leq b\}$$

for all $a \leq r \leq b$. By means of such a family of invariant spaces, one can obtain a triangular representation for an abstraction of V known as a Volterra operator.

Before defining Volterra operators, it is worthwhile to note that, under very general assumptions on k, one can show that V is compact and $(I - \mu V)$ is invertible for all complex μ. The spectrum of V reduces to a single point $\lambda = 0$. Motivated by this, we have

Definition 9.3.1 *A linear operator V on H is called a* **Volterra** *operator if it is compact and has the one-point spectrum $\lambda = 0$.*

Next, we build the terminology that is needed in order to state the required results. From now on, a projection operator will mean an orthogonal projection.

1. For any two projection operators P_1, P_2 on H, we write $P_1 < P_2$ meaning $P_1 H \subset P_2 H$. We write $P_1 \leq P_2$ if either $P_1 < P_2$ or $P_1 = P_2$.
2. A family of orthogonal projections $\pi = \{P\}$ is called a **chain** if, for any distinct $P_1, P_2 \in \pi$, either $P_1 < P_2$ or $P_2 < P_1$.
3. A chain π is **bordered** if $0, I \in \pi$.
4. The closure of a chain π is the set of all operators which are the strong limits of sequences in π. Note that the closure of a chain is again a chain.
5. If a chain coincides with its closure, it is said to be closed.
6. A pair (P^-, P^+) of projection operators in a closed chain π with $P^- < P^+$ is called a **gap** of π if, for any $P \in \pi$, either $P \leq P^-$ or $P \geq P^+$. The dimension of $P^+H - P^-H$ is called the **dimension** of the gap (P^-, P^+).

7. A chain is called **maximal** if it is bordered, closed, and its gaps are one dimensional.
8. A chain π is called an **eigenchain** of a bounded linear operator A on H if $PAP = AP$ for all $P \in \pi$.
9. The **dual** $\pi^{\perp}$ of a chain is a chain consisting of the projections of the form $P^{\perp} = I - P$ for all $P \in \pi$. If π is an eigenchain for an operator A, then the dual chain $\pi^{\perp}$ is an eigenchain for the adjoint operator A^*.

One can define a factorization of a bounded operator A on H along a chain π as follows:

$$A = A_+ D A_-$$

where A_+ and A_- are bounded operators having respectively π and $\pi^{\perp}$ as eigenchains and D commutes with all $P \in \pi$. However, such a factorization is not unique even if one exists. In fact, if B and C are any two bounded invertible operators which commute with all the projections $P \in \pi$, then define $A'_+ := A_+ B$, $A'_- := CA_-$, and $D' = B^{-1} D C^{-1}$ so that $A = A'_+ D' A'_-$. Hence, we are led to the following notion:

Definition 9.3.2 *A* **special factorization** *of a bounded operator A along a chain π is a representation of A given by*

$$A = (I + X_+) D (I + X_-) \tag{9.3.1}$$

where X_+ and X_- are Volterra operators having respectively π and $\pi^{\perp}$ as eigenchains, D commutes with all $P \in \pi$, and $D - I$ is compact.

Since X_+ and X_- are Volterra operators, the factors $(I + X_+)$ and $(I + X_-)$ are invertible. If A is invertible, then D is also invertible. If a bounded invertible operator A admits a special factorization relative to a maximal chain π, then the factorization is unique (see [28], pp. 158–159). If a bounded, self-adjoint invertible operator A admits such a factorization, then it is clear that $X_+^* = X_-$ and $D^* = D$.

If an operator A admits special factorization, then $I - A$ is compact since it can be written as

$$A - I = X_+ + X_- + X_+ X_- + (I + X_+)(D - I)(I + X_-)$$

and X_+, X_-, and $D - I$ are all compact by Theorem 9.1.5. Moreover, if A is invertible, then $I - A^{-1}$ is also a compact operator. Indeed, if

$$I - A = S \quad \text{where } S \text{ is compact}$$

and A^{-1} exists as a bounded operator, then define the operator

$$T := I - A^{-1} = I - (I - S)^{-1}.$$

Therefore, $T = ST - S$, a compact operator. So, in the factorization theorems, we will take A to be of the form $(I - T)^{-1}$ where T is compact. Such a form is a necessary condition for A to have a special factorization.

We need the following notion of integrals along closed chains to state the main result. Let π be a closed chain. A partition β of π is a chain consisting of a finite number of elements $\{P_0 < P_1 < \cdots < P_n\}$ of π such that $P_0 = \min_{P\in\pi} P$ and $P_n = \max_{P\in\pi} P$. Let $F : \pi \to L(H)$ be an operator-valued function. For a partition β as above, define

$$S(\beta) = \sum_{j=1}^{n} F(P_{j-1})\Delta P_j \text{ where } \Delta P_j = P_j - P_{j-1}.$$

An operator A is called the limit in norm of $S(\beta)$, denoted by

$$A = (m)\int_{\pi} F(P)\,dP, \tag{9.3.2}$$

if for any $\epsilon > 0$, there exists a partition $\beta(\epsilon)$ of π such that, for every partition $\beta \supset \beta(\epsilon)$, $\|S(\beta) - A\| < \epsilon$. If the limit of $S(\beta)$ exists, we shall say that the integral (9.3.2) converges. The integral

$$B = (m)\int_{\pi} F(P)\,dP$$

is defined analogously.

The theorem of Gohberg and Krein (see [28] Chapter 4), in the form that we need is stated below without proof. We follow the presentation in [39] in the rest of this section.

Theorem 9.3.1 *Let π be a maximal chain. Then, for every operator $T \in L_2$ such that each of the operators $I - PTP$ and $P \in \pi$, is invertible, the integrals*

$$X_+ = (m)\int_{\pi} (I - PTP)^{-1}PT\,dP \tag{9.3.3}$$

and

$$X_- = (m)\int_{\pi} dP\,TP(I - PTP)^{-1} \tag{9.3.4}$$

converge in norm. The operator $A = (I - T)^{-1}$ has a special factorization (9.3.1) along π with X_+, X_-, $D - I \in L_2$, and

$$D = I + \sum_j \left(P_j^+ - P_j^-\right)\left[\left(I - P_j^+ T P_j^+\right)^{-1} - I\right]\left(P_j^+ - P_j^-\right), \tag{9.3.5}$$

where $\{(P_j^+, P_j^-)\}$ is the set of all gaps in the chain π.

We use Theorem 9.3.1 along with the following lemmas to deduce a useful result.

Lemma 9.3.2 *Let $T \in L_2$. If $I - T$ is self-adjoint, positive, and invertible, then for any projection P, $I - PTP$ is invertible.*

Proof Consider for any $f \in H$,

$$\begin{aligned}((I - PTP)f, f) &= ((I - P)f, f) + ((P^2 - PTP)f, f)\\ &= ((I - P)^2 f, f) + ((I - T)Pf, Pf)\\ &= \|(I - P)f\|^2 + \|(I - T)^{1/2}Pf\|^2\\ &\geq \|(I - P)f\|^2 + c^2\|Pf\|^2\end{aligned}$$

where $c = \|I - T\|$. Thus,

$$((I - PTP)f, f) \geq c_1^2\|f\|^2$$

with $c_1 = \min\{1, c\}$. Thus, $I - PTP$ is invertible. ∎

Lemma 9.3.3 *If a self-adjoint, positive, invertible operator A has a special factorization 9.3.1 along a maximal chain, then the factor D is self-adjoint, positive, and invertible.*

Proof From the hypotheses, it is clear that D is self-adjoint and invertible. To prove positive definiteness of D, consider for any $f \in H$,

$$(Df, f) = ((I + X_+)^{-1}A(I + X_-)^{-1}f, f)$$

Denoting $(I + X_-)^{-1}f$ as g,

$$\begin{aligned}(Df, f) &= ((I + X_+)^{-1}Ag, (I + X_-)g)\\ &= ((I + X_-)^*(I + X_+)^{-1}Ag, g)\\ &= ((I + X_+)(I + X_+)^{-1}Ag, g)\\ &= (Ag, g).\end{aligned}$$

∎

Lemma 9.3.4 *If V is a Volterra operator and in L_2 with π as an eigenchain, then the operator $W = (I + V)^{-1} - I$ is also a Volterra operator in L_2 and has π as an eigenchain.*

Proof The statement $W \in L_2$ follows immediately since

$$W + V + VW = 0 \tag{9.3.6}$$

Since $W + V + WV$ is also equal to zero, V and W commute. Hence, we have (cf. [63], p. 426)

$$r_{V+VW} \leq r_V r_{I+W}$$

where r_V, for instance, denotes the spectral radius of V, that is, the radius of the smallest closed disk centered at 0 which contains the spectrum of V. By assumption, $r_V = 0$. Hence, $r_{V+VW} = 0$, which implies that $V + VW$ is Volterra. From Equation (9.3.6), W is Volterra.

By the definition of W, $I + W$ is the resolvent at 1 of the operator $-V$, so that

$$W = \sum_{j=1}^{\infty} (-1)^j V^j.$$

The Neumann sum on the right side converges in norm. Since π is an eigenchain of V, it easy to verify that $PV^jP = V^jP$ for any $P \in \pi$ and for any j. Hence, $PWP = WP$. ■

From Theorem 9.3.1 and Lemmas 9.3.2 to 9.3.4, we have the following theorem.

Theorem 9.3.5 *Let S be a self-adjoint, positive, invertible operator such that $S = I - T$ and $T \in L_2$. Then S and S^{-1} have the following factorizations along any maximal chain $\pi = \{P\}$:*

$$S = (I + W_-)D^{-1}(I + W_+) \quad and \quad S^{-1} = (I + X_+)D(I + X_-),$$

where

(i) *W_+, W_-, X_+, X_- are Volterra operators in L_2. X_+ and X_- are given by (9.3.3) and (9.3.4), the integral converging in norm, and*

$$I + W_+ = (I + X_+)^{-1} \text{ and } I + W_- = (I + X_-)^{-1}.$$

(ii) *W_+, X_+ have π as an eigenchain whereas W_-, X_- have $\pi^\perp$ as an eigenchain.*

(iii) *$W_+^* = W_-$ and $X_+^* = X_-$.*

(iv) *D is a self-adjoint, positive, invertible operator given by (9.3.5).*

(v) *$D - I \in L_2$ and $DP = PD$ for all $P \in \pi$.*

9.4 Nonanticipative Representations

Consider a mean-square continuous Gaussian process $\{X_t : t \in [0, 1]\}$ on a complete probability space $(\Omega, \mathcal{F}, Q)$. For all $s, t \in [0, 1]$, the covariance between X_s and X_t, will be denoted by $C_Q(s, t)$.

For any $t \in [0, 1]$, let $L(X; t)$ be the closed linear subspace spanned by $\{X_s : 0 \leq s \leq t\}$ in $L^2(\Omega, \mathcal{F}, Q)$. Let π_t denote the orthogonal projection of $L(X; 1)$ to $L(X; t)$. We cannot claim that π is a maximal chain since there could be gaps in the chain. If there is a gap between $L(X; t+) := \cap_{s>t} L(X; s)$ and $L(X; t)$, we can fill the gap by adding a finite number of projections to the chain π and form a maximal chain. The details are left as an exercise. For notational simplicity, we will denote such a maximal chain by π as well.

Let $\mathcal{F}^{X,0}$ denote the σ-field generated by $\{X_t : t \in [0, 1]\}$. Suppose that P is another probability measure on $(\Omega, \mathcal{F}^{X,0})$ such that, on the probability space $(\Omega, \mathcal{F}^{X,0}, P)$, the process $\{X_t : t \in [0, 1]\}$ is a centered Gaussian process which is continuous in the mean-square sense with covariance function $C_P(s, t)$.

In what follows, we will use the notation $\{X_t, P\}$ to denote the Gaussian process $\{X_t : t \in [0, 1]\}$ defined on the space $(\Omega, \mathcal{F}^{X,0}, P)$, with a similar meaning for $\{X_t, Q\}$.

Suppose that P and Q are equivalent in the sense that they are mutually absolutely continuous measures on $\mathcal{F}^{X,0}$.

Definition 9.4.1 *The process $\{X_t, P\}$ has a* **nonanticipative representation** *with respect to $\{X_t, Q\}$ if on $(\Omega, \mathcal{F}^{X,0})$ there exists a centered Gaussian process $\{Y_t, Q\}$ with covariance function C_P that satisfies the following property:*

$$Y_t \in L(X; t) \quad \textit{for each} \quad t \in [0, 1]. \tag{9.4.1}$$

Next, we briefly recall the definition of a reproducing kernel Hilbert space.

Definition 9.4.2 *Let C be a given symmetric, nonnegative kernel on $[0,1] \times [0,1]$. A Hilbert space $K(C)$ of functions f defined on $[0,1]$ is called the* **reproducing kernel Hilbert space** *(RKHS, for short) for C if the following properties are satisfied:*

(i) The function $C(\cdot, t) \in K(C)$, and

(ii) $f(t) = (f, C(\cdot, t))_{K(C)}$ for each $t \in [0, 1]$.

It is well known that there exists a unique RKHS $K(C)$ for a symmetric, nonnegative kernel C. The second requirement in the above definition is known as the *reproducing* property. Instead of writing $K(C_P)$ and $K(C_Q)$, we will simply write K_P and K_Q, respectively. Further, we will use the notation $(\cdot, \cdot)$ for the inner product in $L^2[0,1]$ and $\langle \cdot, \cdot \rangle$ for the inner product in K_Q.

As an example, consider a standard one-dimensional Wiener process $\{X_t, Q\}$. We have

$$\begin{aligned} C_Q(s,t) &= E(X_s X_t) = \min\{s,t\} \\ &= \int_0^1 1_{[0,s]}(u) 1_{[0,t]}(u)\, du. \end{aligned}$$

Thus, $\langle C_Q(\cdot, s), C_Q(\cdot, t) \rangle = \int_0^1 1_{[0,s]}(u) 1_{[0,t]}(u)\, du$, and an isometric isomorphism from K_Q onto $L^2[0,1]$ is thus defined by mapping $C_Q(\cdot, t)$ to $1_{[0,t]}$. In addition, any element $f \in K_Q$ can be represented in the form

$$f(t) = \int_0^t g(u)\, du$$

with $g \in L^2[0,1]$. Thus,

$$K_Q = \left\{ f \mid f(t) = \int_0^t g(s)\, ds, \text{ and } g \in L^2[0,1] \right\}$$

equipped with the inner product

$$\langle f_1, f_2 \rangle = (g_1, g_2).$$

The identification of X_t with $C_Q(\cdot, t)$ can be extended to an isometric isomorphism from $L(X; 1)$ to K_Q. Thus, the spaces K_Q, $L(X; 1)$ and $L^2(0,1)$ are isometrically isomorphic to each other.

In general, if X is a mean-square continuous Gaussian process, the identification of X_t with $C_Q(\cdot, t)$ as above leads to an isometric isomorphism ϕ from $L(X; 1)$ to K_Q. The following relation between $L(X; 1)$ and K_Q is important for us. Define the subspace

$$F(t) = \{f : f(s) = 0, 0 \leq s \leq t\}$$

of K_Q. Let $M(t) = K_Q/F(t)$. Then one can show that $\phi(L(X; t)) = M(t)$. Let $\psi(t)$ denote the orthogonal projection of K_Q to $M(t)$ for all t. Then the chain γ defined by

$$\gamma = \phi\pi\phi^{-1}$$

is a maximal chain in K_Q and contains the chain $\psi(t)$.

The next theorem gives necessary and sufficient conditions for the equivalence of two Gaussian measures [40].

Theorem 9.4.1 *Two Gaussian measures P and Q are equivalent if and only if the covariance C_P defines an operator S on the RKHS $K(Q)$ with the following properties:*

(i) $C_P(\cdot, t) = SC_Q(\cdot, t)$ *for* $0 \leq t \leq 1$.

(ii) *S is a bounded, self-adjoint, positive operator.*

(iii) $S = I - T$ *and* $T \in L_2$.

(iv) $1 \notin \sigma(T)$, *the spectrum of* T.

We will now prove the following theorem.

Theorem 9.4.2 *Every mean-square continuous Gaussian process $\{X_t, P\}$ that is equivalent to a mean-square continuous Gaussian process $\{X_t, Q\}$ has a nonanticipative representation with respect to $\{X_t, Q\}$.*

Proof Consider the maximal chain γ_t for the RKHS K_Q. Applying Theorem 9.3.5 to the operator S defined in Theorem 9.4.1, we obtain

$$S = (I + W_-)\Lambda(I + W_+)$$

where $\Lambda = D^{-1}$. Since D is self-adjoint and positive, so is Λ. Also, Λ has a square root, $\Lambda^{1/2}$, which is self-adjoint and positive. By its construction, $\Lambda^{1/2}$ commutes with P since Λ does.

Define the operator

$$F = \Lambda^{1/2}(I + W_+).$$

Then we have $S = F^*F$. Consider the operator $\tilde{F}$ on $L(X; 1)$ corresponding to F:

$$\tilde{F} = \tilde{\Lambda}^{1/2}(I + \tilde{W}_+).$$

Define $Y_t = \tilde{F}X_t$ for all $t \in [0, 1]$. Since $\tilde{F}$ is a *linear* operator on $L(X; 1)$, $\{Y_t, Q\}$ is a Gaussian process. Furthermore, $E_Q[Y_t] = 0$ and

$$\begin{aligned} E_Q Y_t Y_s &= (\tilde{F}X_s, \tilde{F}X_t) \\ &= \langle FC_Q(\cdot, s), FC_Q(\cdot, t)\rangle \\ &= \langle SC_Q(\cdot, s), C_Q(\cdot, t)\rangle \\ &= \langle C_P(\cdot, s), C_Q(\cdot, t)\rangle \\ &= C_P(s, t) \end{aligned}$$

by the reproducing property of the RKHS K_Q. Thus, $\{Y_t, Q\}$ is a centered Gaussian process with covariance function $C_P(s, t)$. It remains to show that $Y_t \in L(X; t)$. Consider, for each $t \in [0, 1]$,

$$\begin{aligned} Y_t &= \tilde{\Lambda}^{1/2}(I + \tilde{W}_+)X_t \\ &= \tilde{\Lambda}^{1/2}(I + \tilde{W}_+)\pi_t X_t \\ &= \tilde{\Lambda}^{1/2}\pi_t(I + \tilde{W}_+)\pi_t X_t \quad (\pi \text{ is an eigenchain for } \tilde{W}_+) \\ &= \pi_t \tilde{\Lambda}^{1/2}(I + \tilde{W}_+)\pi_t X_t \quad (\tilde{\Lambda}^{1/2} \text{ commutes with } \pi_t) \\ &= \pi_t Y_t. \end{aligned}$$

Thus, $Y_t \in L(X; t)$; and the proof is over. ■

Let $\{X, Q\}$ be a Wiener process. A natural isometric isomorphism α can be defined from $L(X; 1)$ onto $L^2[0, 1]$ such that $\alpha X_t = 1_{[0,t]}$, and

$$\alpha\{L(X; t)\} = \{h \in L^2[0, 1] : h(s) = 0 \text{ a.e. for } t < s \leq 1\}.$$

The chain $\rho = \{P_t : P_t = \alpha\pi_t\alpha^{-1},\ t \in [0, 1]\}$ is maximal in $L^2[0, 1]$, and $P_t \in \rho$ is such that

$$P_t h(s) = h(s)1_{[0,t]}(s). \tag{9.4.2}$$

Proposition 9.4.3 *Let $g(s, t)$ be the kernel in $L^2([0, 1] \times [0, 1])$ for a Hilbert-Schmidt operator G on $L^2[0, 1]$. If the chain ρ is an eigenchain for G, then*

$$g(u, v) = 0 \ \text{a.e. for} \ u > v.$$

Proof From the definition of an eigenchain for G, we have, for each $t \in [0, 1]$,

$$GP_t h = P_t G P_t h.$$

From Equation (9.4.2), we know that $\int_0^t g(u, v)h(v)\,dv = 0$ a.e. for $u > t$. Hence,

$$\int_0^1 \int_0^1 g(u, v)(1 - 1_{[0,t]}(u))1_{[0,t]}(v)h_1(u)h_2(v)\,du\,dv = 0$$

for all $t \in [0,1]$ and $h_1, h_2 \in L^2[0,1]$. Take

$$h_1 = 1_{[a,b]}, \quad h_2 = 1_{[c,d]}$$

with either $b \leq c$ or $a \leq d$. Then

$$\int_0^1 \int_0^1 g(u,v)(1 - 1_{[0,u]}(v))1_{[0,t]}(v)h_1(u)h_2(v)du\, dv = 0.$$

Since the family of functions of $(u,v) \in [0,1] \times [0,1]$ given by $1_{[a,b]}(u)1_{[c,d]}(v)$ with $b \leq c$ or $d \leq a$ is dense in $L^2([0,1] \times [0,1])$, we can conclude that

$$g(u,v)(1 - 1_{[0,u]}(v)) = 0 \text{ a.e. for } u,v \in [0,1].$$

Thus, $g(u,v) = 0$ a.e. for $u > v$. ■

Since there are no gaps in the chain π, $D = I$ so that $\Lambda^{1/2} = I$. Therefore, $\alpha Y_t = \alpha(I + \tilde{W}_+)X_t$. Since the spaces $L(X;1)$ and $L^2[0,1]$ are isometrically isomorphic, there exists an operator $\hat{W}_+$ on $L^2[0,1]$ that corresponds to $\tilde{W}_+$ so that

$$\alpha \tilde{W}_+ X_t = \hat{W}_+ 1_{[0,t]}.$$

We thus obtain

$$\alpha Y_t = (I + \hat{W}_+)1_{[0,t]}.$$

Applying Proposition 9.4.3 to the Hilbert-Schmidt operator $\hat{W}_+$, it follows that

$$\begin{aligned}(I + \hat{W}_+)1_{[0,t]} &= 1_{[0,t]} + \int_0^1 \hat{W}_+(\cdot, v)1_{[0,t]}(v)\, dv \\ &= 1_{[0,t]} + \int_0^t \hat{W}_+(\cdot, v)\, dv\end{aligned}$$

where $\hat{W}_+(u,v)$ denotes the kernel in $L^2([0,1] \times [0,1])$ that corresponds to the operator $\hat{W}_+$. Thus, for all $u \in L^2[0,1]$, we have

$$(\alpha Y_t)(u) = 1_{[0,t]}(u) + \int_0^t \hat{W}_+(u,v)\, dv. \tag{9.4.3}$$

It is worthwhile to recall that the isometric isomorphism from K_Q to $L^2[0,1]$ maps $f \to g$ such that

$$f(t) = \int_0^t g(u)\, du,$$

and the isometric isomorphism α^{-1} from $L^2[0,1]$ to $L(X;1)$ maps

$$g \to \int_0^1 g(u)\,dX_u.$$

Therefore, Equation (9.4.3) yields the nonanticipative representation

$$\begin{aligned} Y_t &= \int_0^1 1_{[0,t]}(u)\,dX_u + \int_0^1 \int_0^t \hat{W}_+(u,v)\,dv\,dX_u \\ &= X_t + \int_0^t \left\{ \int_0^v \hat{W}_+(u,v)dX_u \right\} dv \end{aligned} \tag{9.4.4}$$

since $\hat{W}_+(u,v) = 0$ a.e. for $u > v$.

9.5 Gaussian Solutions of Stochastic Equations

Suppose that we are given a complete probability space $(\Omega, \mathcal{F}, P)$ with a filtration $\{\mathcal{F}_t : 0 \le t \le T\}$. On it, let $\{W_t\}$ and $\{Z_t\}$ be two $\mathcal{F}_t$-adapted processes with the following properties:

P-1 $\{W_t, \mathcal{F}_t\}$ is a Wiener martingale.

P-2 For all $t \in [0,T]$,

$$Z_t = \int_0^t b(s,Z)\,ds + W_t P \text{ a.s.} \tag{9.5.1}$$

where $b : [0,T] \times C[0,T] \to \mathbb{R}$ is a nonanticipative functional.

P-3 $P\{\omega \in \Omega : \int_0^T b^2(s, Z(\omega))\,ds < \infty\} = 1$.

P-4 The probability distribution $\mu_Z = PZ^{-1}$ on $(C, \mathcal{B}_T(C))$ is Gaussian.

Equation (9.5.1) is a functional stochastic differential equation written as an integral equation. Using the results obtained in the last two sections and the Gohberg-Krein factorization (Theorem 9.3.1), we will produce an *explicit* representation for the process Z in terms of W. The process Z is a weak solution of (9.5.1). However, we will show that Z is the unique strong solution of (9.5.1). Several other interesting properties of Z follow from the results of this section. The interplay between stochastic and functional analysis in the context of Gaussian processes is highlighted by this approach.

Theorem 9.5.1 *Suppose that the four properties listed above are satisfied by the processes W and Z. Then*

(i) *Z can be expressed in terms of W by the formula*

$$Z_t = W_t + \int_0^t \left[\int_0^v g(u,v)\, dW_u \right] dv$$

for all t, P-a.s., where G is a square integrable Volterra kernel determined by Theorem 9.3.1.

(ii) *For each t,*

$$\mathcal{F}_t^Z = \mathcal{F}_t^W.$$

Proof

Step 1 Let μ denote the distribution of W. Condition P-3 and the Girsanov theorem imply that $\mu_Z << \mu$. Since both μ and μ_Z are Gaussian measures, they are either equivalent or mutually singular. Therefore, we can conclude that $\mu_Z \equiv \mu$. Condition P-3 can now be written as

$$\mu\{x \in C : \int_0^T b^2(s,x)\, ds < \infty\} = 1. \tag{9.5.2}$$

From Equation (9.5.1), we see that W_t is $\mathcal{F}_t^Z$-adapted so that

$$\mathcal{F}_t^W \subseteq \mathcal{F}_t^Z \subseteq \mathcal{F}_t.$$

Consider the Gaussian process Z on the probability space $(\Omega, \mathcal{F}_T^Z, P)$. W is a standard Wiener process on the same space. We can assume without loss of generality that $EZ_t = 0$ for all t. By what was proved in the previous section, we obtain a nonanticipative representation Y for the process Z with respect to W which reads as follows:

$$Y_t = W_t + \int_0^t \left[\int_0^v g(u,v)\, dW_u \right] dv. \tag{9.5.3}$$

To prove part (i), we need to show that $Y = Z$.

Step 2 In this step, we will prove that W can be expressed in terms of Y. This would amount to an inversion of Equation (9.5.3).

If we denote the covariance function of Z by R, then R determines an operator S by the equivalence of the measures μ_Z and μ. Though S is an operator on the RKHS for the Wiener process, we know that an operator corresponding to S can be determined on $L^2[0,T]$. Since the latter is more useful to us throughout the proof, we will, by a slight abuse in notation, denote the operator on $L^2[0,T]$ as S.

The special factorization allows us to write

$$S = (I + G^*)(I + G) \quad \text{and} \quad S^{-1} = (I + K)(I + K^*),$$

where G and K are Volterra operators on $L^2[0,1]$. The kernels corresponding to G and K are denoted by $g(u,v)$ and $k(u,v)$. Note that both $g(u,v)$ and $k(u,v)$ are zero almost everywhere when $u > v$. Since, for any $f \in L^2[0,1]$,

$$(Kf)(s) = \int_0^1 k(s,t)f(t)\,dt = \int_s^1 k(s,t)f(t)\,dt$$

so that $(KGf)(s) = \int_0^1 \{\int_s^t k(s,u)g(u,t)\,du\} f(t)\,dt$.

Since $I + G = (I+K)^{-1}$, we have the relations

$$g(s,t) + k(s,t) + \int_s^t k(s,u)\,g(u,t)\,du = 0 \tag{9.5.4}$$

$$g(s,t) + k(s,t) + \int_s^t g(s,u)\,k(u,t)\,du = 0 \tag{9.5.5}$$

Define $\phi(t) = \int_0^t G(u,t)\,dW_u$. From Equation (9.5.3) we can obtain

$$dY_t = dW_t + \phi(t)\,dt.$$

Using this we obtain

$$\begin{aligned}
&Y_t + \int_0^t \left[\int_0^v k(u,v)\,dY_u\right] dv \\
&= W_t + \int_0^t \left[\int_0^v g(u,v)\,dW_u\right] dv + \int_0^t \left[\int_0^v k(u,v)\,dW_u + \int_0^v g(u,v)\phi(u)\,du\right] dv \\
&= W_t + \int_0^t \left[\int_0^v g(u,v)\,dW_u\right] dv + \int_0^t \left[\int_0^v k(u,v)\,dW_u\right] dv \\
&\quad + \int_0^t \left[\int_0^v k(u,v)\left(\int_0^u g(s,u)\ dW_s\right) du\right] dv \\
&= W_t + \int_0^t \left[\int_0^v g(u,v)\,dW_u\right] dv + \int_0^t \left[\int_0^v \{k(s,v) + \int_s^v k(u,v)\,g(s,u)\,du\}\,dW_s\right] dv
\end{aligned} \tag{9.5.6}$$

since

$$\int_0^v \int_0^v k(u,v)\,g(s,u)\,dW_s\,du = \int_0^v \int_0^v k(u,v)\,g(s,u)\,du\,dW_s.$$

Using Equation (9.5.4) in (9.5.6), we obtain that $Y_t + \int_0^t \left[\int_0^v k(u,v)\,dY_u\right] dv$ equals

$$W_t + \int_0^t \left[\int_0^v g(u,v)\,dW_u\right] dv - \int_0^t \left[\int_0^v g(u,v)\,dW_u\right] dv = W_t.$$

Thus,

$$W_t = Y_t + \int_0^t \left[\int_0^v k(u, v)\, dY_u \right] dv. \tag{9.5.7}$$

Though the aim of this step is complete, let us observe that Equations (9.5.3) and (9.5.7) imply that, for each t,

$$\mathcal{F}_t^Y = \mathcal{F}_t^W. \tag{9.5.8}$$

Step 3 Setting $\beta(t, Y) = -\int_0^t k(u, t)\, dY_u$, we can write (9.5.7) as

$$Y_t = W_t + \int_0^t \beta(s, Y) ds. \tag{9.5.9}$$

If we are able to show that

$$P\left\{ \int_0^t b(s, Z)\, ds = \int_0^t \beta(s, Z)\, ds \;\; \forall \;\; t \right\} = 1, \tag{9.5.10}$$

then from property P-2, we can write

$$Z_t = \int_0^t \beta(s, Z)\, ds + W_t.$$

Hence, we have

$$W_t = Z_t + \int_0^t \left\{ \int_0^s k(u, s)\, dZ_u \right\} ds. \tag{9.5.11}$$

The above can be inverted by following the procedure used in arriving at (9.5.7) from (9.5.3). We obtain

$$Z_t = W_t + \int_0^t \left\{ \int_0^s g(u, s)\, dW_u \right\} ds. \tag{9.5.12}$$

Comparing (9.5.3) and (9.5.12), it follows that, P-a.s.,

$$Y_t = Z_t \;\; \forall \;\; t.$$

From the above equality and (9.5.9), we can conclude that

$$\mathcal{F}_t^Z = \mathcal{F}_t^W$$

for all t. Thus, all that remains to be proved is Equation (9.5.10):

$$P\left\{ \int_0^t b(s, Z)\, ds = \int_0^t \beta(s, Z)\, ds \;\; \forall t \right\} = 1.$$

Step 4 Consider

$$E_\mu\left[\int_0^T \beta^2(t,x)\,dt\right] = E\left[\int_0^T \beta^2(t,W)\,dt\right]$$
$$= \int_0^T \int_0^t k^2(u,t)\,du\,dt < \infty. \tag{9.5.13}$$

Therefore, we have

$$\mu\left\{x \in C : \int_0^T \beta(t,x)\,dt < \infty\right\} = 1. \tag{9.5.14}$$

Now, we make a basic observation that the distributions of Z and its nonanticipative representation Y are the same. That is, $\mu_Y = \mu_Z$. We already know that $\mu_Z \equiv \mu$ so that $\mu_Y \equiv \mu$. From this and (9.5.14) we obtain

$$P\left\{\omega : \int_0^T \beta^2(t,Y(\omega))\,dt < \infty\right\} = 1. \tag{9.5.15}$$

From the Girsanov theorem, the Radon-Nikodym derivative $\frac{d\mu_Z}{d\mu}$ relative to $\mathcal{B}_T$ is given by

$$\exp\left\{\int_0^T b(s,x)\,dx_s - \frac{1}{2}\int_0^T b^2(s,x)\,ds\right\}.$$

Since $\mu_Y = \mu_Z$, it follows from (9.5.9) that $\frac{d\mu_Z}{d\mu}$ relative to $\mathcal{B}_T$ is also given by

$$\exp\left\{\int_0^T \beta(s,x)\,dx_s - \frac{1}{2}\int_0^T \beta^2(s,x)\,ds\right\}.$$

For each t, if we take the conditional expectation $E(\frac{d\mu_Z}{d\mu} \mid \mathcal{B}_{t+})$, we obtain that, μ a.s.,

$$\exp\left\{\int_0^t b(s,x)\,dx_s - \frac{1}{2}\int_0^t b^2(s,x)\,ds\right\} = \exp\left\{\int_0^t \beta(s,x)\,dx_s - \frac{1}{2}\int_0^t \beta^2(s,x)\,ds\right\} \tag{9.5.16}$$

by the uniqueness of the Radon-Nikodym derivative. From (9.5.16), it follows that, μ a.s.,

$$\int_0^t [b(s,x) - \beta(s,x)]\,dx_s = \frac{1}{2}\int_0^t [b^2(s,x) - \beta^2(s,x)]\,ds.$$

The left side is a continuous martingale, whereas the right side is a process with paths of finite variation. Hence, for all t,

$$\int_0^t b(s,x)\,dx_s = \int_0^t \beta(s,x)\,dx_s \tag{9.5.17}$$

$$\int_0^t b^2(s,x) = \int_0^t \beta^2(s,x)\,ds \tag{9.5.18}$$

μ a.s. Equations (9.5.18) and (9.5.13) yield

$$\int_0^T E_\mu\left(b^2(t,x)\right)\,dt < \infty. \tag{9.5.19}$$

Now, from Equations (9.5.17) and (9.5.18), we obtain

$$\int_0^T E_\mu[b(s,x) - \beta(s,x)]^2\,ds = E_\mu\left(\left(\int_0^T [b(s,x) - \beta(s,x)]\,dx_s\right)^2\right) = 0. \tag{9.5.20}$$

Therefore, if we define the set

$$\Gamma = \{(s,x) \in [0,T] \times C : b(s,x) \neq \beta(s,x)\},$$

then $\lambda \times \mu(\Gamma) = 0$, where λ is the Lebesgue measure. Since $\mu_Z \equiv \mu$, we have, for all $s \in [0,T]$,

$$P\{\omega \in \Omega : b(s,Z(\omega)) = \beta(s,Z(\omega))\} = 1. \tag{9.5.21}$$

From Equation (9.5.20), we can infer that

$$E_\mu\left[\int_0^t \beta(s,x)\,ds - \int_0^t b(s,x)\,ds\right]^2 \leq T\int_0^T E_\mu[b(s,x) - \beta(s,x)]^2\,ds = 0.$$

Again, by using $\mu_Z \equiv \mu$, we can conclude that, for all t,

$$\int_0^t \beta(s,x)\,ds = \int_0^t b(s,x)\,ds \quad \mu_Z\text{-a.s.} \tag{9.5.22}$$

In other words, Equation (9.5.10) follows, and the proof is over. ■

Theorem 9.5.2 *Let b be a nonanticipative functional such that*

$$\mu\{x \in C : \int_0^T b^2(t,x)\,dx < \infty\} = 1. \tag{9.5.23}$$

(i) Then the stochastic equation

$$dZ_t = b(t,Z)\,dt + dW_t \tag{9.5.24}$$

has a weak solution that is Gaussian if and only if b is of the form

$$b(t,x) = \int_0^t g(u,t)\, dx_u \tag{9.5.25}$$

for almost every (t,x), with respect to the product measure $\lambda \times \mu$, where $g(u,t)$ belongs to $L^2([0,T] \times [0,T])$ and $g(u,t) = 0$ a.e. for $u \geq t$.

(ii) If a Gaussian solution exists, then a unique Gaussian strong solution exists.

Proof

Step 1 Suppose that b is given by (9.5.25). Consider the canonical Wiener process x_t on the Wiener space $(C, \mathcal{B}_T, \mu)$. Let $\tilde{G}$ be the Volterra operator determined by the kernel g on the linear space $L(x; T)$. Define the process y_t on $(C, \mathcal{B}_T, \mu)$ by

$$y_t(x) = (I + \tilde{G})x_t.$$

Let G denote the operator on $L^2[0,T]$ that corresponds to $\tilde{G}$. Under μ, y is Gaussian with zero mean function and covariance R, where the operator S determined by R on $L^2[0,T]$ is given by $S = (I + G^*)(I + G)$. Since G is a Volterra operator, S has all the properties required to ensure the equivalence of μ and the distribution of y, denoted by μ_y. Thus, μ_y is the required unique weak solution of Equation (9.5.24).

The existence of a Gaussian weak solution of (9.5.24) implies the required form of b. Indeed, this follows from the proof of Theorem 9.5.1. Thus, part (i) has been proved.

Step 2 To prove part (ii), let $(\Omega, \mathcal{F}_T, P)$ be a probability space, and let $(W_t, \mathcal{F}_t)$ be a Wiener martingale defined on it. If ν is the Gaussian weak solution of (9.5.24), then ν is equivalent to μ; ν is determined by its covariance function, which we denote by R. Without loss of generality, we can assume that its mean function is zero.

R defines a nonanticipative representation Y_t of ν with respect to μ on the given space $(\Omega, \mathcal{F}_T, P)$. It is given by (9.5.9) where $\beta(t, Y) = -\int_0^t k(u,t)\, dY_u$. By (9.5.8) and (9.5.9), Y_t can be taken to be jointly measurable in (t, ω), and adapted to $\mathcal{F}_t$. Hence Y is a strong solution of (9.5.24).

Finally, let X be any strong solution of (9.5.24) with $\mu_X = \nu$. By assumption (9.5.23), condition P-3 is satisfied. Therefore, as in the proof of Theorem 9.5.1, we obtain $X_t = Y_t$ for all t, P-a.s. Note that the nonanticipative representation Y_t is unique since the Gohberg-Krein factorization of S is unique. Therefore, the strong solution X is unique. ∎

Example 9.5.3 *On the time interval $[0,T]$, a process $X := \{X_t\}$ is called a Brownian bridge if it is a Gaussian process with mean zero and covariance $E(X_t X_s) = (t \wedge s) - \frac{ts}{T}$. We will obtain a stochastic differential equation for X.*

If W is a Wiener process, then $X_t = W_t - \frac{t}{T} W_T$ for $0 \leq t \leq T$ is a Brownian bridge. The process depends on W_T at all times $t > 0$. To obtain a nonanticipative representation *of X, we need to find a kernel* $g(t,u)$ *satisfying*

$$\int_0^{t\wedge s} g(t,u)\,g(s,u)\,du = t\wedge s - \frac{ts}{T}.$$

Suppose $g(t,u)$ *is of the form* $f(t)\,g(u)$. *The above equation becomes*

$$f(t)f(s)\int_0^{t\wedge s} g^2(u)\,du = t\wedge s - \frac{ts}{T}. \tag{9.5.26}$$

Let us assume that $f(0)\,g^2(0)$ *is nonzero and denote it as* c^{-1}. *With* $s < t$ *in* (9.5.26), *differentiate it with respect to s to obtain*

$$f(t)f'(s)\int_0^{s} g^2(u)\,du + f(t)f(s)\,g^2(s) = \frac{T-t}{T}.$$

For $s = 0$, *the above equation becomes* $f(t) = c(T-t)/T$. *Using it in* (9.5.26), *we have*

$$c^2\frac{(T-t)(T-s)}{T^2}\int_0^{t\wedge s} g^2(u)\,du = t\wedge s - \frac{ts}{T},$$

which reduces to

$$c^2\frac{(T-s)}{T}\int_0^{s} g^2(u)\,du = s$$

when $s < t < T$. *Hence,* $g(s) = \frac{T}{c(T-s)}$, *and we can write*

$$g(t,s) = c\left(\frac{T-t}{T}\right)\frac{T}{c(T-s)} = \frac{T-t}{T-s}.$$

Hence,

$$Y_t = \int_0^t \frac{T-t}{T-s}\,dW_s \quad \textit{if}\ t < T,$$

and $Y_T := 0$ *is the nonanticipative representation of the Brownian bridge. As is easily seen,* Y_t *solves*

$$dY_t = dW_t - \frac{Y_t}{T-t}\,dt.$$

Exercises

1. Prove Theorem 9.2.1.
2. Show that the space $L_2(H)$ of Hilbert-Schmidt operators on H, equipped with the inner product

$$\langle T, S\rangle = \sum_n \langle Te_n, Se_n\rangle,$$

is a Hilbert space.
3. Show that if A is a bounded operator on H and B, a trace class operator, then AB is a trace class operator, and

$$\|AB\|_1 \leq \|A\| \, \|B\|_1.$$

Likewise, BA is in $L_1(H)$ and $\|BA\|_1 \leq \|B\|_1 \|A\|$.

10 Jump Markov Processes

Poisson processes form one of the earliest examples of jump processes, and form the building blocks for more general jump Markov processes. Jump processes arise in a variety of disciplines such as biology, epidemiology, queueing theory, and mathematical finance. Stochastic analysis for cadlag semimartingales (which are not necessarily Markov processes) is well developed and actively studied. The lectures of P. A. Meyer [54], and the books by Metivier [52] and Protter [60] offer excellent accounts of the theory.

Markov jump processes form a large and useful subclass of cadlag semimartingales. For instance, birth and death processes, and more generally, Lévy processes, that is, processes with independent increments, are Markov jump processes. The book by Itô [33] provides an impressive and complete description of Lévy processes. Since Markov jump processes possess more structure, one can study the weak convergence of a sequence of such processes as well as the martingale problems associated with the sequence and their limit (cf. [36] and [24]).

In this chapter, we present a simple introduction to jump processes, starting with Poisson and compound Poisson processes. The processes are shown to have the strong Markov property and their generators are identified. Stochastic integration and the Itô formula are derived for cadlag semimartingales. Sufficient conditions are given under which the processes do not explode. Finally, diffusion approximation is explained by constructing a sequence of jump processes.

10.1 Definitions and Basic Results

An integer-valued process $N := \{N_t\}$, defined for all $t \geq 0$, is known as a **Poisson process** if

(i) $N_0 = 0$ a.s.

(ii) The process N has independent increments; that is, for any n, the random variables $N_{t_j} - N_{s_j}$ for $j = 1, 2, \ldots, n$ are independent if $[s_j, t_j)$ are disjoint intervals.

(iii) There exists a number $\lambda > 0$ such that $N(t+s) - N(s)$ has a Poisson distribution with parameter λt for all positive t and $s \geq 0$.

The sample paths of N are right-continuous step functions whose jumps are of size one. Thus, Poisson processes are examples of counting processes and are useful in modeling the number of arrivals in a queue or the number of hits registered in a Geiger counter by a radioactive material. By the independence of increments, it follows that, for any $0 = t_0 < t_1 < \cdots < t_n$, and non negative integers $k_1 \leq k_2 \cdots \leq k_n$,

$$
\begin{aligned}
&P\{N_{t_n} = k_n \mid N_{t_j} = k_j,\ j = 1, 2, \ldots, n-1\} \\
&\quad = P\{N_{t_n} - N_{t_{n-1}} = k_n - k_{n-1} \mid N_{t_j} - N_{t_{j-1}} = k_j - k_{j-1}, j = 1, \ldots, n-1\} \\
&\quad = P\{N_{t_n} - N_{t_{n-1}} = k_n - k_{n-1}\},
\end{aligned}
$$

so that N has the Markov property. In fact, it is a strong Markov process. To show this, define the σ-fields $\mathcal{F}_t = \sigma(N_s : s \leq t)$ for all $t \geq 0$. Let τ be an $\mathcal{F}_t$-stopping time taking a countable number of values $\{t_j\}$. If $F \in \mathcal{F}_\tau$, then $F_j := F \cap \{\tau = t_j\} \in \mathcal{F}_{t_j}$. By independence of increments,

$$P[(N_{t+\tau} - N_\tau = k) \cap F_j] = P[(N_{t+t_j} - N_{t_j} = k)]\, P(F_j)$$

for each j. Adding over j,

$$P[(N_{t+\tau} - N_\tau = k) \cap F] = P[(N_{t+\tau} - N_\tau = k)]P(F) = e^{-\lambda t}\frac{(\lambda t)^k}{k!} P(F).$$

Thus, one can infer that $N(\cdot + \tau) - N(\tau)$ is a Poisson process (with parameter λ) which is independent of $\mathcal{F}_\tau$. Given any stopping time τ, note that $\tau = \lim_{j\to\infty} \tau_j$ where

$$\tau_j = \frac{[2^j \tau] + 1}{2^j}.$$

Each τ_j is a countably valued stopping time, and hence $N(\cdot + \tau_j) - N(\tau_j)$ is a Poisson process, independent of $\mathcal{F}_{\tau_j}$. Since $\tau_j \downarrow \tau$, $N(\cdot + \tau) - N(\tau)$ and the σ-field $\mathcal{F}_\tau$ are independent using the right continuity of paths of N.

From the strong Markov property of N, it is quite simple to prove that the waiting times between successive jumps for N are independent exponential random variables with parameter λ. These variables are known as inter-arrival times. We will denote them as $\tau_j, j \geq 1$. Consider

$$P[\tau_1 > t] = P[N_t = 0] = e^{-\lambda t}$$

so that τ_1 is an exponential random variable with parameter λ. With $N(\cdot + \tau_1) - N(\tau_1)$ being a Poisson process independent of τ_1, we have

$$P[\tau_2 > t] = P[N_{t+\tau_1} - N_{\tau_1} = 0] = e^{-\lambda t},$$

and τ_2 is independent of τ_1. The same argument can be used for each inter-arrival time. From the discussion, we have proved the following:

Proposition 10.1.1 *Let $N := \{N_t\}$ be a Poisson process with parameter $\lambda > 0$. Then N is a strong Markov process, and the inter-arrival times, denoted by $\{\tau_j\}$, are independent and identically distributed random variables, each exponentially distributed with parameter λ.*

It is quite clear that $M_t := N_t - \lambda t$ is an $\mathcal{F}_t$-martingale. Since λt is a continuous function of t, it is a natural process. We call the process λt the **compensator** of N.

Using the independence of increments, and the first two moments of a Poisson random variable, it is a simple exercise to show that $M_t^2 - \lambda t$ is an $\mathcal{F}_t$-martingale. Thus, the Meyer process for the martingale M is given by $\langle M \rangle_t = \lambda t$.

Definition 10.1.1 *Let $\{Y_j\}$ be a sequence of independent and identically distributed random variables, each with distribution μ. Suppose that N is an independent Poisson process with parameter λ. Then the process $S_t = Y_1 + \cdots + Y_{N_t}$ is called a* **compound Poisson process.**

This is an extension of a Poisson process, since the size of a jump is not necessarily one. Rather, each jump is distributed according to the measure μ. However, the jump times are determined, as before, by N. These are simple examples of what are known as jump Markov processes. In fact, we can do a little better. Instead of taking Y_j's to be iid random variables, let us take Y to be a Markov chain.

A sequence of random variables $\{Y_j\}$ defined on a probability space $(\Omega, \mathcal{F}, P)$ is called a time-homogeneous **Markov chain** with state space $\mathbb{R}^1$, initial distribution ν, and one-step transition probability measure Π if

(i) each Y_j is an $\mathbb{R}^1$-valued random variable,
(ii) $P(Y_0 \in A) = \nu(A)$ for all $A \in \mathcal{B}$, and
(iii) $P\{Y_{j+1} \in B \mid Y_j, Y_{j-1}, \ldots, Y_0\} = \Pi(Y_j, B)$ for all Borel sets B.

The state space for the Markov chain need not necessarily be $\mathbb{R}^1$. It can be any set S equipped with a σ-field. For any $k \in \mathbb{N}$ and Borel B, define the k-step transition probability measure inductively by

$$\Pi^k(x, B) = \int_{\mathbb{R}^1} \Pi(z, B)\, \Pi^{k-1}(x, dz).$$

Definition 10.1.2 *Let $\{Y_j\}$ be a time-homogeneous Markov chain as described above. Let N be an independent Poisson process with parameter $\lambda > 0$. The continuous-time process $X_t := Y_{N_t}$ is called a* **Poisson jump Markov process.**

We will first show that X has the Markov property. Let $\mathcal{F}_s$ denote $\mathcal{F}_s^X$. The class of sets

$$\mathcal{A} = \{A_1 \cap A_2 \cap \{N_s = k\} \text{ where } k \in \mathbb{N},\ A_1 \in \mathcal{F}_s^N,\ A_2 \in \sigma(Y_i : i \le k)\}$$

generates $\mathcal{F}_s$. By independence of increments, for any Borel set B,

$$
\begin{aligned}
P[X_{t+s} \in B \mid \mathcal{F}_s] &= P[Y_{N_{t+s}-N_s+N_s} \in B \mid \mathcal{F}_s] \\
&= \sum_{j=0}^{\infty} e^{-\lambda t} \frac{(\lambda t)^j}{j!} P[Y_{j+N_s} \in B \mid \mathcal{F}_s]. \qquad (10.1.1)
\end{aligned}
$$

We will show that

$$
P[Y_{j+N_s} \in B \mid \mathcal{F}_s] = \Pi^j(X_s, B). \qquad (10.1.2)
$$

Let us call a typical set $A_1 \cap A_2 \cap \{N_s = k\} \in \mathcal{A}$ as F_k. To prove the equality (10.1.2), consider

$$
\begin{aligned}
P\left[(Y_{j+N_s} \in B) \cap F_k\right] &= P\left[A_1 \cap \{N_s = k\}\right] P\left[\{Y_{j+k} \in B\} \cap A_2\right] \\
&= P\left[A_1 \cap \{N_s = k\}\right] \int_{A_2} \Pi^j(Y_k(\omega), B)\, P(d\omega) \\
&= \int_{F_k} \Pi^j(X_s(\omega), B)\, P(d\omega)
\end{aligned}
$$

by using the Markov property of Y, and independence of Y and N. Equations (10.1.1) and (10.1.2) yield

$$
P[X_{t+s} \in B \mid \mathcal{F}_s] = \sum_{j=0}^{\infty} e^{-\lambda t} \frac{(\lambda t)^j}{j!} \Pi^j(X_s, B)
$$

so that X has the Markov property and the transition probabilities for this Markov process are given by

$$
p(t, x, B) = P_x\{X_t \in B\} = \sum_{j=0}^{\infty} e^{-\lambda t} \frac{(\lambda t)^j}{j!} \Pi^j(x, B). \qquad (10.1.3)
$$

The strong Markov property of X can be proved easily. In fact, let τ be an $\mathcal{F}_t$-stopping time that takes values in a countable set $\{t_j\}$. Take any $A \in \mathcal{F}_\tau$, and Borel set B. If C_j denotes the event $\{\tau = t_j\}$, then $A \cap C_j \in \mathcal{F}_{t_j}$ and

$$
P_x[\{X_{\tau+t} \in B\} \cap A \cap C_j] = \int_{A \cap C_j} p(t, X_{t_j}, B)\, dP_x
$$

for each j. Summing up over j,

$$
P_x[\{X_{\tau+t} \in B\} \cap A] = \int_A p(t, X_\tau, B)\, dP_x.
$$

For a general stopping time τ, the same equality holds by approximating τ by a decreasing sequence of countably valued stopping times. Thus,

$$P_x[\{X_{\tau+t} \in B\} \mid \mathcal{F}_\tau] = p(t, X_\tau, B) \text{ P}_x\text{-a.s.}$$

which is the strong Markov property.

Using the transition probability $p(t, x, dy)$ given by (10.1.3), define the semigroup

$$(S_t f)(x) = \int f(y)\, p(t, x, dy)$$

for all bounded, Borel-measurable functions f. Then,

$$(S_t f)(x) - f(x) = (e^{-\lambda t} - 1)f(x) + e^{-\lambda t}\lambda t \int f(y)\, \Pi(x, dy) + o(t)$$

where $o(t)/t$ goes to zero uniformly in x. Therefore,

$$\frac{(S_t f)(x) - f(x)}{t} \to \lambda \int [f(y) - f(x)]\, \Pi(x, dy)$$

uniformly in x, and, hence, the infinitesimal generator for the process X is given by

$$(Lf)(x) = \lambda \int [f(y) - f(x)]\, \Pi(x, dy).$$

The above discussion on the process X is collected and given in the following result:

Proposition 10.1.2 *Let X be a jump Markov process with a constant jump intensity λ. Then X is a strong Markov process with stationary transition probabilities given by*

$$p(t, x, B) = \sum_{j=0}^{\infty} e^{-\lambda t} \frac{(\lambda t)^j}{j!} \Pi^j(x, B).$$

The infinitesimal generator for the semigroup $(S_t f)(x) = \int f(y)\, p(t, x, dy)$ is given by L where

$$Lf(x) = \lambda \int [f(y) - f(x)]\, \Pi(x, dy). \tag{10.1.4}$$

Remark 10.1.1 The generator L for the jump Markov process X is a bounded operator. Therefore, we can expand the semigroup in powers of L as

$$S_t = \sum_{j=0}^{\infty} \frac{t^j}{j!} L^j.$$

Theorem 10.1.3 *Let X be a jump Markov process with jump intensity λ. For any $x \in \mathbb{R}^1$, let P_x be the distribution of X with $X_0 = x$, and transition probabilities $p(t, x, dy)$ given by Equation (10.1.3). Then, for any bounded, Borel-measurable function f on $\mathbb{R}^1$,*

$$M_f(t) = f(X_t) - f(x) - \int_0^t (Lf)(X_s)\, ds \tag{10.1.5}$$

is a $\mathcal{F}_t$-martingale with respect to P_x.

Proof Since L is the infinitesimal generator for the semigroup S_t, it follows that

$$S_t f - f = \int_0^t S_s Lf\, ds.$$

Using the Markov property of X, for any $0 \le s \le t$,

$$\begin{aligned} E_x[M_f(t) - M_f(s) \mid \mathcal{F}_s] &= E_x\left[f(X_t) - f(X_s) - \int_s^t (Lf)(X_r)\, dr \mid \mathcal{F}_s\right] \\ &= (S_{t-s}f)(X_s) - f(X_s) - \int_s^t S_{r-s}(Lf)(X_s)\, dr \\ &= 0. \end{aligned}$$

■

In any interval $[0, T]$, there are only a finite number of jumps for the process X. For all $R > 0$, define $\tau_R = \inf\{t : |X_t| > R\}$. Then $\tau_R \uparrow \infty$ a.s. as $R \uparrow \infty$. Let us assume that $\Pi(x, \{y : |y - x| > l\}) = 0$. If f is bounded and Borel measurable with $f(y) = y$ for $y \in [-R - l, R + l]$, then

$$M_f(t \wedge \tau_R) = X_{t\wedge\tau_R} - x - \int_0^{t\wedge\tau_R} (Lf)(X_s)\, ds.$$

We have thus obtained a *semimartingale decomposition* of the jump process X:

$$X_t = x + \int_0^t (Lf)(X_s)\, ds + M_f(t).$$

In this section, we have taken the state space of the jump process X to be $\mathbb{R}^1$. However, the proofs of all the results remain the same if we replace $\mathbb{R}^1$ by $\mathbb{R}^d$ or, in general, by a Polish space S.

10.2 Stochastic Calculus for Processes with Jumps

Consider a Poisson process $\{N_t\}$ with parameter λ. The process $M_t := N_t - \lambda t$ is a martingale, and its quadratic variation process is N_t. We denote the quadratic variation by $[M]_t$. Though it is an increasing, adapted process, it is not a natural process in the sense of Chapter 3. However, we know that $M_t^2 - \lambda t$ is a martingale, and λt, being a continuous function of t, is natural. For this reason, we denote λt by $\langle M\rangle_t$. This example shows that there are two processes, $[M]_t$ and $\langle M\rangle_t$, associated with a square integrable cadlag martingale M.

We now construct stochastic integrals with respect to a square integrable cadlag martingale M. Consider a Riemann-Stieltjes integral $\int_0^t f(s)\, dg(s)$ where f is bounded and g is right continuous. Then f is Riemann-Stieltjes integrable provided that f is left continuous at all discontinuity points of g. In view of this result, we start with adapted left continuous processes as integrands to build stochastic integrals with respect to a square-integrable cadlag martingale M. Let us fix any interval $[0, T]$.

Definition 10.2.1 *A process $\{H_t\}$ is called a simple predictable process if it can be expressed in the form*

$$H_t = \sum_{n=0}^{k-1} H_n 1_{(t_n, t_{n+1}]}(t),$$

where $0 \leq t_0 < t_1 < \cdots < t_k = T$ and, for each $n \in \mathbb{N}$, $H_n \in \mathcal{F}_{t_n}$. In addition, H_n is bounded uniformly.

For all $t \leq T$, define the integral

$$\int_0^t H_s\, dM_s = \sum_{n=0}^{\infty} H_n(M_{t\wedge t_{n+1}} - M_{t\wedge t_n}).$$

Let us call the above integral $(H \cdot M)_t$. It is clear that the integral $H \cdot M$ is linear in H and is a martingale.

From Chapter 3, we know that there exists a process $\langle M\rangle_t$ such that for any $s < t \leq T$,

$$E[(M_t - M_s)^2 | \mathcal{F}_s] = E[\langle M\rangle_t - \langle M\rangle_s | \mathcal{F}_s].$$

Using this, it is a simple exercise to show the isometry

$$E[(H \cdot M)_t^2] = E[\int_0^t |H_s|^2\, d\langle M\rangle_s]. \tag{10.2.1}$$

Before we extend this isometry in an abstract manner, let us define a predictable process to gain insight into the class of integrands that we are about to encounter.

Definition 10.2.2 *Let $\mathcal{P}$ denote the smallest σ-field generated by the class of all adapted left-continuous processes. A stochastic process X is called* **predictable** *if the map $(t, \omega) \to X_t(\omega)$ is $\mathcal{P}$-measurable on $[0, T] \times \Omega$.*

Although this definition may seem abstract, note that the process X_{t-} is a predictable process whenever X_t is a cadlag process. This observation is quite useful to us in many applications.

Let H be a bounded, predictable process. Then we can find a sequence $\{H^{(n)}\}$ of simple predictable processes such that

$$\lim_{n\to\infty} E\left[\int_0^T \left(H_s^{(n)} - H_s\right)^2 d\langle M\rangle_s\right] = 0.$$

Hence, by the isometry (10.2.1), we can define, for all $t \in [0, T]$,

$$(H \cdot M)_t = \lim_{n\to\infty} \left(H^{(n)} \cdot M\right)_t \quad \text{in } L^2(P).$$

The integral is well defined and retains the martingale property and the isometry (10.2.1). The boundedness of H can be replaced by the requirement that $E[\int_0^T H_s^2\, d\langle M\rangle_s] < \infty$. Thus, the stochastic integral $H \cdot M$ is defined for all predictable integrands H that satisfy $E[\int_0^T H_s^2\, d\langle M\rangle_s] < \infty$. The integral retains linearity in H, the martingale property, and the isometry (10.2.1).

The class of integrands can be extended to the class of predictable processes H satisfying

$$\int_0^T H_s^2\, d\langle M\rangle_s < \infty \text{ a.s.}$$

However, $H \cdot M$ for this larger class of integrands will only be a local martingale. If $\{\tau_n\}$ is a localizing sequence of stopping times, then the isometry holds for the process $(H \cdot M)_{t\wedge\tau_n}$.

Next, we proceed to establish the Itô formula for functions of a cadlag semimartingale X as in [59]. If the semimartingale decomposition is $X_t = X_0 + M_t + A_t$, then we write $[X]_t^c$ for the continuous part of the quadratic variation of X. In fact, $[X]_t^c = \langle M\rangle_t$.

Theorem 10.2.1 *Let X be a semimartingale, and f, a real-valued function in $C^2(\mathbb{R})$. Then $f(X)$ is a semimartingale, and*

$$f(X_t) = f(X_0) + \int_{0+}^t f'(X_{s-})\, dX_s + \frac{1}{2}\int_{0+}^t f''(X_{s-})\, d[X]_s^c + \sum_{0<s\leq t} \{f(X_s) - f(X_{s-}) - f'(X_{s-})\Delta X_s\} \tag{10.2.2}$$

Proof Consider a refining sequence of partitions π_n of the interval $[0, t]$. Let

$$\pi_n = \{t_0^n = 0 \leq t_1^n \leq t_2^n \leq \cdots \leq t_{k_n}^n = t\}.$$

We can write

$$f(X_t) - f(X_0) = \sum_{i=0}^{k_n} [f(X_{t_{i+1}^n}) - f(X_{t_i^n})].$$

We know that a.s.,

$$\sum_{s\leq t}(\Delta X_s)^2 \leq [X]_t < \infty.$$

For any $\delta > 0$, we can split $\sum_{s\leq t}(\Delta X_s)^2$ into two parts such that the first sums up all jumps of magnitude, say, $>\delta$, and the second is the addition of all jumps of magnitude $\leq\delta$. The first is a finite sum, whereas the second is possibly an infinite sum. Given an $\epsilon > 0$, we can choose δ such that the second sum is less than ϵ^2. Let A denote the jump times used for the first sum, and B, for the second sum. Define

$$I_n = \{i : A \cap (t_i^n, t_{i+1}^n] \neq \emptyset\}$$

and

$$J_n = \{i : A \cap (t_i^n, t_{i+1}^n] = \emptyset\}.$$

Then, we can write

$$f(X_t) - f(X_0) = \sum_{I_n}[f(X_{t_{i+1}^n}) - f(X_{t_i^n})] + \sum_{J_n}[f(X_{t_{i+1}^n}) - f(X_{t_i^n})].$$

It is clear that

$$\lim_{n\to\infty}\sum_{I_n}[f(X_{t_{i+1}^n}) - f(X_{t_i^n})] = \sum_{A}[f(X_s) - f(X_{s-})].$$

Using the Taylor formula, we can write

$$\begin{aligned}
\sum_{J_n}[f(X_{t_{i+1}^n}) - f(X_{t_i^n})] &= \sum_{J_n} f'(X_{t_i^n})(X_{t_{i+1}^n} - X_{t_i^n}) + \frac{1}{2}\sum_{J_n} f''(X_{t_i^n})(X_{t_{i+1}^n} - X_{t_i^n})^2 \\
&\quad + \sum_{J_n} R(X_{t_i^n}, X_{t_{i+1}^n}) \\
&= \sum f'(X_{t_i^n})(X_{t_{i+1}^n} - X_{t_i^n}) + \frac{1}{2}\sum f''(X_{t_i^n})(X_{t_{i+1}^n} - X_{t_i^n})^2 \\
&\quad - \sum_{I_n} f'(X_{t_i^n})(X_{t_{i+1}^n} - X_{t_i^n}) - \frac{1}{2}\sum_{I_n} f''(X_{t_i^n})(X_{t_{i+1}^n} - X_{t_i^n})^2 \\
&\quad + \sum_{J_n} R(X_{t_i^n}, X_{t_{i+1}^n}).
\end{aligned}$$

As $n \to \infty$, the first two terms converge to $\int_0^t f'(X_{s-})\, dX_s$ and $\frac{1}{2}\int_0^t f''(X_{s-})\, d[X]_s$, respectively. Likewise,

$$\sum_{I_n} f'(X_{t_i^n})(X_{t_{i+1}^n} - X_{t_i^n}) \to \sum_{A} f'(X_{s-})\Delta X_s,$$

and

$$\frac{1}{2}\sum_{I_n} f''(X_{t_i^n})(X_{t_{i+1}^n} - X_{t_i^n})^2 \to \sum_A f''(X_{s-})(\Delta X_s)^2.$$

For $R > 0$, define $\tau_R = t \wedge \inf\{s \geq 0 : |X_s| \geq R\}$. The process $X1_{[0,\tau_R)}$ is bounded by R. Since $\tau_R \uparrow t$ a.s. as $R \to \infty$, we can assume without loss of generality that the semimartingale X itself is bounded by R for all $s \leq t$. Then f'' can be viewed as a uniformly continuous. Using the right continuity of X, we can argue that

$$\limsup_{n\to\infty} \sum_{J_n} R(X_{t_i^n}, X_{t_{i+1}^n}) \leq r(\epsilon+)[X]_t,$$

where $r(|x-y|)(x-y)^2$ is the remainder estimate in approximating $f(y)$ by the second order Taylor polynomial around x. Hence, r that appears above is an increasing function which tends to zero as $\epsilon \to 0$.

Thus, all that we need to show is the convergence of

$$\sum_A \left[f(X_s) - f(X_{s-}) - f'(X_{s-})\Delta X_s - \frac{1}{2}f''(X_{s-})(\Delta X_s)^2\right] \quad \text{to}$$

$$\sum_{0\leq s\leq t} \left[f(X_s) - f(X_{s-}) - f'(X_{s-})\Delta X_s - \frac{1}{2}f''(X_{s-})(\Delta X_s)^2\right]$$

as $\epsilon \to 0$. The convergence holds if this limit series converges absolutely. To prove this, we can, by the use of stopping times, assume that $|X_t| \leq k$ for a fixed constant $k > 0$. Then f'' restricted to $[-k, k]$ is uniformly bounded by a constant K. Hence, a.s.,

$$\sum_{0\leq s\leq t} |f(X_s) - f(X_{s-}) - f'(X_{s-})\Delta X_s| \leq K \sum_{0\leq s\leq t} (\Delta X_s)^2 \leq K[X]_t < \infty,$$

and $\sum_{0\leq s\leq t}[\frac{1}{2}|f''(X_{s-})|(\Delta X_s)^2] \leq K[X]_t$. The required absolute convergence is thus shown. The formula is thus established by collecting the terms that accrue in the limit. ■

10.3 Jump Markov Processes

In Section 10.1, the waiting time parameter λ for the process X was a positive constant. However, in many applications, it is a function of the present value of X. For instance, consider an integer-valued process X known as a *birth and death process*. Suppose the initial population X_0 is j. At the time of its first jump, the process has either a birth or a death, so that the population size becomes either $j + 1$ or $j - 1$. Births and deaths are assumed to be independent. The birth rate will be proportional to j if each individual gives birth at a rate λ independently of others. Likewise, the death rate will be proportional to

j. If $j = 0$, let us assume there is only birth. Then the population, at the time of a jump, moves as follows:

$$j \to j+1 \quad \text{with rate } \lambda j, \text{ and}$$
$$j \to j-1 \quad \text{with rate } \mu j.$$

In general, the birth rate can be taken as a function λ_j of j, and the death rate, μ_j. At the time of its first jump from state j, we have

$$P_j\{X_\tau = j+1\} = \frac{\lambda_j}{\lambda_j + \mu_j}, \quad \text{and}$$
$$P_j\{X_\tau = j-1\} = \frac{\mu_j}{\lambda_j + \mu_j}.$$

The waiting time parameter is $\lambda_j + \mu_j$.

To define a jump Markov process, let Y be a Markov chain with one-step transition probability measures given by $\Pi(x, dy)$ and initial distribution given by ν. Let $Z = \{Z_j\}$ be an independent sequence of iid exponential variables with parameter 1. If λ is a given positive function, define

$$\tau_j = \frac{Z_j}{\lambda(Y_{j-1})} \quad \text{for all } j \geq 1.$$

Define $\sigma_n = \sum_{j=1}^{n} \tau_j$ for all n, and $S_0 \equiv 0$.

Definition 10.3.1 *Let Y and Z be as above. The continuous-time process defined by*

$$X_t = Y_n \quad \text{if} \quad \sigma_n \leq t < \sigma_{n+1}$$

is called a **jump Markov process** *with initial distribution ν, jump measure $\Pi(x, dy)$, and waiting time parameter $\lambda(x)$.*

In other words, if we define N_t as the counting process with inter-arrival times specified by $\{\tau_j\}$, then $X_t = Y_{N_t}$. Such processes are also known as *pure jump* Markov processes. A nice sub class of such processes is furnished by those with a *bounded* waiting time function $\lambda(x)$. For such a process, the infinitesimal generator takes the form:

$$(Lf)(x) = \lambda(x) \int [f(y) - f(x)] \, \Pi(x, dy). \qquad (10.3.1)$$

The generator can be reduced to a generator of the simpler type given by equation (10.1.4). Indeed, define a constant $\lambda > \sup \lambda(x)$, and

$$\mu(x, B) = \frac{\lambda(x)}{\lambda} \Pi(x, B) + \left(1 - \frac{\lambda(x)}{\lambda}\right) 1_B(x).$$

Then $\mu(x,\cdot)$ is a probability measure. However, $\mu(x,\cdot)$ is not a genuine jump distribution, since $\mu(x,\{x\})$ can be strictly positive. Keeping this in mind, the infinitesimal generator of X can be written

$$(Lf)(x) = \lambda \int [f(y) - f(x)]\,\mu(x,dy).$$

When λ is not a bounded function, X is still a strong Markov process, and its generator is given by

$$Lf(x) = \lambda(x) \int [f(y) - f(x)]\,\Pi(x,dy).$$

However, L is no longer a bounded operator. In this situation, the sequence of stopping times σ_n need not increase to ∞ a.s., but may converge to a time σ, known as the **explosion time** for X. On the set$\{\sigma < \infty\}$, the process X has infinite number of jumps in a finite time, and for this reason, σ is called as explosion time. Note that the process X stops at σ. We will study conditions that guarantee non-explosion of the process X, that is $\sigma = \infty$ a.s.

Proposition 10.3.1 *Let X be a jump Markov process as described above. Then, non-explosion of X is equivalent to the condition*

$$\sum_{j=1}^{\infty} \frac{1}{\lambda(Y_{j-1})} = \infty.$$

Proof Define $\mathcal{F}^Y := \vee_t \mathcal{F}_t^Y$, that is, the smallest σ-field containing $\cup_t \mathcal{F}_t$, and $A_j := \dfrac{1}{\lambda(Y_{j-1})}$. Then, for any $u \geq 0$,

$$E\left\{e^{-u\tau_j} \mid \mathcal{F}^Y\right\} = E\left\{e^{-uA_jZ_j} \mid \mathcal{F}^Y\right\} = \frac{1}{1+uA_j}.$$

By independence of Z_j,

$$E\left\{e^{-u\sigma_n} \mid \mathcal{F}^Y\right\} = \prod_{j=1}^{n} \frac{1}{1+uA_j}.$$

Letting $n \to \infty$, we obtain

$$E\{e^{-u\sigma} \mid \mathcal{F}^Y\} = \prod_{j=1}^{\infty} \frac{1}{1+uA_j} = \exp\left\{-\sum_j \log(1+uA_j)\right\}$$

a.s., by using the dominated convergence theorem. The infinite series converges if and only if $\sum_{j=1}^{\infty} A_j < \infty$. Allowing $u \to 0$,

$$P\left[\sigma < \infty \mid \mathcal{F}^Y\right] = 1_{\{\sum_{j=1}^{\infty} A_j < \infty\}} \text{ a.s.}$$

Thus, $\sigma = \infty$ a.s. if and only if $\sum_{j=1}^{\infty} A_j = \infty$ a.s. ■

If the function λ is bounded, then it follows easily from the above proposition that $\sigma = \infty$ a.s. The necessary and sufficient condition for non-explosion of the process X depends on Y and, hence, is not deterministic.

When L is unbounded, the domain of L is difficult to ascertain. However, suppose that λ is continuous, and the map $x \to \Pi(x, dy)$ is continuous in the sense that if $x_n \to x$, then the sequence of probability measures $\Pi(x_n, dy)$ converges weakly to $\Pi(x, dy)$. Then L maps bounded continuous functions into continuous functions.

If $f \in C_b(\mathbb{R}^1)$ such that $\sup_x |Lf(x)| < \infty$, then, as in Section 10.1, one can show that

$$M_f(t) = f(X_t) - X(0) - \int_0^t Lf(X_s)\, ds$$

is a $\mathcal{F}_t^X$-martingale with respect to P_x.

In the following, we obtain a sufficient condition for non-explosion of X. Let $\mathcal{F}_t$ denote the σ-field generated by $\{X_{s\wedge\sigma} : 0 \leq s \leq t\}$.

Lemma 10.3.2 *Let X be a jump Markov process with waiting time parameter $\lambda(x)$. Suppose that, for each $N > 0$, there exists a constant K_N such that*

$$\sup_{|x|\leq N} \lambda(x) = K_N < \infty. \tag{10.3.2}$$

Then, for every $T > 0$ and $N > 0$, we have

$$P_x\left\{\sup_{0\leq t\leq\sigma} |X_t| \leq N, \sigma \leq T\right\} = 0.$$

In particular, $P_x\{\sup_{0\leq t\leq\sigma} |X_t| < \infty, \sigma < \infty\} = 0$.

Proof Let A denote the event $\{\sup_{0\leq t\leq\sigma} |X_t| \leq N, \sigma \leq T\}$, and

$$A_n := \left\{\sup_{0\leq t\leq\sigma_n} |X_t| \leq N, \sigma_n \leq T\right\}.$$

Then A_n is a decreasing sequence, and $A \subset \cap_n A_n$. Hence,

$$P_x(A) \leq \lim_{n\to\infty} P_x(A_n). \tag{10.3.3}$$

Consider

$$\begin{aligned} P_x(A_{n+1}) &\leq P_x\left\{\sup_{0\leq t\leq\sigma_n}|X_t|\leq N,\ \sigma_{n+1}\leq T\right\} \\ &= E_x\left[P_x\left\{\sup_{0\leq t\leq\sigma_n}|X_t|\leq N,\ \sigma_{n+1}\leq T\,|\,\mathcal{F}_{\sigma_n}\right\}\right] \\ &= E_x\left[1_{A_n}E_{X_{\sigma_n}}\left(1_{\{\tau_1\leq T-\sigma_n\}}\right)\right] \quad \text{(by the strong Markov property)} \\ &\leq P_x(A_n)(1-e^{-K_N T}). \end{aligned}$$

By iteration of the above argument, we have $P_x(A_{n+1}) \leq (1-e^{-K_N T})^{n+1}$. From this and (10.3.3), $P_x(A) = 0$.

The last statement in the lemma follows from

$$\left\{\sup_{0\leq t\leq\sigma}|X_t|<\infty,\ \sigma<\infty\right\} = \cup_{N=1}^{\infty}\cup_{k=1}^{\infty}\left\{\sup_{0\leq t\leq\sigma}|X_t|\leq N,\ \sigma\leq k\right\}. \quad \blacksquare$$

Theorem 10.3.3 *Let X be a jump Markov process such that there exists a finite $J > 0$ such that $\Pi(x, \{y : |y-x| > J\}) = 0$. Further, assume that the function λ satisfies (10.3.2). Define the functions*

$$b(x) = \lambda(x)\int_{\mathbb{R}^1}(y-x)\,\Pi(x,dy) \tag{10.3.4}$$

$$a(x) = \lambda(x)\int_{\mathbb{R}^1}(y-x)^2\,\Pi(x,dy). \tag{10.3.5}$$

Suppose that the following condition holds for a constant K:

$$|b(x)|^2 + a(x) \leq K(1+|x|^2). \tag{10.3.6}$$

Then there is no explosion for X, and the process $M_t = X_t - x - \int_0^t b(X_s)\,ds$ is a square integrable martingale with respect to P_x with

$$\langle M\rangle_t = \int_0^t a(X_s)\,ds.$$

Proof

Step 1 Define the stopping time $T_R = \inf\{t : |X_{t\wedge\sigma}| > R\} \wedge \sigma$. By the boundedness of jumps,

$$1_{\{t<\sigma\}}|X_{t\wedge T_R}| < R + J.$$

Let f be a bounded, continuous function such that

$$f(x) = \begin{cases} x & \text{if } |x| < R+J, \\ 0 & \text{if } |x| > R+2J. \end{cases}$$

Then, for $|x| > R + 3J$, we have $\int [f(y) - f(x)]\,\Pi(x, dy) = 0$ by the choice of f. Using this and the boundedness of $\lambda(x)$ on $|x| \le R + 3J$, we have

$$\sup_x |Lf(x)| = \sup_{\{x:|x|\le R+3J\}} |Lf(x)| < \infty.$$

Hence, $M_f(t) = 1_{\{t<\sigma\}} f(X_t) - f(x) - \int_0^t 1_{\{t<\sigma\}} Lf(X_s)\,ds$ is a P_x-martingale.
For $|x| < R$, we have

$$Lf(x) = \lambda(x) \int (y - x)\,\Pi(x, dy) = b(x).$$

Therefore, by Doob's optional sampling theorem,

$$M_{t\wedge T_R} = 1_{\{t\wedge T_R<\sigma\}} X_{t\wedge T_R} - x - \int_0^{t\wedge T_R} 1_{\{s<\sigma\}} b(X_s)\,ds \tag{10.3.7}$$

is a P_x-martingale.

Step 2 Consider the function $\phi(x) = f^2(x)$. As in Step 1, we have $\sup_x |L\phi(x)| < \infty$. For $|x| < R$, we have $L\phi(x) = 2xb(x) + a(x)$ since we can write

$$y^2 - x^2 = 2x(y - x) + (y - x)^2.$$

Therefore, as in Step 1,

$$Z_{t\wedge T_R} = 1_{\{t\wedge T_R<\sigma\}} X^2_{t\wedge T_R} - x^2 - \int_0^{t\wedge T_R} 1_{\{s<\sigma\}} [2X_s\, b(X_s)\,ds + a(X_s)]\,ds \tag{10.3.8}$$

is a P_x-martingale. Using the Itô formula for the function ϕ and the semimartingale $1_{\{t\wedge T_R<\sigma\}} X_{t\wedge T_R}$, we obtain from (10.3.7) that

$$\begin{aligned} 1_{\{t\wedge T_R<\sigma\}} X^2_{t\wedge T_R} &= x^2 + 2\int_0^{t\wedge T_R} 1_{\{s<\sigma\}} X_s\, b(X_s)\,ds \\ &\quad + 2\int_0^{t\wedge T_R} 1_{\{s<\sigma\}} X_s\,dM_s + \langle M\rangle_{t\wedge T_R}. \end{aligned} \tag{10.3.9}$$

Comparing with (10.3.8), it follows that

$$\langle M\rangle_{t\wedge T_R} = \int_0^{t\wedge T_R} 1_{\{s<\sigma\}} a(X_s)\,ds. \tag{10.3.10}$$

Step 3 Fix any $t \le T$. Taking expectation in (10.3.9), we obtain

$$E\left[1_{\{t\wedge T_R<\sigma\}} X^2_{t\wedge T_R}\right] = x^2 + E\left(\int_0^{t\wedge T_R} 1_{\{s<\sigma\}} [2X_s\, b(X_s) + a(X_s)]\,ds\right). \tag{10.3.11}$$

We can bound $2xb(x)$ by $x^2 + b(x)^2$. Using this, and the hypothesis (10.3.6) in Equation (10.3.9),

$$E\left[1_{\{t\wedge T_R<\sigma\}} X^2_{t\wedge T_R}\right] \le x^2 + KT + (K+1)\left(\int_0^t E 1_{\{s\wedge T_R<\sigma\}} X^2_{s\wedge T_R}\,ds\right).$$

By the Gronwall inequality, it follows that

$$\sup_{t\leq T} E\left[1_{\{t\wedge T_R<\sigma\}}X^2_{t\wedge T_R}\right] \leq (x^2 + KT)e^{(K+1)T}.$$

Allowing $R \uparrow \infty$,

$$\sup_{t\leq T} E\left[1_{\{t<\sigma\}}X^2_t\right] \leq (x^2 + KT)e^{(K+1)T}. \tag{10.3.12}$$

Upon squaring both sides of Equation (10.3.7), one easily obtains

$$M^2_{t\wedge T_R} \leq 3\left[1_{\{t\wedge T_R<\sigma\}}X^2_{t\wedge T_R} + x^2 + T\int_0^{t\wedge T_R} 1_{\{s<\sigma\}}b^2(X_s)\,ds\right].$$

Using hypothesis (10.3.6), we have

$$M^2_{t\wedge T_R} \leq 3\left[1_{\{t\wedge T_R<\sigma\}}X^2_{t\wedge T_R} + x^2 + TK\int_0^{t\wedge T_R} 1_{\{s<\sigma\}}(1 + X^2_s)\,ds\right].$$

If we take expectations in the above inequality and use the estimate (10.3.12), we obtain the existence of a finite constant C_T such that

$$E(M^2_{t\wedge T_R}) \leq C_T.$$

Since C_T is independent of R,

$$\sup_{t\leq T} E(1_{\{t<\sigma\}}M^2_t) \leq C_T. \tag{10.3.13}$$

We already know that $M_{t\wedge T_R}$ is a locally square integrable martingale. The estimate (10.3.13) implies that $1_{\{t<\sigma\}}M_t$ is indeed a square integrable martingale. Hence Equation (10.3.10) yields

$$\langle M\rangle_t = \int_0^t 1_{\{t<\sigma\}}a(X_s)\,ds$$

where

$$M_t = 1_{\{t<\sigma\}}X_t - x - \int_0^t 1_{\{s<\sigma\}}b(X_s)\,ds.$$

Therefore,

$$E\left[\sup_{t<T\wedge\sigma} X^2_t\right] \leq x^2 + 2\int_0^t K\left[1 + E1_{\{s<\sigma\}}X^2_s\right]\,ds + E\langle M\rangle_T < \infty.$$

Thus, $E[\sup_{t<T\wedge\sigma} X^2_t] < \infty$. Now, Lemma 10.3.2 allows us to infer that $\sigma > T$ a.s., and T is arbitrary. Thus, $\sigma = \infty$ a.s. ∎

If the jump Markov process X in Theorem 10.3.3 is $\mathbb{R}^d$ valued, then an analogous result can be proved by making a few simple changes to the above proof. We carefully indicate below all the required changes.

Define

$$b(x) = \lambda(x) \int_{\mathbb{R}^d} (y - x)\, \Pi(x, dy)$$

so that $b(x)$ is a d-dimensional vector, and define $a(x)$ as a $d \times d$-matrix with its elements given by

$$a_{ij}(x) = \lambda(x) \int_{\mathbb{R}^d} (y_i - x_i)(y_j - x_j)\, \Pi(x, dy).$$

Fix any $\theta \in \mathbb{R}^d$. Then (X_t, θ) is a real-valued jump Markov process. In Step 1 of the above proof, define the function f as

$$f(x) = \begin{cases} (x, \theta) & \text{if } |x| < R + J, \\ 0 & \text{if } |x| > R + 2J. \end{cases}$$

Then $Lf(x) = \lambda(x) \int_{\mathbb{R}^d} (y - x, \theta)\, \Pi(x, dy) = (b(x), \theta)$ if $|x| < R$. Hence, we can argue exactly as in Step 1 to show that

$$M_{t\wedge T_R} = 1_{\{t\wedge T_R<\sigma\}}(X_{t\wedge T_R}, \theta) - (x, \theta) - \int_0^{t\wedge T_R} 1_{\{s<\sigma\}}(b(X_s), \theta)\, ds$$

is a P_x-martingale.

In Step 2, if $|x| < R + J$, then $L\phi(x) = 2(b(x), \theta)(x, \theta) + (\theta, a(x)\theta)$. Working with (X_t, θ) in the place of X_t, and arguing as before, we obtain

$$\langle M\rangle_{t\wedge T_R} = \int_0^{t\wedge T_R} 1_{\{s<\sigma\}}(\theta, a(X_s)\theta)\, ds.$$

Likewise in Step 3, one obtains by a Gronwall argument that

$$\sup_{t\le T} E[1_{\{t<\sigma\}}(X_t, \theta)^2] \le |\theta|^2(|x|^2 + KT)e^{(K+1)T}.$$

Hence, $\sup_{t\le T} E(1_{\{t<\sigma\}} M_t^2) \le |\theta|^2 C_T$. In the present case, the last inequality in Step 3 reads as:

$$E[\sup_{\theta\le 1} \sup_{t<T\wedge\sigma} (X_t, \theta)^2] \le |x|^2 + 2\int_0^t K[1 + E1_{\{s<\sigma\}} X_s^2]\, ds + E\langle M\rangle_T < \infty.$$

Hence, $E[\sup_{t<T\wedge\sigma} |X_t|^2] < \infty$, which implies by Lemma 10.3.2 that $\sigma = \infty$ a.s. We thus have the following result:

Theorem 10.3.4 *Let X be an $\mathbb{R}^d$-valued jump Markov process such that there exists a finite $J > 0$ such that $\Pi(x, \{y : |y - x| > J\}) = 0$. Suppose that, for each $N > 0$, there exists a constant K_N such that*

$$\sup_{|x|\leq N} \lambda(x) = K_N < \infty.$$

Define the functions

$$b(x) = \lambda(x) \int_{\mathbb{R}^1} (y - x)\, \Pi(x, dy)$$

$$a_{ij}(x) = \lambda(x) \int_{\mathbb{R}^1} (y_i - x_i)(y_j - x_j) \Pi(x, dy) \;\; \forall\, i, j = 1, 2, \ldots, d.$$

Suppose that there exists a constant K such that $|b(x)|^2 + a(x) \leq K(1 + |x|^2)$.

Then there is no explosion for X, and the process $M_t = X_t - x - \int_0^t b(X_s)\, ds$ is a square integrable martingale with respect to P_x with

$$\langle M \rangle_t = \int_0^t a(X_s)\, ds.$$

10.4 Diffusion Approximation

In order to discuss weak convergence of cadlag processes, we need to consider their path space, that is, the space of all $\mathbb{R}^d$-valued functions that are right continuous with left limits. We denote as $D([0, 1]; \mathbb{R}^d)$ the space of all functions f defined on $[0, 1]$ such that f is right continuous with left limits on $(0, 1)$, $f(0) = f(0+)$, and $f(1) = f(1-)$. The domain can be replaced by any interval $[0, T]$. One can equip $D([0, 1]; \mathbb{R}^d)$ by a certain metric known as the Skorohod metric. The topology given by this metric is known as the Skorohod topology. From now on, $D := D([0, 1]; \mathbb{R}^d)$ will always be equipped with the Skorohod topology. One can show that D is a complete separable metric space. Such facts are not immediate, and the proofs are lengthy. Of great importance to us is the following:

Definition 10.4.1 *A sequence $\{f_n\}$ converges to f in D if there exists a strictly increasing sequence $\{\lambda_n\}$ of maps from $[0, 1]$ to $[0, 1]$ such that*

$$\lim_{n\to\infty} \sup_{0\leq t\leq 1} \left[|\lambda_n(t) - t| + |\lambda_n^{-1}(t) - t|\right] = 0,$$

and

$$\lim_{n\to\infty} \sup_{0\leq t\leq 1} |f_n(\lambda_n(t)) - f(t)| = 0.$$

We state below a useful criterion for tightness of a sequence of processes with paths in D. Its proof and several results on the Skorohod space can be found in [36] and [24].

Theorem 10.4.1 (Aldous criterion) *Let $\{X_n\}$ be a sequence of processes with paths in D. Suppose that, for each rational $t \in [0, 1]$, the family of random variables $\{X_n(t)\}$ is tight. Then $\{X_n\}$ is tight in D if the following condition is satisfied:*

For every sequence (T_n, δ_n) where each T_n is a stopping time such that $T_n \le 1$, and $\delta_n > 0$, $\delta_n \to 0$, we have $|X_n(T_n + \delta_n) - X_n(T_n)| \to 0$ in probability as $n \to \infty$.

Suppose we are given a bounded, continuous function $b : \mathbb{R}^d \to \mathbb{R}^d$ and a continuous function $a : \mathbb{R}^d \to \mathbb{R}^{d\times d}$ such that $a(x)$ is symmetric and uniformly elliptic. Let

$$\operatorname{tr} a(x) = \sum_{i=1}^{d} a_{ii}(x) \le K(1 + |x|^2).$$

Define the operator A on $C_0^2(\mathbb{R}^d)$ by

$$Af(x) = \sum_{i=1}^{d}\sum_{j=1}^{d} a_{ij}(x)\frac{\partial^2 f}{\partial x_i \partial x_j}(x) + \sum_{i=1} b_i(x)\frac{\partial f}{\partial x_i}(x). \tag{10.4.1}$$

Our aim in this section is to prove the existence of a solution P_x to the (A, δ_x)-martingale problem for all $x \in \mathbb{R}^d$ by forming an approximating sequence of jump Markov processes and identifying the limit of the sequence.

Let $\{e_k(x) : k = 1, \ldots, d\}$ be the orthonormal eigenvectors of the matrix $a(x)$. For any given $\epsilon > 0$, define a function $\lambda(x)$ by

$$\lambda(x) = \frac{\operatorname{tr} a(x)}{\epsilon^2}.$$

As we shall see, the function λ will serve as the waiting time parameter. As ϵ becomes smaller, the jumps become more frequent. Define two functions $p_k(x)$ and $q_k(x)$ as the solution of

$$\lambda(x)\epsilon^2(p_k(x) + q_k(x)) = (e_k(x), a(x)e_k(x))$$

and

$$\lambda(x)\epsilon(p_k(x) - q_k(x)) = (b(x), e_k(x))$$

where the inner product refers to that of $\mathbb{R}^d$.

Since $\sum_{k=1}(e_k(x), a(x)e_k(x)) = \operatorname{tr} a(x)$, the first of the two equations implies that for all x,

$$\sum_{k=1}^{d}(p_k(x) + q_k(x)) = 1.$$

Solving for $p_k(x)$, one obtains $2\lambda(x)\epsilon^2 p_k(x) = (e_k(x), a(x)e_k(x)) + \epsilon(b(x), e_k(x))$, which can be rewritten as

$$p_k(x) = \frac{(e_k(x), a(x)e_k(x))}{\operatorname{tr} a} + \epsilon\frac{(b(x), e_k(x))}{\operatorname{tr} a}.$$

By boundedness of the function b and uniform ellipticity of a, it follows that $p_k(x)$ is positive for all x when ϵ is small enough. A similar statement holds for $q_k(x)$. We will choose and fix such an ϵ. Thus, we have ensured that ϵ doesn't depend on x.

With $\epsilon, e_k, \lambda, p_k$, and q_k defined as above, let X denote a jump Markov process with waiting time parameter given by $\lambda(x)$ and jump distribution specified by

$$\begin{aligned}\Pi(x, x + \epsilon e_k(x)) &= p_k(x), \text{ and}\\ \Pi(x, x - \epsilon e_k(x)) &= q_k(x).\end{aligned}$$

Later, we would vary ϵ, and denote X as X^ϵ. By the definition of λ, and the bound on tr a, we have

$$0 \leq \lambda(x) \leq \frac{K}{\epsilon^2}(1 + |x|^2).$$

Also, $\Pi(x, \{y : |x - y| > \epsilon\}) = 0$. Thus, the hypotheses of Theorem 10.3.4 are satisfied so that

$$M_t = X_t - x - \int_0^t b(X_s)\, ds$$

is a P_x-martingale with $\langle M\rangle_t = \int_0^t a(X_s)\, ds$. For any $\theta \in \mathbb{R}^d$, let $X_\theta(t)$ denote (X_t, θ), and $M_\theta(t) = (M_t, \theta)$. For any $f \in C_0^2(\mathbb{R}^d)$, apply the Itô formula to $f((X_\theta(t))$ to obtain

$$\begin{aligned}f(X_\theta(t)) = f((x,\theta)) &+ \int_0^t f'(X_\theta(s-))(b(X_s),\theta)\, ds + \int_0^t f'(X_\theta(s-))dM_\theta(s)\\ &+ \sum_{s\in[0,t]} \{f(X_\theta(s)) - f(X_\theta(s-)) - f'(X_\theta(s-))(\Delta X_s,\theta)\}\end{aligned} \tag{10.4.2}$$

Note that the process $\sum_{s\in[0,t]} f''(X_\theta(s-))(\Delta X_s,\theta)^2$ can also be written as $\int_0^t f''(X_\theta(s-))\, d[M_\theta]_s$. Since $\langle M_\theta\rangle_t = \int_0^t (\theta, a(X_s)\theta)\, ds$, it follows that

$$\sum_{s\in[0,t]} f''(X_\theta(s-))(\Delta X_s,\theta)^2 - \int_0^t f''(X_\theta(s-))(\theta, a(X_s)\theta)\, ds$$

is a martingale. Let us call it $\xi_\theta(t)$. Using Equation (10.4.2) and the process $\xi_\theta(t)$, we obtain

$$\begin{aligned}\int_0^t f'(X_\theta(s-))\, dM_\theta(s) + \frac{1}{2}\xi_\theta(t) = f(X_\theta(t)) - f((x,\theta)) &- \int_0^t f'(X_\theta(s-))(b(X_s),\theta)\, ds\\ &-\frac{1}{2}\int_0^t f''(X_\theta(s-))(\theta, a(X_s)\theta)\, ds\\ &- \sum_{s\in[0,t]} \Bigg\{ \Delta f(X_\theta(\cdot))_s - f'(X_\theta(s-))(\Delta X_s,\theta)\\ &-\frac{1}{2} f''(X_\theta(s-))(\Delta X_s,\theta)^2 \Bigg\}.\end{aligned} \tag{10.4.3}$$

where $\Delta f(X_\theta(\cdot))_s = f(X_\theta(s)) - f(X_\theta(s-))$. We will call the last term on the right side of Equation (10.4.3) as the remainder term $R_\theta(t)$. Thus, the process

$$N_\theta^f(t) := f(X_\theta(t)) - f((x,\theta)) - \int_0^t f'(X_\theta(s-))(b(X_s),\theta)\, ds - \int_0^t \frac{1}{2} f''(X_\theta(s-))(\theta, a(X_s)\theta)\, ds + R_\theta(t) \tag{10.4.4}$$

is a P_x-martingale. Since f is bounded along with its first two derivatives, each summand in $R_\theta(t)$ can be bounded by $C_1[1 + |\theta|\epsilon + |\theta|^2\epsilon^2]$ for a finite constant $C_1 > 0$. The sum in $R_\theta(t)$ should be over those jump times $s \le t$ such that $X_\theta(s-)$ lies in the support of f. By our hypotheses, $\lambda(x) \le \frac{K}{\epsilon^2}(1 + |x|^2)$, and λ is therefore bounded on compacts. Hence, in the time interval $[0, t]$, the expected number of terms in $R_\theta(t)$ is bounded by $k\epsilon^2 t$ for a constant $k > 0$. Thus,

$$E|R_\theta(t)| < C(\theta, f)t\epsilon. \tag{10.4.5}$$

So far, we have suppressed the dependence of X_t on ϵ for notational convenience. From now on, we will write X_t as X_t^ϵ. The probability distribution of the process X_t^ϵ will be denoted by the measure P^ϵ on the space $D([0,\infty) : \mathbb{R}^d)$. We will denote a generic element of the space $D([0,\infty) : \mathbb{R}^d)$ by z. The canonical process is denoted by Z so that $Z_t(z) = z(t)$. Let $\mathcal{F}_t$ be the right-continuous filtration generated by Z till time t.

We will prove that (i) the sequence of probability measures $\{P^\epsilon\}$ converges weakly to a probability measure P, (ii) the support of P is contained in $C([0,\infty) : \mathbb{R}^d)$, and (iii) P is a solution of the following martingale problem:

1. $P(z : z_0 = x) = 1$, and
2. For all $f \in C_0^2(\mathbb{R}^1)$, and $\theta \in \mathbb{R}^d$, the process

$$H_f^\theta(t) = f(Z_\theta(t)) - f((x,\theta)) - \int_0^t f'(Z_\theta(s-))(b(Z_s),\theta)\, ds - \int_0^t \frac{1}{2} f''(Z_\theta(s-))(\theta, a(Z_s)\theta)\, ds$$

is an $\mathcal{F}_t$-martingale with respect to P.

We start with the following lemma.

Lemma 10.4.2 *Consider the jump Markov process $X = \{X_t\}$ described above. For any given $T > 0$, there exists a corresponding constant k, which does not depend on ϵ, such that*

$$E[\sup_{0\le t\le T} |X_t|^2] \le k[T + |x|^2]e^{kT}. \tag{10.4.6}$$

Proof Let $X_i(t), M_i(t)$ denote the ith coordinate of X, M respectively. By the Itô formula,

$$|X_t|^2 = |x|^2 - 2\sum_{i=1}^d \int_0^t X_i(s)\, b_i(X_s)\, ds + 2\sum_{i=1}^d \int_0^t X_i(s)\, dM_i(s) + [M]_t. \tag{10.4.7}$$

We will write $\sum_{i=1}^d X_i(s)\, b_i(X_s)$ as $(X_s, b(X_s))$, and $\sum_{i=1}^d \int_0^t X_i(s)\, dM_i(s)$ as $\int_0^t X_s\, dM_s$. Since

$$|(X_s, b(X_s))| \le |X_s||b(X_s)| \le \frac{1}{2}[|X_s|^2 + |b(X_s)|^2]$$

and b is bounded by a constant c, we obtain

$$|X_t|^2 \le |x|^2 + \int_0^t |X_s|^2\, ds + c^2 t + 2\int_0^t X(s)\, dM(s) + [M]_t.$$

Let $\{\tau_n\}$ be a localizing sequence of stopping times so that the process $\int_0^{t\wedge\tau_n} X_s\, dM_s$ is a martingale for each n. Then,

$$\begin{aligned} E[|X_{t\wedge\tau_n}|^2] &\le |x|^2 + E\int_0^{t\wedge\tau_n} |X_s|^2\, ds + c^2 t + E[\langle M\rangle_{t\wedge\tau_n}] && (10.4.8)\\ &\le |x|^2 + (c^2 + K)t + (1+K)E\int_0^{t\wedge\tau_n} |X_s|^2\, ds \\ &\le |x|^2 + (c^2 + K)t + (1+K)\int_0^t E[|X_{s\wedge\tau_n}|^2]\, ds. && (10.4.9)\end{aligned}$$

The Gronwall inequality yields

$$E[|X_{t\wedge\tau_n}|^2] \le [|x|^2 + (c^2 + K)t]e^{1+K)t}. \quad (10.4.10)$$

Allowing $n \to \infty$, by the Fatou lemma, we obtain

$$E[|X_t|^2] \le [|x|^2 + (c^2 + K)t]e^{1+K)t}. \quad (10.4.11)$$

Likewise, by using the Itô formula for the real-valued semimartingale $|X_t|^2$, one obtains

$$\begin{aligned} |X_t|^4 = |x|^4 - 4\sum_{i=1}^d \int_0^t |X_s|^2 X_i(s)\, b_i(X_s)\, ds + 4\sum_{i=1}^d \int_0^t |X_s|^2 X_i(s)\, dM_i(s) \\ + 2\int_0^t |X_s|^2 [M]_s + 4\sum_{i=1}^d \sum_{j=1}^d \int_0^t X_i(s) X_j(s)\, d[M_i, M_j]_s. \end{aligned}$$

By localization and taking expectation as before,

$$\begin{aligned} E|X_{t\wedge\tau_n}|^4 \le |x|^4 + 2\int_0^t E\left[|X_{s\wedge\tau_n}|^4\right] ds + 2c^2 \int_0^t E\left[|X_{s\wedge\tau_n}|^2\right] ds \\ + 6\int_0^t E\left[|X_{s\wedge\tau_n}|^2 \operatorname{tr}\, a(X_{s\wedge\tau_n})\right] ds. \end{aligned}$$

Again, we use (10.4.10) and the condition on tr a to bound $E|X_{t\wedge\tau_n}|^4$. A Gronwall argument yields

$$E[|X_t|^4] \leq C_1[|x|^4 + t]e^{C_2 t}$$

where C_1, C_2 are suitable constants.

From the Doob inequality, one obtains

$$\begin{aligned} E\left\{\sup_{t\leq T}\left|\int_0^t X(s)\, dM(s)\right|^2\right\} &\leq 4\int_0^T E\left(\sum_{i=1}^d\sum_{j=1}^d \int_0^t X_i(s)X_j(s)\, d[M_i, M_j]_s\right) \\ &\leq 4\int_0^T E\left(|X_s|^2 \text{ tr } a(X_s)\right) ds \\ &\leq 4K\int_0^T E\left(|X_s|^2 + |X_s|^4\right) ds. \end{aligned} \tag{10.4.12}$$

Upon taking supremum over $[0, T]$ termwise in (10.4.7), taking expectation, and using (10.4.12), the required bound is obtained for a constant $k > 0$. ■

Remembering that the process $\{X_t\}$ depends on $\epsilon > 0$, let us set $\epsilon = \frac{1}{n}$ and denote the resulting probability distribution on $D([0, T] : \mathbb{R}^d)$ as P^n.

Lemma 10.4.3 *The sequence $\{P^n\}$ is tight in the space of probability measures on $D([0, T] : \mathbb{R}^d)$. Every limit point P of $\{P^n\}$ has the property that $P[C([0, T] : \mathbb{R}^d)] = 1$.*

Proof We will show tightness by the Aldous criterion. Fix any $t \in [0, T]$, and consider the marginals $\{P_t^n\}$. The previous lemma says that

$$\sup_n E_{P^n}\{\sup_{s\leq T} |Z_s|^2\} \leq k[T + |x|^2]e^{kT}.$$

Therefore, given any $\epsilon > 0$, there exists an R such that $P^n\{|Z_t| \geq R\} < \epsilon$ for all n; that is, $\{P_t^n\}$ is tight.

Next, take any stopping time $\tau < T$ and $\delta > 0$, and consider

$$\begin{aligned} E_{P^n}\left(\int_\tau^{\tau+\delta} \text{tr } a(Z(s))\, ds\right) &\leq E_{P^n}\left\{E\left(\int_0^\delta K(1 + |Z_s|^2)\, ds \mid Z_\tau\right)\right\} \\ &\leq K\delta + \int_0^\delta E_{P^n}(|Z_s|^2)\, ds \\ &\leq C^*\delta \end{aligned}$$

where C^* is the bound in (10.4.11). From this, tightness of $\{P^n\}$ follows.

Let P be the limit of any weakly convergent subsequence of $\{P^n\}$. We will denote the subsequence by the full sequence $\{P^n\}$ for simplicity of notation. For any $\delta > 0$, the set

$$\left\{z : \sup_{0\leq t\leq T} |\Delta z(t)| \leq \delta\right\}$$

is a closed set in $D([0, T] : \mathbb{R}^d)$, and therefore,

$$P\left\{z : \sup_{0\leq t\leq T} |\Delta z(t)| \leq \delta\right\} \geq \limsup_{n\to\infty} P^n \left\{z : \sup_{0\leq t\leq T} |\Delta z(t)| \leq \delta\right\} = 1.$$

Thus, $P\{C([0, T] : \mathbb{R}^d)\} = 1$. ■

Theorem 10.4.4 *The sequence of probability measures $\{P^n\}$ converges weakly to a solution P of the martingale problem for (A, δ_x), where A is given by Equation (10.4.1). Moreover,*

$$E_P[\sup_{0\leq t\leq T} |Z_t|^2] < \infty.$$

Proof Let $\{P^{n_j}\}$ denote a weakly convergent subsequence, and as before we will call it $\{P^n\}$. Since $\sup_n E_{P^n}\{\sup_{s\leq T} |Z_s|^2\} \leq k[T + |x|^2]e^{kT}$, we have, by the Fatou lemma, that $E_P[\sup_{0\leq t\leq T} |Z_t|^2] < \infty$.

Since $P_n \to P$ weakly, by the Skorohod representation theorem, there exists a sequence $\{\xi_n\}$ of $\mathbb{R}^d$-valued random variables and an $\mathbb{R}^d$-valued random variable ξ on some probability space $(\Omega', \mathcal{F}', (\mathcal{F}'_t), P')$ such that the law of ξ_n equals P_n and that of ξ equals P, and $\xi_n \to \xi$ a.s. as $n \to \infty$. Define $\xi_n^\theta(t) = (\xi_n(t), \theta)$ for any $\theta \in \mathbb{R}^d$ and all $n \in \mathbb{N}$. Let $\xi^\theta(t)$ be similarly defined.

To prove the convergence of the martingale problems, we will fix any $0 \leq s < t \leq T$, and any bounded random variable ψ which is $\mathcal{F}'_s$-measurable. For any real-valued function $f \in C_0^2(\mathbb{R}^d)$, define

$$\begin{aligned} M_n(f) &= f(\xi_n^\theta(t)) - f(\xi_n^\theta(s)) - \int_s^t f'(\xi_n^\theta(u))(\theta, b(u, \xi_n(u)))\, du \\ &\quad - \frac{1}{2}\int_s^t f''(\xi_n^\theta(u))(\theta, a((\xi_n(u))\theta)\, du - (R_\theta(t) - R_\theta(s)) \end{aligned}$$

where $R_\theta(t)$ is the the remainder term given in Equation (10.4.3). Let

$$\begin{aligned} M(f) &= f(\xi^\theta(t)) - f(\xi^\theta(s)) - \int_s^t f'(\xi^\theta(u))(\theta, b(u, \xi(u)))\, du \\ &\quad - \frac{1}{2}\int_s^t f''(\xi^\theta(u))(\theta, a((\xi(u))\theta)\, du. \end{aligned}$$

We need to prove that $\lim_{n\to\infty} E_{P'}\{[M_n(f) - M(f)]\psi\} = 0$.

By tightness of P_n, given any $\epsilon > 0$, there exists a compact set F in $D([0,T] : \mathbb{R}^d)$ such that

$$P'(\xi_n \in F^c) < \epsilon \quad \text{for all } n \text{, and} \quad P'(\xi \in F^c) < \epsilon.$$

Therefore, for any $a > 0$, and any $0 \le s < t \le T$

$$\begin{aligned}
&\limsup_{n\to\infty} P'\left\{\left|\int_s^u \left[a(s,\xi_n(u)) - a(u,\xi(u))\right] du\right| > a\right\} \\
&\le 2\epsilon + \limsup_{n\to\infty} \frac{1}{a} E_{P'}\left|\int_0^T \left[a(u,\xi_n(u)) - a(u,\xi(u))\right] 1_F(\xi_n) 1_F(\xi)\, du\right. \\
&\le 2\epsilon.
\end{aligned}$$

With ϵ being arbitrary, $\int_0^T [a(u,\xi_n(u)) - a(u,\xi(u))]du \to 0$ in probability. Likewise, $\int_0^T [(\xi_n(u), b(\xi_n(u))) - (\xi(u), b(\xi(u)))]du \to 0$ in probability. Since $R_\theta(t) - R_\theta(s)$ has already been shown to converge to 0 a.s., $\xi_n(t) \to \xi(t)$, and $\xi_n(s) \to \xi(s)$, we obtain

$$M_n(f) \to M(f) \quad \text{in probability as} \quad n \to \infty.$$

From the proof of Lemma 10.4.2, we have

$$\sup_n E|M_n(f)|^2 < \infty$$

so that $\{M_n(f)\}$ is uniformly integrable. Hence, $\lim_{n\to\infty} E[|(M_n(f) - M(f))\psi|] = 0$. For each n, we know that $E[M_n(f)\psi] = 0$. Therefore, $E[M(f)\psi] = 0$. In other words, the process

$$f(\xi_\theta(t)) - f((x,\theta)) - \int_0^t f'(\xi_\theta(s-))(b(\xi_s),\theta)\, ds - \int_0^t \frac{1}{2} f''(\xi_\theta(s-))(\theta, a(\xi_s)\theta)\, ds$$

is a P'-martingale, which finishes the proof. ■

Exercises

1. Let N_t be a Poisson process with parameter $\lambda > 0$. Show that $\lim_{t\to s} E[(N_t - N_s)^2] = 0$ though N has discontinuous sample paths.
2. Let N_t be a Poisson process with parameter $\lambda > 0$, and $\mathcal{F}_t = \sigma(N_s : 0 \le s \le t)$.
   (i) Show that the process $M_t := N_t - \lambda t$ is an $\mathcal{F}_t$-martingale.
   (ii) Show that $M_t^2 - \lambda t$ is an $\mathcal{F}_t$-martingale.

3. Let $S_t = \sum_{j=1}^{N_t} Y_j$ denote the compound Poisson process defined in Section 10.1. Let $m_1 = EY_1$ and $m_2 = E(Y_1^2)$ exist. Suppose that $\mathcal{F}_t = \sigma(S_s : 0 \leq s \leq t)$.
   (i) Show that $S_t - \lambda m_1 t$ is an $\mathcal{F}_t$-martingale.
   (ii) Show that $(S_t - \lambda m_1 t)^2 - \lambda m_2 t$ is an $\mathcal{F}_t$-martingale.
4. Let V be an adapted cadlag process with paths of finite variation on compacts. Show that

$$[V]_t = \sum_{s \leq t} (\Delta V_s)^2;$$

that is, V has pure jump quadratic variation.
5. Let M be a cadlag martingale with $M_0 = 0$. Prove that the infinite product

$$X_t := \Pi_{0<s\leq t}(1 + \Delta M_s) \exp\{-\Delta M_s\}$$

is convergent and that X is a finite variation process. Hint: It suffices to consider the product only for s such that $|\Delta M_s| < 1/2$. Take the logarithm, and use the inequality $|\log(1 + x) - x| \leq x^2$ to prove the absolute convergence of the series.
6. Let M and X be as in the previous problem. Define $Y_t = M_t - \frac{1}{2}[M]_t^c$. Use the Itô formula with the function $f(x, y) = xe^y$ to show that $Z_t = f(X_t, Y_t)$ solves the integral equation

$$Z_t = 1 + \int_0^t Z_{s-}\, dM_s.$$

Z is known as the stochastic exponential of M when M is a cadlag martingale. In fact, writing it explicitly, we have

$$Z_t = \exp\left\{M_t - \frac{1}{2}[M]_t\right\} \Pi_{0<s\leq t}(1 + \Delta M_s) \exp\left\{-\Delta M_s + \frac{1}{2}(\Delta M_s)^2\right\}.$$

It is denoted as $\mathcal{E}(M)$.
7. If M and N are two cadlag martingales, show that

$$\mathcal{E}(M)\mathcal{E}(N) = \mathcal{E}(M + N) + [M, N]).$$

11 Invariant Measures and Ergodicity

Consider a dynamical system with state space E. If $x \in E$ is the initial state of the system, let us denote the state of the system at time t as $T_t x$, where $T_t : E \to E$ is a measure-preserving transformation. The equality $T_t(T_s x) = T_{t+s}x$ holds in several systems. Let ϕ be an observable real-valued function on E. The time taken to obtain observations is much longer than the natural time scale for molecular interactions. Therefore, we end up with measurements of the average

$$\frac{1}{t}\int_0^t \phi(T_s x)\, ds$$

instead of $\phi(x)$. Boltzmann hypothesized that if orbits wandered all over the the state space, then the time averages would converge, as $t \to \infty$, to a phase (state) space average. Later, Gibbs identified the conditions under which the time averages converged to a phase-space mean.

The Kolmogorov strong law of large numbers is an ergodic theorem for any given sequence $\{X_n\}$ of independent, identically distributed random variables with finite mean. As an ergodic result, it is set in discrete time with the transformation T being the translation $Tx_n = x_{n+1}$ where $\{x_n\}$ is a sample path of the iid sequence. Stationary stochastic processes provide a natural context to obtain generalizations of the strong law of large numbers. In physical applications, an a.e. limit is required in an ergodic theorem, and the strong law of large numbers fits in perfectly. However, the very next instance of an ergodic theorem that we encounter in probability theory is the one for positive-recurrent, irreducible, aperiodic Markov chains, where the limit is not in the almost sure sense. There is a mathematical reason for switching from a.e. convergence to other modes of convergence such as $L^2(P)$ convergence. The use of Hilbert space methods to prove these limit theorems, though mathematically elegant, prevents us from making a.s. conclusions.

In 1931, Birkhoff proved an important result known as the ergodic theorem while von Neumann established the mean ergodic theorem. Since then, several improvements, extensions, and refinements have been made of these results. The ensuing theory is one of the major topics in ergodic theory. The avid reader can consult books on ergodic theory by Halmos, Sinai, Petersen, and Walters, to name a few.

In this chapter, our aim is to provide an elementary introduction to ergodic behavior of solutions of stochastic differential equations. We start with simple examples and prove a mean ergodic theorem for one-dimensional stochastic differential equations. This is a result of Skorohod and can be extended to more general one-dimensional stochastic differential equations. Next, the Krylov-Bogolyubov criterion that ensures the existence of invariant measures for diffusion processes and, in more generality, for time-homogeneous Markov processes is proved. A simple criterion for the uniqueness of invariant measures is proved as well. Next, we introduce the notions of ergodic measures, strong Feller property and irreducibility. A useful theorem of Doob is proved that states that the strong Feller property, and irreducibility imply ergodicity of an invariant measure. Examples are given to illustrate the ideas involved in establishing the ergodic behavior of stochastic systems.

11.1 Introduction

Suppose that $\{X_t\}$ is an $\mathbb{R}^d$-valued stochastic process. Let μ be a probability measure on $(\mathbb{R}^d, \mathcal{B}(\mathbb{R}^d))$ with the following property for all $t \geq 0$:

$$\mathcal{L}(X_0) = \mu \text{ implies } \mathcal{L}(X_t) = \mu$$

where $\mathcal{L}$ denotes probability law or distribution. Then we call μ an invariant measure or a stationary measure for the process $\{X_t\}$.

Example 11.1.1 *Consider the Ornstein-Uhlenbeck process*

$$X_t = \xi - \alpha \int_0^t X_s \, ds + W_t$$

for all $t \geq 0$. Here, $\alpha > 0$ is a constant, and the initial variable ξ is independent of the Wiener process W_t. We know that the solution is given by

$$X_t = \xi e^{-\alpha t} + \int_0^t e^{-\alpha(t-s)} \, dW_s.$$

Let ξ be in $L^2(P)$. Then, X_t has Gaussian distribution with expectation $EX_t = e^{-\alpha t}E(\xi)$ and variance

$$V(X_t) = e^{-2\alpha t}V(\xi) + \frac{1 - e^{-2\alpha t}}{2\alpha}.$$

Therefore, if we choose ξ to be a $N(0, \frac{1}{2\alpha})$ random variable, then X_t has a $N(0, \frac{1}{2\alpha})$ distribution for all $t \geq 0$.

Thus, the $N(0, \frac{1}{2\alpha})$ distribution is an invariant measure for $\{X_t\}$. The invariant measure being absolutely continuous with respect to the Lebesgue measure, we have the invariant density function

$$f(x) = \sqrt{\frac{\alpha}{\pi}}\, e^{-\alpha x^2} \quad \text{for} \quad -\infty < x < \infty.$$

Example 11.1.2 *For the one-dimensional Wiener process X, an invariant measure does not exist. To see this, assume that there exists an invariant measure μ; that is, if X_0 is a random variable with distribution μ, then the distribution of X_t is μ as well. Hence, for any $n \in \mathbb{N}$,*

$$\begin{aligned}\mu\{[-n,n]\} &= \frac{1}{\sqrt{2\pi t}} \int_{[-n,n]} \int e^{-\frac{(x-y)^2}{2t}}\, d\mu(x)\, dy \\ &\le \frac{1}{\sqrt{2\pi t}} 2n\end{aligned}$$

Allowing $t \to \infty$, $\mu\{[-n,n]\} = 0$. Hence $\mu = 0$, a contradiction, since μ is a probability measure.

Example 11.1.3 *Consider the one-dimensional stochastic differential equation*

$$dX_t = b(X_t)\, dt + dW_t$$

with $X_0 = \xi$, which is independent of the Wiener process W_t. By the Itô formula, for any $\phi \in C_b^2(\mathbb{R})$, we have

$$\phi(X_t) = \phi(X_0) + \int_0^t \phi'(X_s)\, dW_s + \int_0^t L\phi(X_s)\, ds$$

where $L\phi(x) = \frac{1}{2}\phi''(x) + b(x)\phi'(x)$. Suppose that an invariant measure μ for X_t exists. Using it, take expectation on both sides to infer that

$$\int L\phi(x)\, \mu(dx) = 0.$$

If μ has a smooth enough density function f, and if b is differentiable, then by an integration by parts, we obtain

$$\int \phi(x) \left\{ \frac{1}{2} f''(x) - (bf)'(x) \right\} dx = 0;$$

that is, $\frac{d}{dx}(\frac{1}{2}f'(x) - bf(x)) = 0$. Hence, we infer that $\frac{1}{2}f' - (bf) = 0$, since f is a density function. The form of the density of an invariant measure is therefore given by

$$f(x) = C \exp\{2 \int_0^x b(y)\, dy\}.$$

Suppose that an invariant measure μ exists for a process $\{X_t\}$. Then we say that μ is an ergodic measure for $\{X_t\}$ if for all $\phi \in L^2(\mu)$,

$$\lim_{t\to\infty} \frac{1}{t} \int_0^t \phi(X_s)\, ds = \int \phi\, d\mu \quad \text{in } L^2(\mu). \tag{11.1.1}$$

Here, note that we have defined an ergodic measure in the mean-square sense. We will prove, in the next section, the existence of an ergodic measure for one-dimensional diffusions under appropriate hypotheses. Our proof will avoid the use of Markov property so that it can be used in more generality.

11.2 Ergodicity for One-dimensional Diffusions

Let us consider the one-dimensional stochastic differential equation

$$dX_t = b(X_t)\, dt + \sigma(X_t)\, dW_t \tag{11.2.1}$$

with $X_0 = x \in \mathbb{R}^1$, where b and σ are Lipschitz continuous functions and W is a standard one-dimensional Wiener process. We often write the solution as X_t^x to indicate the solution was started in state x.

Let σ be positive valued and satisfy

$$\int_{\mathbb{R}^1} \frac{1}{\sigma(x)^2} dx < \infty.$$

Define the function

$$h(x) = \int_0^x \exp\left\{-\int_0^z \frac{2b(u)}{\sigma^2(u)}\, du\right\} dz. \tag{11.2.2}$$

Let $\tilde{\sigma}(x) = h'(h^{-1}(x))\sigma(h^{-1}(x))$. We will work under the following hypotheses:

Hypotheses H

H.1 $\lim_{x\to\infty} h(x) = \infty$ and $\lim_{x\to-\infty} h(x) = -\infty$.

H.2 $\tilde{\sigma} > 0$ and Lipschitz continuous with $\int_{\mathbb{R}^1} \frac{1}{\tilde{\sigma}(x)^2}\, dx < \infty$.

Proposition 11.2.1 *Assume hypotheses H. If $\{X_t^x\}$ and $\{X_t^y\}$ are two solutions of the stochastic differential equation (11.2.1) started at x and y respectively, then*

$$\lim_{t\to\infty} \frac{1}{t} E \int_0^t \{g(X_s^x) - g(X_s^y)\}\, ds = 0 \tag{11.2.3}$$

for all g, bounded and Borel measurable.

Proof Applying the Itô formula for the function h given by (11.2.2), we obtain

$$h(X_t^x) = h(x) + \int_0^t h'(X_s^x)\,\sigma(X_s^x)\,dW_s.$$

Let Y_t^x denote $h(X_t^x)$, and we can write the above equation as

$$Y_t^x = h(x) + \int_0^t \tilde{\sigma}(h(X_s^x))\,dW_s.$$

Let ϕ be a bounded, Borel-measurable function. Then, by the occupation density formula, we have

$$\frac{1}{t}\left|\int_0^t E\{\phi(X_s^x) - \phi(X_s^y)\}\,ds\right| = \frac{1}{t}\left|E\int_{-\infty}^{\infty} \frac{\phi(a)}{\tilde{\sigma}^2(a)}\left\{L_t^a(Y^x) - L_t^a(Y^y)\right\}\,da\right| \tag{11.2.4}$$

where L_t^a denotes the local time at a in the interval $[0, t]$. Using the Fubini theorem and Tanaka formula, the right side of (11.2.4) becomes

$$= \frac{1}{t}\left|E\int_{-\infty}^{\infty} \frac{\phi(a)}{\tilde{\sigma}^2(a)}\left\{(E|Y_t^x - a| - |h(x) - a|) - (E|Y_t^y - a| - |h(y) - a|)\right\}\,da\right|.$$

Let $x > y$ so that $h(x) > h(y)$. Then, $Y_s^x \geq Y_s^y$ for all s. For, if τ denotes the first time that that the two processes are equal, then $Y_{\tau+t}^x = Y_{\tau+t}^y$ for all $t \geq 0$, since they are both solutions of the same equation started at the common value Y_τ^x. Therefore,

$$\begin{aligned}\frac{1}{t}\left|\int_0^t E\{\phi(X_s^x) - \phi(X_s^y)\}\,ds\right| &\leq \frac{1}{t}\int_{-\infty}^{\infty} \frac{\phi(a)}{\tilde{\sigma}^2(a)}\left\{(E(Y_t^x - Y_t^y) + (h(x) - h(y))\right\}\,da \\ &= 2((h(x) - h(y))\frac{1}{t}\int_{-\infty}^{\infty} \frac{\phi(a)}{\tilde{\sigma}^2(a)}\,da,\end{aligned}$$

which tends to 0 as $t \to \infty$. ■

Proposition 11.2.1 shows that the time averages $\lim_{t\to\infty} \frac{1}{t} E\int_0^t g(X_s^x)\,ds$ forget the initial state x as t becomes large. We proceed to find the limit of these time averages. From now on, we will use the notation

$$\beta(a) := \frac{1}{h'(a)\sigma^2(a)}, \quad \text{and } D^{-1} = \int_{-\infty}^{\infty} \frac{1}{\tilde{\sigma}^2(a)}\,da = \int_{-\infty}^{\infty} \beta(a)\,da. \tag{11.2.5}$$

Proposition 11.2.2 *Assume hypotheses H. If $\{X_t^x\}$ is the solution of the stochastic differential equation (11.2.1) started at x, then*

$$\liminf_{t\to\infty} \frac{1}{t} E\left\{\int_0^t g(X_s^x)\,ds\right\} = \frac{\int_{\mathbb{R}^1} g(a)\beta(a)\,da}{\int_{\mathbb{R}^1} \beta(a)\,da} \tag{11.2.6}$$

for all nonnegative, bounded, Borel-measurable functions g.

Proof As before, let $Y_t^x = h(X_t^x)$ so that by Tanaka's formula, one obtains

$$EL_t^a(Y^x) = E|Y_t^x - a| - |h(x) - a|.$$

Therefore,

$$E|Y_t^x - h(x)| - 2|h(x) - a| \leq EL_t^a(Y^x) \leq E|Y_t^x - h(x)|. \tag{11.2.7}$$

As a consequence,

$$\liminf_{t\to\infty} \frac{1}{t} E|Y_t^x - h(x)| = \liminf_{t\to\infty} \frac{1}{t} EL_t^a(Y^x).$$

Let ϕ be any nonnegative, bounded, Borel-measurable function. Let us define

$$\psi(a) := \frac{\phi(a)}{\tilde{\sigma}^2(a)}.$$

Consider

$$\begin{aligned}\int_{\mathbb{R}^1} \liminf_{t\to\infty} \left(\frac{1}{t} E|Y_t^x - h(x)|\right) \psi(a)\, da &= \int_{\mathbb{R}^1} \liminf_{t\to\infty} \left(\frac{1}{t} EL_t^a(Y^x)\right) \psi(a)\, da \\ &\leq \liminf_{t\to\infty} \int_{\mathbb{R}^1} \left(\frac{1}{t} EL_t^a(Y^x)\right) \psi(a)\, da\end{aligned}$$

by the Fatou lemma; continuing,

$$\leq \liminf_{t\to\infty} \frac{1}{t} E|Y_t^x - h(x)| \int_{\mathbb{R}^1} \psi(a)\, da \tag{11.2.8}$$

by using (11.2.7). From (11.2.8), we infer that

$$\begin{aligned}\liminf_{t\to\infty} \left(\frac{1}{t} E|Y_t^x - h(x)|\right) \int_{\mathbb{R}^1} \psi(a)\, da &= \liminf_{t\to\infty} \frac{1}{t} \int_{\mathbb{R}^1} EL_t^a(Y^x)\, \psi(a)\, da \\ &= \liminf_{t\to\infty} \frac{1}{t} E \int_0^t \phi(Y_s^x)\, ds\end{aligned} \tag{11.2.9}$$

Setting $\phi \equiv 1$ in (11.2.9), we obtain

$$\liminf_{t\to\infty} \left(\frac{1}{t} E|Y_t^x - h(x)|\right) = \left[\int_{\mathbb{R}^1} \frac{1}{\tilde{\sigma}^2(a)}\, da\right]^{-1}.$$

Using this in (11.2.9), it follows that

$$\liminf_{t\to\infty} \frac{1}{t} E \int_0^t \phi(Y_s^x)\, ds = \left[\int_{\mathbb{R}^1} \frac{1}{\tilde{\sigma}^2(a)}\, da\right]^{-1} \int_{\mathbb{R}^1} \psi(a)\, da.$$

The proof is completed upon taking $\phi = g(h^{-1})$ and noting that

$$\int_{\mathbb{R}^1} \frac{1}{\tilde{\sigma}^2(a)}\, da = \int_{\mathbb{R}^1} \beta(a)\, da.$$

■

Proceeding in a similar manner, one can obtain

$$\limsup_{t\to\infty} \frac{1}{t} E\left\{\int_0^t g(X_s^x)\, ds\right\} = \frac{\int_{\mathbb{R}^1} g(a)\beta(a)\, da}{\int_{\mathbb{R}^1} \beta(a)\, da} \tag{11.2.10}$$

for all bounded, nonnegative functions g with compact support. Equations (11.2.6) and (11.2.10) show that for all such functions g,

$$\lim_{t\to\infty} \frac{1}{t} E\left\{\int_0^t g(X_s^x)\, ds\right\} = \frac{\int_{\mathbb{R}^1} g(a)\beta(a)\, da}{\int_{\mathbb{R}^1} \beta(a)\, da}. \tag{11.2.11}$$

The same method of proof also yields

Lemma 11.2.3 *Assume hypotheses H. Let $\{X_t^x\}$ be the solution of the stochastic differential equation (11.2.1) started at x. If g is a bounded, nonnegative function with compact support, then*

$$\liminf_{t\to\infty} \frac{2}{t^2} E\left\{\int_0^t g(X_s^x) s\, ds\right\} = \frac{\int_{\mathbb{R}^1} g(a)\beta(a)\, da}{\int_{\mathbb{R}^1} \beta(a)\, da}. \tag{11.2.12}$$

Lemma 11.2.4 *Assume hypotheses H. Let $\{X_t^x\}$ be the solution of the stochastic differential equation (11.2.1) started at x, and $Y_t^x = h(X_t^x)$. Then, for any two bounded, nonnegative functions ϕ_1, ϕ_2 with compact support,*

$$\liminf_{t\to\infty} \frac{1}{t^2} E \int_0^t \phi_2(Y_u^x) \int_u^t \phi_1(Y_s^x)\, ds\, du$$
$$= \liminf_{t\to\infty} \frac{1}{t} E \int_{\mathbb{R}^1} \frac{\phi_1(b)}{\tilde{\sigma}(b)} \int_0^t \phi_2(Y_u^x) \left\{|Y_t^x - b| - Y_u^x - b|\right\}\, du\, db. \tag{11.2.13}$$

Proof Using the occupation density formula, we can write

$$\frac{1}{t^2} E \int_0^t \phi_2(Y_u^x) \int_u^t \phi_1(Y_s^x)\, ds\, du = \frac{1}{t^2} E \int_0^t \phi_2(Y_u^x) \int_{\mathbb{R}^1} \{L_t^b(Y^x) - L_u^b(Y^x)\} \frac{\phi_1(b)}{\tilde{\sigma}^2(b)}\, db\, du. \tag{11.2.14}$$

Applying the Tanaka formula,

$$\liminf_{t\to\infty}\frac{1}{t^2}E\int_0^t \phi_2(Y_u^x)\int_u^t \phi_1(Y_s^x)\,ds\,du = \frac{1}{t^2}E\left[\int_{\mathbb{R}^1}\int_0^t \phi_2(Y_u^x)\Big\{|Y_t^x-b|-|Y_u^x-b| - \int_u^t \operatorname{sgn}(Y_v^x-b)\tilde{\sigma}(Y_v^x)\,dW_v\Big\}\frac{\phi_1(b)}{\tilde{\sigma}^2(b)}\,db\right]. \tag{11.2.15}$$

Consider the expression

$$\frac{1}{t^2}\int_{\mathbb{R}^1}\int_0^t \phi_2(Y_u^x)\left\{\int_u^t \operatorname{sgn}(Y_v^x-b)\tilde{\sigma}(Y_v^x)\,dW_v\right\}\frac{\phi_1(b)}{\tilde{\sigma}^2(b)}\,db.$$

It can be written as

$$\frac{1}{t^2}\int_{\mathbb{R}^1}\frac{\phi_1(b)}{\tilde{\sigma}^2(b)}\int_0^t \phi_2(Y_u^x)(M_t-M_u)\,du\,db \tag{11.2.16}$$

with $M_s = \int_0^s \operatorname{sgn}(Y_v^x-b)\tilde{\sigma}(Y_v^x)\,dW_v$ for all s.

Define $K_s = \int_0^s \phi_2(Y_u^x)\,du$. Then (11.2.16) can be written as

$$\begin{aligned}
&= \frac{1}{t^2}\int_{\mathbb{R}^1}\frac{\phi_1(b)}{\tilde{\sigma}^2(b)}\int_0^t (M_t-M_u)\,dK_u\,db\\
&= \frac{1}{t^2}\int_{\mathbb{R}^1}\frac{\phi_1(b)}{\tilde{\sigma}^2(b)}\int_0^t K_u\,dM_u\,db
\end{aligned}$$

by stochastic integration by parts. The process $\int_0^t K_u\,dM_u$, being a martingale, has its expectation zero. Therefore, the last term on the right side of (11.2.15) is zero. With this observation, Equation (11.2.15) yields the result. ∎

The statement of the above lemma holds if lim inf is replaced by lim sup.

Theorem 11.2.5 (L^2 ergodic theorem) *Under the hypotheses H, if $\{X_t^x\}$ denotes the solution of the stochastic differential equation (11.2.1), then as $t\to\infty$,*

$$\frac{1}{t}E\left\{\int_0^t g(X_s^x)\,ds\right\} \xrightarrow{L^2(P)} D\int_{\mathbb{R}^1} g(a)\beta(a)\,da \tag{11.2.17}$$

for all bounded, Borel-measurable functions g with β and D given by Equation (11.2.5).

Proof

Step 1 Let ϕ be a nonnegative, bounded, Borel-measurable function with compact support. In this step, we will prove that

$$\limsup_{t\to\infty} E\left(\frac{1}{t}\int_0^t \phi(Y_s^x)\,ds\right)^2 \le \left(D\int_{\mathbb{R}^1}\frac{\phi(a)}{\tilde{\sigma}^2(a)}\,da\right)^2. \tag{11.2.18}$$

Consider

$$\limsup_{t\to\infty} E\left(\frac{1}{t}\int_0^t \phi(Y_s^x)\,ds\right)^2 = \limsup_{t\to\infty}\frac{2}{t^2}E\int_0^t \phi(Y_u^x)\int_u^t \phi(Y_s^x)\,ds\,du$$
$$= \limsup_{t\to\infty}\frac{2}{t^2}E\int_{\mathbb{R}^1}\frac{\phi(b)}{\tilde{\sigma}^2(b)}\int_0^t \phi(Y_u^x)\{Y_t^x - b| - |Y_u^x - b|\}\,du\,db$$

by Lemma 11.2.4 with $\phi_1 = \phi_2 = \phi$. We would like to factor out $\int_{\mathbb{R}^1}\frac{\phi(b)}{\tilde{\sigma}^2(b)}\,db$ from the right side of the above equation. For this purpose, multiply both sides by $\int_{-M}^{M}\frac{dv}{\tilde{\sigma}^2(v)}\,dv$. Then,

$$\limsup_{t\to\infty} E\int_{-M}^{M}\frac{dv}{\tilde{\sigma}^2(v)}\,dv\left(\frac{1}{t}\int_0^t \phi(Y_s^x)\,ds\right)^2$$
$$= \limsup_{t\to\infty}\frac{2}{t^2}E\int_{-M}^{M}\frac{1}{\tilde{\sigma}^2(v)}\int_{\mathbb{R}^1}\frac{\phi(b)}{\tilde{\sigma}^2(b)}\int_0^t \phi(Y_u^x)\{Y_t^x - v| - |Y_u^x - v|\}\,du\,db\,dv$$

by using the fact that ϕ is bounded with compact support; continuing,

$$= \left(\int_{\mathbb{R}^1}\frac{\phi(b)}{\tilde{\sigma}^2(b)}\,db\right)\limsup_{t\to\infty}\frac{2}{t^2}E\int_{-M}^{M}\frac{1}{\tilde{\sigma}^2(v)}\int_0^t \phi(Y_u^x)\{Y_t^x - v| - |Y_u^x - v|\}\,du\,dv$$
$$= \left(\int_{\mathbb{R}^1}\frac{\phi(b)}{\tilde{\sigma}^2(b)}\,db\right)\limsup_{t\to\infty}\frac{2}{t^2}E\int_0^t \phi(Y_u^x)\int_u^t 1_{[-M,M]}(Y_s^x)\,ds\,du$$

by Lemma 11.2.4 with $\phi_1 = 1_{[-M,M]}$ and $\phi_2 = \phi$;

$$\leq \left(\int_{\mathbb{R}^1}\frac{\phi(b)}{\tilde{\sigma}^2(b)}\,db\right)\limsup_{t\to\infty}\frac{2}{t^2}E\int_0^t \phi(Y_u^x)(t-u)\,du$$

by using nonnegativity of ϕ. The above expression equals

$$D\left[\int_{\mathbb{R}^1}\frac{\phi(b)}{\tilde{\sigma}^2(b)}\,db\right]^2$$

by (11.2.11) and Lemma 11.2.3. Thus, we have proved that

$$\left(\int_{-M}^{M}\frac{dv}{\tilde{\sigma}^2(v)}\,dv\right)\limsup_{t\to\infty} E\left(\frac{1}{t}\int_0^t \phi(Y_s^x)\,ds\right)^2 \leq D\left[\int_{\mathbb{R}^1}\frac{\phi(b)}{\tilde{\sigma}^2(b)}\,db\right]^2$$

for all $M > 0$. Letting $M \to \infty$, we obtain the inequality (11.2.18).

Step 2 Using Jensen's inequality in (11.2.18), we obtain

$$\begin{aligned}\left(D\int_{\mathbb{R}^1}\frac{\phi(a)}{\tilde{\sigma}^2(a)}\,da\right)^2 &\geq \limsup_{t\to\infty} E\left(\frac{1}{t}\int_0^t \phi(Y_s^x)\,ds\right)^2\\ &\geq \liminf_{t\to\infty} E\left(\frac{1}{t}\int_0^t \phi(Y_s^x)\,ds\right)^2\\ &\geq \liminf_{t\to\infty}\left[E\left(\frac{1}{t}\int_0^t \phi(Y_s^x)\,ds\right)\right]^2\\ &= \left(D\int_{\mathbb{R}^1}\frac{\phi(a)}{\tilde{\sigma}^2(a)}\,da\right)^2\end{aligned}$$

by Proposition 11.2.2. Thus,

$$\lim_{t\to\infty}\left[E\left(\frac{1}{t}\int_0^t \phi(Y_s^x)\,ds\right)\right]^2 = \left(D\int_{\mathbb{R}^1}\frac{\phi(a)}{\tilde{\sigma}^2(a)}\,da\right)^2.$$

Taking $\phi = g(h^{-1})$, the above equality and Proposition 11.2.5 imply the required $L^2(P)$ convergence for all bounded, Borel-measurable $g \geq 0$ with compact support.

The general case follows upon first proving the result for bounded, Borel-measurable $g \geq 0$ by using the function $g_m = g1_{[-m,m]}$ for any given $m > 0$ and then letting $m \to \infty$. If g is not necessarily nonnegative, we can argue by splitting g into its positive and negative parts. ■

11.3 Invariant Measures for d-dimensional Diffusions

Following the method used in Example 11.1.3, we will show the existence and uniqueness of invariant measures for a class of d-dimensional stochastic differential equations. Consider

$$X_t = X_0 + \int_0^t b(X_s)\,ds + W_t$$

where b is a vector with each component b_i being a C^∞ function mapping $\mathbb{R}^d$ to $\mathbb{R}^1$. We will assume that b is Lipschitz continuous. W is a d-dimensional Wiener process, and X_0 is the initial random variable, independent of W. We know that the differential operator that corresponds to X_t is given by

$$Lf = \frac{\Delta}{2}f + b\cdot\nabla f \quad \text{for } f \in C^2(\mathbb{R}^d).$$

The formal adjoint of L is L^*, so that for all $g \in C^2(\mathbb{R}^d)$,

$$L^* g = \frac{\Delta}{2} g - \nabla \cdot (bg).$$

We constructed an invariant measure in Example 11.1.3 by solving $L^* g = 0$, which forms the motivation for the following theorem due to Varadhan.

Theorem 11.3.1 *Suppose that there exists a nonnegative $C^2(\mathbb{R}^d)$ function g such that $L^* g = 0$ and $\int g(x)\, dx = 1$. Then the Borel measure μ defined by $\mu(B) = \int_B g(x)\, dx$ is an invariant measure for $\{X_t\}$.*

Proof If $X_0 = x$, we will call the corresponding solution X_t^x. Let G be a bounded open set in $\mathbb{R}^d$ with a smooth boundary. Let a non negative function $f \in C_0^\infty(G)$ be given, and $\tau = \inf\{t \geq 0 : X_t^x \notin G\}$. Then we know that $u(t,x) = E\{f(X_{t\wedge\tau}^x)\}$ is a solution of the initial-boundary value problem:

$$\begin{aligned} \frac{\partial u}{\partial t} &= Lu && \text{on } (0,\infty) \times G, \\ u(0,x) &= f(x) && \text{on } \{0\} \times \bar{G}, \text{ and} \\ u(t,x) &= 0 && \text{on the boundary of } G. \end{aligned}$$

With the smoothness of f and b, the solution is smooth and unique. By the probabilistic representation of u, we have $u(t,x) \geq 0$ for all $x \in G$. By the definition of g, and integration by parts, one obtains

$$\frac{\partial}{\partial t} \int_G u(t,x)\, g(x)\, dx = \frac{1}{2} \int_{\partial G} g(x) \frac{\partial u}{\partial n}(t,x)\, d\sigma$$

where n is the outer normal to G, and ∂G denotes the boundary of G. Since $u \geq 0$ in G, and $u = 0$ on ∂G, we get $\frac{\partial u}{\partial n}(t,x) \leq 0$ for all $t > 0$. Therefore,

$$\frac{\partial}{\partial t} \int_G u(t,x)\, g(x)\, dx \leq 0.$$

Consequently,

$$\int_G u(t,x) g(x) dx \leq \int_G u(0,x)\, g(x)\, dx = \int f(x)\, g(x)\, dx.$$

Letting G to increase to $\mathbb{R}^d$, by Fatou's lemma, we obtain

$$E\{f(X_t^x)\} \leq \int f(x)\, g(x)\, dx. \tag{11.3.1}$$

Define a measure ν by

$$\nu(B) = \int_B P\{X_t^x \in B\}\, d\mu(x)$$

for all Borel sets B in $\mathbb{R}^d$. Take any $\phi \in C_0^\infty(\mathbb{R}^d)$ such that $\phi \geq 0$. Then,

$$\int \phi(y)\, d\nu(y) = \int E(\phi(X_t^x))\, d\mu(x)$$
$$\leq \int \phi(x)\, g(x)\, dx = \int \phi(x)\, d\mu(x)$$

by using (11.3.1). Thus, for all nonnegative $\phi \in C_0^\infty(\mathbb{R}^d)$,

$$\int \phi(y)\, d\nu(y) \leq \int \phi(x)\, d\mu(x).$$

Therefore, $\nu(B) \leq \mu(B)$ for all bounded Borel sets B, and, hence, for all Borel sets. Since μ, ν are probability measures, it follows that they are equal. In other words, μ is an invariant measure. ■

Remark 11.3.1 The converse of Theorem 11.3.1 holds as well; that is, if μ is an invariant measure for $\{X_t\}$, then there exists a C^2 function g such that $g \geq 0$ and $\int g(x)dx = 1$, with $L^*g = 0$. The measure μ is given by

$$\mu(B) = \int_B g(x)\, dx.$$

for all Borel sets B. The proof is left as an exercise.

Theorem 11.3.2 *Assume the hypotheses of the previous theorem. Suppose there are two invariant measures μ_1 and μ_2 for $\{X_t\}$. Then, $\mu_1 = \mu_2$.*

Proof We know that there exist smooth functions g_1 and g_2 such that $L^*g_i = 0$, and, for any Borel set B, $\mu_1(B) = \int_B g_1(x)\, dx$ and $\mu_2(B) = \int_B g_2(x)\, dx$. It suffices to show that $g_1 = g_2$ a.e.

Define $h = g_1 - g_2$. Since μ_1 and μ_2 are invariant measures,

$$\int p(t, x, B)\, d\mu_i(x) = \mu_i(B)$$

for any B, Borel; that is,

$$\int p(t, x, B)\, g_i(x)\, dx = \int_B g_i(x)\, dx.$$

Hence, $\int p(t,x,B)\, h(x)\, dx = \int_B h(x)\, dx$ for all Borel sets B, so that

$$\int p(t,x,y)\, h(x)\, dx = h(y) \quad \text{a.e.}$$

With h^+ denoting the positive part of h, the above equality implies

$$\begin{aligned}\int h^+(y)\, dy &= \int \left(\int h(x)\, p(t,x,y)\, dx \right)^+ dy \\ &\leq \int \int h^+(x)\, p(t,x,y)\, dx\, dy \\ &= \int h^+(x)\, dx \int p(t,x,y)\, dy \\ &= \int h^+(x)\, dx.\end{aligned}$$

Thus, $\int (\int h(x)\, p(t,x,y)\, dx)^+\, dy = \int \int h^+(x)\, p(t,x,y)\, dx\, dy$. As a consequence,

$$\left(\int h(x)\, p(t,x,y)\, dx \right)^+ = \int h^+(x)\, p(t,x,y)\, dx \quad \text{for a.e. } y \tag{11.3.2}$$

since $(\int h(x)\, p(t,x,y)\, dx)^+ \leq \int h^+(x)\, p(t,x,y)\, dx$. With $p(t,x,y) > 0$ for almost every x, y, we obtain from (11.3.2) that $h = h^+$ a.e. Likewise, by considering the negative part, h^-, we can obtain $h = h^-$ a.e. Therefore, $h = 0$ a.e., and the proof is over. ■

11.4 Existence and Uniqueness of Invariant Measures

In the context of $\mathbb{R}^d$-valued (and more general) Markov processes, we can write the definition of invariant measures using the semigroup associated with the processes. Suppose that $\{X_t^x : x \in \mathbb{R}^d,\ t \geq 0\}$ is a family of time-homogeneous, $\mathbb{R}^d$-valued Markov processes indexed by x, with continuous paths and $X_0^x = x$. Define

$$P_x(t,B) = P\{X_t^x \in B\},$$

the *transition* probability measure for all $t \geq 0$, $x \in \mathbb{R}^d$, and $B \in \mathcal{B}(\mathbb{R}^d)$. Then, $P(t,B)$ is a Borel-measurable function for all fixed t and B. The Markov property yields the Chapman-Kolmogorov equation:

$$P_x(t+s,B) = \int_{\mathbb{R}^d} P_x(t,dy) P_y(s,B). \tag{11.4.1}$$

For any given probability measure ν on $(\mathbb{R}^d, \mathcal{B}(\mathbb{R}^d))$, one can define a probability measure

$$P_\nu(t,B) := \int_{\mathbb{R}^d} P_x(t,B)\,\nu(dx).$$

If ξ is a random variable with distribution ν and is independent of $\{X^x : x \in \mathbb{R}^d\}$, then one can construct a continuous, time-homogeneous Markov process, denoted X^ξ, such that $X_0^\xi = \xi$, and for any $t > 0$, $P_\nu(t,\cdot)$ is the distribution of X_t^ξ.

One can define a semigroup of linear operators, S_t, on $B_b(\mathbb{R}^d)$ as follows:

$$S_t\phi(x) = E\phi(X_t^x) = \int_{\mathbb{R}^d} \phi(y)P_x(t,dy). \tag{11.4.2}$$

The semigroup, S_t, is a strongly continuous Markov semigroup. Indeed, S_t is a positivity-preserving contraction, and $S_t 1 = 1$.

Definition 11.4.1 *A probability measure μ on $(\mathbb{R}^d, \mathcal{B}(\mathbb{R}^d))$ is called an* **invariant measure** *or a* **stationary measure** *for the Markov process X if*

$$\mu(B) = \int_{\mathbb{R}^d} P_x(t,B)\,d\mu(x).$$

for all $B \in \mathcal{B}(\mathbb{R}^d)$, or equivalently, if for all $\phi \in C_b(\mathbb{R}^d)$, and all $t \geq 0$,

$$\int_{\mathbb{R}^d} \phi(x)\,d\mu(x) = \int_{\mathbb{R}^d} (S_t\phi)(x)\,d\mu(x). \tag{11.4.3}$$

In other words, μ is an invariant measure; if the probability distribution of the initial random variable is μ, then the probability distribution of X_t^ξ is also μ.

The process that we are interested in is $X := \{X_t\}$, an $\mathbb{R}^d$-valued diffusion process, with its differential operator A given by

$$Af(x) = \sum_{i=1}^d b_i(x)\frac{\partial f(x)}{\partial x_i} + \sum_{i,j=1}^d a_{ij}(x)\frac{\partial^2 f(x)}{\partial x_i \partial x_j}$$

for all $f \in C_b^2(\mathbb{R}^d)$. We know that X is a time-homogeneous Markov process with transition probability measure denoted by $P(x,t,B) = P_x(X_t \in B)$ for all $B \in \mathcal{B}$, where $\mathcal{B}$ is the Borel s-field in $\mathbb{R}^d$. From now on, an integral sign will stand for integration over $\mathbb{R}^d$ unless specific limits are given.

Definition 11.4.2 *The Markov semigroup S_t is said to have the* **Feller property** *if S_t maps $C_b(\mathbb{R}^d)$ to $C_b(\mathbb{R}^d)$. The semigroup possesses the* **strong Feller property** *if S_t maps $B_b(\mathbb{R}^d)$ to $C_b(\mathbb{R}^d)$.*

We know that, under standard hypotheses, the diffusion process X is Feller. However, we explicitly state it in the following results.

Proposition 11.4.1 *Suppose that S_t has the Feller property. Let μ_t denote $P_\nu(t, \cdot)$, the distribution of X_t^ξ. If μ_t converges weakly to μ as $t \to \infty$, then μ is an invariant measure for S_t.*

Proof With ν being the distribution of the initial variable ξ, we can write, for any $\phi \in C_b(\mathbb{R}^d)$,

$$\int_E \phi(x)\, d\mu_t(x) = \int (S_t\phi)(x)\, d\nu(x) \to \int \phi(x)\, d\mu(x)$$

as $t \to \infty$. Therefore, for any fixed $s > 0$,

$$\int_E (S_{t+s}\phi)(x)\, d\nu(x) \to \int \phi(x)\, d\mu(x) \tag{11.4.4}$$

as $t \to \infty$. On the other hand,

$$\int_E (S_{t+s}\phi)(x)\, d\nu(x) = \int (S_t(S_s\phi))(x)\, d\nu(x) \to \int (S_s\phi)(x)\, d\mu(x) \tag{11.4.5}$$

as $t \to \infty$, since $S_s\phi \in C_b(\mathbb{R}^d)$ by the Feller property. From (11.4.4) and (11.4.5), it follows that

$$\int (S_s\phi)(x)\, d\mu(x) = \int \phi(x)\, d\mu(x)$$

for all $s > 0$. In other words, μ is an invariant measure. ■

First, we present a result on the existence of invariant measures due to Krylov and Bogolyubov.

Theorem 11.4.2 *Consider the diffusion process X described above. Let B_R denote the closed R-ball in $\mathbb{R}^d$. Suppose that, for some $x \in \mathbb{R}^d$, there exists a sequence of times $t_n \uparrow \infty$ such that*

$$\lim_{R\to\infty} \sup_n \frac{1}{t_n} \int_0^{t_n} P_x\{X_t \in B_R^c\}\, dt = 0. \tag{11.4.6}$$

Then there exists an invariant measure μ for the process X.

Proof For all n, define a Borel measure μ_n on $\mathbb{R}^d$ by

$$\mu_n(B) = \frac{1}{t_n}\int_0^{t_n} P_x\{X_s \in B\}\,ds = \frac{1}{t_n}\int_0^{t_n} P(x,s,B)\,ds.$$

By the hypothesis (11.4.6), $\{\mu_n\}$ is a tight sequence of probability measures. Hence there exists a subsequence $\{\mu_{n_j}\}$ which converges weakly to a probability measure μ. We will show that μ is an invariant measure for X. In what follows, we will call this subsequence $\{\mu_n\}$ for notational simplicity.

Recall that X is a Feller process so that its semigroup S_t maps C_b to C_b. Hence, for any bounded continuous function f on $\mathbb{R}^d$,

$$\int (S_t f)(y)\mu(dy) = \lim_{n\to\infty}\int (S_t f)(y)\mu_n(dy). \tag{11.4.7}$$

We will consider $\int (S_t f)(y)\mu_n(dy)$ for any fixed n. By the Chapman-Kolmogorov equation and the Fubini theorem,

$$\begin{aligned}
\int (S_t f)(y)\mu_n(dy) &= \int \frac{1}{t_n}\int_0^{t_n} (S_t f)(y)P(x,s,dy)\,ds \\
&= \frac{1}{t_n}\int_0^{t_n}\int\int f(z)P(y,t,dz)P(x,s,dy)\,ds \\
&= \frac{1}{t_n}\int_0^{t_n}\int f(z)P(x,t+s,dz)\,ds \\
&= \frac{1}{t_n}\int_0^{t_n} (S_{t+s} f)(x)\,ds.
\end{aligned}$$

The last expression can be written as $\frac{1}{t_n}\int_t^{t_n+t}(S_u f)(x)\,du$ which can be split as

$$\frac{1}{t_n}\left[\int_0^{t_n}(S_u f)(x)\,du + \int_{t_n}^{t_n+t}(S_u f)(x)\,du - \int_0^{t}(S_u f)(x)\,du\right].$$

Using this expression in the right side of Equation (11.4.7), we obtain

$$\begin{aligned}
\int (S_t f)(y)\mu(dy) &= \lim_{n\to\infty}\frac{1}{t_n}\left[\int_0^{t_n}(S_u f)(x)\,du\right. \\
&= \lim_{n\to\infty}\int f(y)\mu_n(dy) \\
&= \int f(y)\mu(dy)
\end{aligned}$$

by the weak convergence of μ_n. Thus, μ satisfies Equation (11.4.2) and is therefore an invariant measure. ∎

As before, we will denote the diffusion process X with $X_0 = x$ as $\{X_t^x\}$. Thus, we have a family of diffusion processes on a complete probability space $(\Omega, \mathcal{F}, P)$ equipped with a filtration $(\mathcal{F}_t : t \geq 0)$ that satisfies the usual conditions. The next result is on the uniqueness of invariant measures.

Theorem 11.4.3 *Let B_R denote the closed R-ball in $\mathbb{R}^d$. Suppose that the following condition holds: for any given positive numbers ϵ, δ, and R, there exists a corresponding number $S > 0$ such that*

$$\frac{1}{T}\int_0^T P\{|X_t^x - X_t^y| \geq \delta\}\, dt < \epsilon \tag{11.4.8}$$

for all $x, y \in B_R$ and $T > S$. If an invariant measure exists for X, then it is unique.

Proof Suppose there exists two invariant measures μ and ν for the diffusion X. We will show that

$$\int f(x)\mu(dx) = \int f(x)\nu(dx)$$

for all f, bounded and uniformly continuous. As before, for any $T > 0$ and $x \in \mathbb{R}^d$, define the Borel probability measure

$$\mu_T^x(B) = \frac{1}{T}\int_0^T P(x, t, B)\, dt$$

and the measure ν_T^x analogously. Consider

$$\begin{aligned}\left|\int f(z)\mu(dz) - \int f(z)\nu(dz)\right| &= \left|\int\int f(z)[\mu_T^x(dz)\mu(dx) - \nu_T^y(dz)\,\nu(dy)]\right| \\ &\leq \int\int\left|\int f(z)[\mu_T^x(dz) - \nu_T^y(dz)]\right|\mu(dx)\,\nu(dy).\end{aligned} \tag{11.4.9}$$

Let us denote $\int f(z)[\mu_T^x(dz) - \nu_T^y(dz)]$ as $g(x, y)$. Given any $\epsilon > 0$, choose an R large enough so that $\mu(B_R^c) + \nu(B_R^c) < \epsilon$. Then, the inequality (11.4.9) allows us to conclude that

$$\left|\int f(z)\mu(dz) - \int f(z)\,\nu(dz)\right| \leq \int_{B_R \times B_R} |g(x, y)|\,\mu(dx)\,\nu(dy) + (4\epsilon + 2\epsilon^2)\|f\|_\infty. \tag{11.4.10}$$

Finally, consider

$$
\begin{aligned}
\int_{B_R \times B_R} |g(x,y)|\, \mu(dx)\, \nu(dy) &\leq \int_{B_R \times B_R} \frac{1}{T} \int_0^T E|f(X_t^x) - f(X_t^y)|\, dt\}\, \mu(dx)\, \nu(dy) \\
&\leq 2\|f\|_\infty \sup_{x,y \in B_R} \frac{1}{T} \int_0^T P\{|X_t^x - X_t^y| \geq \delta\}\, dt \\
&\quad + \sup_{\substack{x,y \in B_R \\ |x-y| < \delta}} |f(x) - f(y)| \\
&< 2\|f\|_\infty \epsilon + \epsilon
\end{aligned}
\tag{11.4.11}
$$

if $T > S$, by the hypothesis (11.4.8) and the uniform continuity of f. From the estimates (11.4.10) and (11.4.11), we obtain the uniqueness of invariant measures. ■

Next, we give a stronger condition, which is easy to verify, that ensures the uniqueness of invariant measures.

Proposition 11.4.4 *Suppose that S_t has the Feller property. Let μ_t^x denote $P_x(t, \cdot)$, the distribution of X_t^x. If μ_t^x converges weakly to μ as $t \to \infty$ for all $x \in \mathbb{R}^d$, then μ is the unique invariant measure for S_t.*

Proof For any $\phi \in C_b(\mathbb{R}^d)$,

$$
S_t\phi(x) = \int \phi(y)\, d\mu_t^x(y) \to \int \phi(y)\, d\mu(y).
$$

Therefore, for any invariant measure ν, we have

$$
\begin{aligned}
\int \phi(x)\, d\nu(x) &= \int S_t\phi(x)\, d\nu(x) \quad \text{for all } t \\
&= \lim_{t \to \infty} \int S_t\phi(x)\, d\nu(x) \\
&= \int \int \phi(y)\, d\mu(y)\, d\nu(x)
\end{aligned}
$$

by the bounded convergence theorem. Thus, for all $\phi \in C_b(\mathbb{R}^d)$, we have

$$
\int \phi(x)\, d\nu(x) = \int \phi(y)\, d\mu(y).
$$

Hence, μ is the unique invariant measure. ■

11.5 Ergodic Measures

Definition 11.5.1 *Let μ be an invariant Borel measure on $\mathbb{R}^d$ for S_t. A Borel set A is called an* **invariant set** *with respect to the semigroup S_t if for all $t \geq 0$,*

$$S_t 1_A = 1_A \quad \mu \text{ a.s.}$$

The measure μ is called **ergodic** *if $\mu(A)$ is either 0 or 1 for any invariant set A.*

The following result is standard, and is stated without proof. One can find a proof in the book by Da Prato and Zabczyk.

Proposition 11.5.1 *If μ is an ergodic measure for the Markov family $\{X_t^x\}$, then for any $\phi \in L^2(E, \mu)$,*

$$\lim_{t\to\infty} \frac{1}{t} \int_0^t S_s\phi \, ds = \int \phi \, d\mu \tag{11.5.1}$$

where the limit is taken in the $L^2(\mu)$ sense.

Proposition 11.5.2 *If μ and ν are two ergodic measures with respect to S_t, and if $\mu \neq \nu$, then μ and ν are singular.*

Proof Let A be such that $\mu(A) \neq \nu(A)$. By the above proposition, there exists $t_n \uparrow \infty$ such that

$$\lim_{n\to\infty} \frac{1}{t_n} \int_0^{t_n} (S_t 1_A)(x) \, dt = \mu(A), \quad \mu \text{ almost all } x$$

and

$$\lim_{n\to\infty} \frac{1}{t_n} \int_0^{t_n} (S_t 1_A)(x) \, dt = \nu(A), \quad \nu \text{ almost all } x.$$

Let the two almost sure sets be denoted B and C respectively. Then, $B \cap C = \emptyset$ and $\mu(B) = \nu(C) = 1$, which completes the proof. ■

Theorem 11.5.3 *If μ is the unique invariant measure for the family $\{X^x\}$, then μ is ergodic.*

Proof Suppose that μ is not ergodic. Then, there exists a Borel set A such that $0 < \mu(A) < 1$ and

$$S_t 1_A = 1_A \quad \mu \text{ a.s.} \tag{11.5.2}$$

Define another Borel measure ν by

$$\nu(B) = \frac{\mu(A \cap B)}{\mu(A)}.$$

We will show that ν is an invariant measure. Consider

$$\int P_x(t,B)\,d\nu(x) = \frac{1}{\mu(A)}\int_A P_x(t,B)\,d\mu(x)$$
$$= \frac{1}{\mu(A)}\left\{\int_A P_x(t,A\cap B)d\mu(x) + \int_A P_x(t,A^c\cap B)\,d\mu(x)\right\}. \tag{11.5.3}$$

From $S_t 1 = 1$ and (11.5.2). it follows that

$$P_x(t,A\cap B) \leq P_x(t,A) = 0 \quad \mu \;\text{ a.s. on } \; A^c$$

and

$$P_x(t,A^c\cap B) \leq P_x(t,A^c) = 0 \quad \mu \;\text{ a.s. on } \; A.$$

Hence, equation (11.5.3) can be written as

$$\int P_x(t,B)\,d\nu(x) = \frac{1}{\mu(A)}\int P_x(t,A\cap B)\,d\mu(x).$$

Using the invariance of μ on the right side,

$$\int P_x(t,B)\,d\nu(x) = \frac{\mu(A\cap B)}{\mu(A)} = \nu(B).$$

Thus ν is also an invariant measure, a contradiction to the uniqueness of invariant measures. ■

Definition 11.5.2 *A Markov family X^x is called* **irreducible** *if for all $t > 0$, any nonempty open set G, and any x, we have*

$$P\{X_t^x \in G\} > 0;$$

that is, $P_x(t,G) > 0$.

Proposition 11.5.4 *If the semigroup S_t is irreducible and strongly Feller, then the probability measures $\{P_x(t,\cdot)\}$ are mutually equivalent for each $t > 0$.*

Proof Pick any $t, s > 0$ Assume that $P_{x_0}(t+s,B) > 0$ for some x_0 and Borel set B. By the Chapman-Kolmogorov equation,

$$P_{x_0}(t+s,B) = \int P_{x_0}(t,dy)P_y(s,B).$$

Therefore, there exists a y_0 such that $P_{y_0}(s, B) > 0$. By the strong Feller property, $P_y(s, B)$ is a continuous function of y. Hence, there exists a neighborhood of y_0, say, $B_r(y_0)$, such that $P_y(s, B) > 0$ for all $y \in B_r(y_0)$. So, for any arbitrary x,

$$\begin{aligned} P_x(t + s, B) &= \int_E P_x(t, dy)P_y(s, B) \\ &\geq \int_{B_r(y_0)} P_x(t, dy)P_y(s, B) > 0, \end{aligned}$$

since $P_x\{t, B_r(y_0)\} > 0$ by irreducibility. ■

The main result that we need is the following theorem due to Doob.

Theorem 11.5.5 *Let μ be an invariant measure for the Markov family $\{X^x\}$. If the corresponding semigroup S_t is irreducible and strongly Feller, then μ is the unique invariant measure, and hence, ergodic.*

Proof Let A be an invariant set so that

$$S_t 1_A = 1_A \quad \text{for all} \quad t > 0. \tag{11.5.4}$$

Suppose that $\mu(A) > 0$. We have to show that $\mu(A) = 1$. Since $\mu(A) > 0$, there exists an $x_0 \in A$ such that $P_{x_0}(t, A) = 1$. By the previous proposition, the family of probability measures $P_x(t, \cdot)$ indexed by x are equivalent, so that $P_x(t, A) = 1$ for all x. Using, in addition, the invariance of μ, it follows that

$$\mu(A) = \int P_x(t, A) d\mu(x) = 1.$$

Thus, μ is ergodic. If ν is another invariant measure, then repeating the above arguments, ν is also ergodic. Being invariant measures, both μ and ν are equivalent to $P_x(t, \cdot)$ for any x and t. Therefore, μ and ν are equivalent, and by Proposition 11.2.9, we can conclude that $\mu = \nu$. ■

We will use the result of Doob to prove the existence of an ergodic measure for d-dimensional diffusions. Consider a d-dimensional stochastic differential equation

$$dX_t = b(X_t)\, dt + \sigma(X_t)\, dW_t$$

with X_0 specified. The matrix σ is a $d \times d$ matrix-valued function, and W is a d-dimensional Wiener process. If $X_0 = x \in \mathbb{R}^d$, then we will denote the solution by X_t^x.

Theorem 11.5.6 *Let the coefficients b and σ be in $C_b^2(\mathbb{R}^d)$. Suppose that σ is invertible and $\sup_x \|\sigma^{-1}(x)\| = K < \infty$. Then the solution $\{X_t\}$ is irreducible and has the strong Feller property.*

Proof

Step 1 Fix any function $g \in C_b^2(\mathbb{R}^d)$. Define $v(t,x) = Eg(X_t^x)$. By Theorem 8.6.6 in Chapter 8, v is twice continuously differentiable in x with bounded derivatives. By applying the Itô formula to the function of s, x given by $v(t-s,x)$, and Remark 8.6.3, we obtain

$$g(X_t^x) = v(t,x) + \sum_{i,r=1}^{d} \int_0^t \sigma_{ir}(X_s^x) D_i v(t-s, X_s^x)\, dW_r. \tag{11.5.5}$$

We will write the second term on the right side of (11.5.5) as $\int_0^t \nabla v(t-s, X_s^x)\, \sigma(X_s^x)\, dW_s$.

Note that $\nabla_x X_s^x$ is a $d \times d$ matrix. For any $h \in \mathbb{R}^d$, let $D_h X_s^x$ denote $\langle \nabla_x X_s^x, h\rangle$. Multiply both sides of Equation (11.5.5) by $\int_0^t \sigma^{-1}(X_s^x)\, D_h\, X_s^x\, dW_s$, and take expectation to obtain

$$\begin{aligned}
E[g(X_t^x) \int_0^t \sigma^{-1}(X_s^x)\langle \nabla_x X_s^x, h\rangle dW_s] &= E\int_0^t (\sigma^*(X(s,x))\nabla v(t-s, X_s^x), \sigma^{-1}(X_s^x) D_h X_s^x)\, ds \\
&= E\int_0^t (\nabla v(t-s, X_s^x), D_h X_s^x)\, ds \\
&= \int_0^t (\nabla_x E[v(t-s, X_s^x), h)]\, ds \\
&= \int_0^t (\nabla_x (S_s S_{t-s} g(x), h)\, ds \\
&= t(\nabla_x v(t,x), h) = t D_h v(t,x).
\end{aligned}$$

Thus, we have proved that

$$D_h v(t,x) = \frac{1}{t} E[g(X_t^x) \int_0^t \sigma^{-1}(X_s^x)\langle \nabla_x X_s^x, h\rangle\, dW_s. \tag{11.5.6}$$

Step 2 First, take any $g \in C_b^2(\mathbb{R}^d)$. From Step 1, we have

$$\begin{aligned}
|D_h v(t,x)| &\le \frac{1}{t} \|g\| E\left[\left|\int_0^t \sigma^{-1}(X_s^x)\langle \nabla_x X_s^x, h\rangle dW_s\right|\right] \\
&\le \frac{K}{t} \|g\| \left(E\int_0^t |D_h X_s^x|^2\, ds\right)^{1/2}.
\end{aligned}$$

From Theorem 8.6.5 in Chapter 8, we know that for any fixed $T > 0$, there exist constants C_1 and C_2 such that

$$\left(E\int_0^t |D_h X_s^x|^2\, ds\right) \le C_1 t |h|^2 e^{2C_2 t}.$$

Therefore, we can conclude that

$$|D_h v(t,x)| \le \sqrt{\frac{KC_1}{t}}|h|e^{C_2 t}\|g\|_{L^2}.$$

Therefore,

$$|S_t g(x) - S_t g(y)| \le \sup_h |D_h v(t,x)(y-x)|,$$

and there exists a constant C_t such that $|S_t g(x) - S_t g(y)| \le C_t\|g\|_{L^2}|x-y|$. This inequality continues to hold for any bounded, Borel-measurable g on $\mathbb{R}^d$ by a routine approximation. We have thus proved that X has the strong Feller property.

Irreducibility follows easily by using the Girsanov transformation to remove the drift and using the fact that σ^{-1} is uniformly bounded.

■

Exercises

1. For the one-dimensional diffusion considered in Section 11.2, prove that

$$\limsup_{t\to\infty} \frac{1}{t}E\left\{\int_0^t g(X_s^x)\,ds\right\} = \frac{\int_{\mathbb{R}^1} g(a)\beta(a)\,da}{\int_{\mathbb{R}^1}\beta(a)\,da}$$

for all bounded, nonnegative functions g with compact support.

2. Complete the details of Step 2 in the proof of Theorem 11.2.5.
3. Prove the converse of Theorem 11.3.1 stated in Remark 11.3.1.
4. Prove that $p(t,x,y) > 0$ for almost every x, y, as claimed at the end of the proof of Theorem 11.3.2.

12 Large Deviations Principle for Diffusions

The theory of large deviations is the study of exponential decay of probabilities (of rare events) with respect to an associated parameter. Rare events refer to extreme events such as an overload on a system or transition from one equilibrium state to another, etc. and occur with a small probability. To understand rare events and the associated parameter, consider the simple case of a random sample $X_1, X_2, \ldots$ from a common $N(0, 1)$ distribution. The sequence of sample averages $\{\bar{X}_n\}$ (where $\bar{X}_n = \frac{1}{n}\sum_{i=1}^{n} X_i$) converges a.s. to 0 by the strong law of large numbers. Therefore, for any fixed $\delta > 0$, the event $A_n := \{|\bar{X}_n| \geq \delta\}$ occurs with a small probability. Hence, we call A_n a rare event. The associated parameter is the sample size n. As $n \to \infty$, one can show that $P(A_n)$ converges to 0 and is approximately given by

$$P(A_n) = \exp\left\{-\frac{n\delta^2}{2}\right\}$$

for large n. It is this exponential rate of decay, in this case $\frac{\delta^2}{2}$, which the theory of large deviations aims to find in more complex systems.

The origin of large deviations goes back to Boltzmann in statistical mechanics, which led Gibbs to formulate macrostates. Large deviations results in several important contexts were proved by [9], [66], [55], and [67], to name a few. [27] developed a beautiful theory to study large deviations for random dynamical systems perturbed by small noise. [17] studied certain function space integrals and their asymptotics, which led [17] to formulate a general theory of large deviations. We refer the avid reader to the book by [11] where one can find a wide range of applications of the theory.

In this chapter, we will focus our attention on large deviations for the small noise asymptotics of diffusion processes. We will use the weak convergence approach of Ellis and Dupuis which hinges on the equivalence of the large deviations principle (LDP) and the Laplace-Varadhan principle. In Section 12.1, the definition of LDP, Cramér's theorem, and a few examples are presented. The concept of relative entropy is introduced, and the equivalence of LDP and the Laplace-Varadhan principle is proved in the next section. A variational representation theorem for positive functionals of a Brownian motion is proved in Section 12.3. Using this representation, the Laplace-Varadhan principle is proved for a large family of d-dimensional diffusion processes which arise as pathwise unique, strong solutions of stochastic differential equations.

12.1 Definitions and Basic Results

Let $\{X_n\}$ be a sequence of random variables defined on $(\Omega, \mathcal{F}, P)$, with values in a complete separable metric space E. For instance, one can take E to be $\mathbb{R}^d$, equipped with the Euclidean norm.

Definition 12.1.1 *A function $I : E \to [0, \infty]$ is called a* **rate function** *if I is a lower semi continuous function. The function I is called a* **good rate** *function if, in addition, I has compact level sets, that is, the set $\{x : I(x) \leq c\}$ is compact in E for any finite number c. We will use the notation*

$$I(A) := \inf_{A} I(x).$$

Definition 12.1.2 *Let I be a good rate function on E. The sequence $\{X_n\}$ is said to satisfy the LDP on E with rate I if the following inequalities are satisfied:*

(a) for each closed subset F in E,

$$\limsup_{n\to\infty} \frac{1}{n} \log P\{X_n \in F\} \leq -I(F). \tag{12.1.1}$$

The inequality (12.1.1) is known as the **upper bound** *for the LDP.*

(b) For each open subset G in E,

$$\liminf_{n\to\infty} \frac{1}{n} \log P\{X_n \in G\} \geq -I(G). \tag{12.1.2}$$

The inequality (12.1.2) is called the **lower bound** *for the LDP.*

It is clear that, if A is Borel with its interior denoted by A^0 and its closure by $\bar{A}$, and if $I(A^0) = I(\mathrm{cl}\, A)$, then

$$\lim_{n\to\infty} \frac{1}{n} \log P\{X_n \in A\} = -I(A).$$

We will explain the existence of a rate function in the context of a sequence $X_1, X_2, \ldots$ of iid random variables with a finite moment generating function (MGF) denoted by $M(t)$. Let us assume that $E(X_1) = 0$. Then, setting $\bar{X}_n = \frac{1}{n}\sum_{i=1}^{n} X_i$, we have for any $\delta > 0$ and any$t > 0$,

$$P\{\bar{X}_n \geq \delta\} = \int 1_{\{\sum_{i=1}^n X_i \geq n\delta\}} dP \leq e^{-nt\delta} E e^{t\sum_{i=1}^n X_i}$$
$$= e^{-nt\delta}\{M(t)\}^n.$$

Taking logarithm,

$$\log P\{\bar{X}_n \geq \delta\} \leq -nt\delta + n\log M(t).$$

The inequality holds for any $t > 0$. Therefore,

$$\limsup_{n\to\infty} \frac{1}{n}\log P\{\bar{X}_n \geq \delta\} \leq -\sup_{t>0}(t\delta - \log M(t))$$
$$= -\sup_{t\geq 0}(t\delta - \log M(t))$$

by the continuity of $t\delta - \log M(t)$ at $t = 0$. Note that $\log M(t) \geq E(tX_1) = 0$ by the Jensen inequality and our assumption that $E(X_1) = 0$. From this observation, it follows that if $t < 0$, then $t\delta - \log M(t) < 0$. Therefore,

$$\sup_{t\geq 0}(t\delta - \log M(t)) = \sup_t(t\delta - \log M(t)).$$

We have considered only the upper bound for the LDP so far. It leads to a reasonable guess for the rate function, namely, $I(\delta) = \sup_t(t\delta - \log M(t))$.

That this is so is the content of Cramér's theorem:

Theorem 12.1.1 *Let $\bar{X}_n = \frac{1}{n}\sum_{i=1}^n X_i$ where $\{X_i\}$ is a sequence of iid d-dimensional r.v.s. with distribution μ and a finite moment generating function. Then $\bar{X}$ satisfies LDP with rate*

$$I(x) = \sup_\alpha \left\{ < \alpha, x > -\log \int_{\mathbb{R}} e^{<\alpha,y>}\, \mu(dy). \right\} \tag{12.1.3}$$

The proof is quite standard and is available in several books. For instance, the reader is referred to [73] or [11].

Example 12.1.2

1. *If $\mu = p\delta_1 + q\delta_0$, the right side of Equation (12.1.3) is easily computed and*

$$I(x) = x\log(x/p) + (1-x)\log\left(\frac{1-x}{q}\right)$$

if $x \in [0,1]$. Otherwise $I(x) = \infty$.

2. *If μ is the Poisson distribution with parameter λ, then*

$$I(x) = \log\frac{x}{\lambda} - x + \lambda$$

if $x \geq 0$. Otherwise, $I(x) = \infty$.

3. *If* μ *is the Normal*(m, σ^2) *distribution, then* $I(x) = \dfrac{(x-m)^2}{2\sigma^2}$.
4. *4. If* μ *is the exponential*(λ) *distribution, one obtains* $I(x) = \lambda x - \log(\lambda x) - 1$ *if* $x > 0$. *Otherwise,* $I(x) = \infty$.

12.2 Large Deviations and Laplace-Varadhan Principle

In this section we start with the definition of the Laplace-Varadhan principle and prove the theorem of Varadhan and its converse, which together will establish the equivalence of LDP and the Laplace-Varadhan principle. The method of Laplace states that, for any $h \in C_b([0,1])$,

$$\lim_{n\to\infty} \frac{1}{n} \log \int_0^1 e^{-nh(x)}\, dx = -\min_{[0,1]} h(x). \tag{12.2.1}$$

The proof of (12.2.1) is straightforward. The analogous statement for functions of random variables is contained in the following statement:

Theorem 12.2.1 (Varadhan) *Let* $\{X_n\}$ *be a sequence of random variables taking values in the space E. Let* $\{X_n\}$ *satisfy LDP on E with a good rate function I. Then, for all* $h \in C_b(E)$,

$$\lim_{n\to\infty} 1/n \log E\{e^{-nh(X_n)}\} = -\inf_{x\in E}\{h(x) + I(x)\}. \tag{12.2.2}$$

Before we prove the theorem, let us define the Laplace-Varadhan principle.

Definition 12.2.1 *A sequence of E-valued random variables* $\{X_n\}$ *is said to satisfy the* **Laplace-Varadhan principle** *with a good rate function I if Equation* (12.2.2) *holds for all* $h \in C_b(E)$. *The inequality*

$$\limsup_{n\to\infty} \frac{1}{n} \log E\left\{e^{-nh(X_n)}\right\} \le -\inf_{x\in E}\left\{h(x) + I(x)\right\} \tag{12.2.3}$$

is known as the **upper bound** *for the Laplace-Varadhan principle, and the inequality*

$$\liminf_{n\to\infty} \frac{1}{n} \log E\left\{e^{-nh(X_n)}\right\} \ge -\inf_{x\in E}\{h(x) + I(x)\} \tag{12.2.4}$$

is known as the **lower bound** *for the Laplace-Varadhan principle.*

Proof

Step 1 The Laplace-Varadhan principle follows if we prove both the upper and lower bounds, (12.2.3) and (12.2.4), for the Laplace-Varadhan principle. In this step, we will show the upper bound for the Laplace-Varadhan principle.

Let $\|h\| = M$. For any $k \in \mathbb{N}$, and $1 \le j \le 2k$, consider the closed sets

$$F_{k,j} = \left\{x \in E : -M + (j-1)\frac{M}{k} \le -h(x) \le -M + j\frac{M}{k}\right\}.$$

The large deviation upper bound is used to obtain the following estimate:

$$\begin{aligned}\limsup_{n\to\infty} \frac{1}{n} \log E\left[e^{-nh(X_n)}\right] &= \limsup_{n\to\infty} \frac{1}{n} \log\left[\sum_{j=1}^{2k} \int_{F_{k,j}} e^{-nh(x)} P_n(dx)\right] \\ &\leq \limsup \frac{1}{n} \log\left[\sum_{j=1}^{2k} \exp\left\{n\left(-M + \frac{jM}{k}\right)\right\} P_n(F_{k,j})\right] \\ &\leq \limsup_{n\to\infty} \frac{1}{n} \log\left[2k \max_j \left\{\exp\{n\left(-M + \frac{jM}{k}\right)\right\} P_n(F_{k,j})\}\right] \\ &= \limsup_{n\to\infty} \frac{1}{n} \max_j \left\{-nM + \frac{jMn}{k} + \log P_n(F_{k,j})\right\} \\ &\leq \max_j \sup_{x\in F_{k,j}} \left\{-M + \frac{jM}{k} - I(x)\right\}.\end{aligned}$$

Using the definition of $F_{k,j}$ in the above bound,

$$\begin{aligned}\limsup_{n\to\infty} \frac{1}{n} \log E[e^{-nh(X_n)}] &\leq \max_j \sup_{x\in F_{k,j}} \{-h(x) - I(x)\} + \frac{2M}{k} \\ &\leq \sup_x \{-h(x) - I(x)\} + \frac{2M}{k}.\end{aligned}$$

Allowing $k \to \infty$, we obtain the upper bound (12.2.3).

Step 2 In this step, we prove the lower bound:

$$\liminf_{n\to\infty} \frac{1}{n} \log E\{e^{-nh(X_n)}\} \geq -\inf_{x\in E}\{h(x) + I(x)\}. \tag{12.2.5}$$

Given any $x \in E$ and $\epsilon > 0$, we apply the LDP lower bound to the open set

$$G = \{y : h(y) < h(x) + \epsilon\}$$

in the following calculation:

$$\begin{aligned}\liminf_{n\to\infty} \frac{1}{n} \log E(e^{-nh(X_n)}) &\geq \liminf_{n\to\infty} \frac{1}{n} \log E\left(1_G(X_n)e^{-nh(X_n)}\right) \\ &\geq \liminf_{n\to\infty} \frac{1}{n} \log E\left(1_G(X_n)e^{-n(h(x)+\epsilon)}\right) \\ &= \liminf_{n\to\infty} \frac{1}{n} \left[\log E\left(I_G(X_n)\right) - n(h(x) + \epsilon)\right] \\ &\geq -I(G) - h(x) - \epsilon.\end{aligned}$$

By the arbitrariness of x and ϵ, the lower bound is obtained. The theorem follows from the bounds (12.2.3) and (12.2.5). ■

The converse of Varadhan's theorem is the content of the following result:

Theorem 12.2.2 *If I is a rate function on E and the limit*

$$\lim_{n\to\infty} 1/n \log Ee^{-nh(X_n)} = -\inf_E\{I(x) + h(x)\}$$

holds for all $h \in C_b(E)$, then $\{X_n\}$ satisfies the large deviation principle with rate function I.

Proof

Step 1 Given a closed set F, define the lower semicontinuous function

$$f(x) = \begin{cases} 0 & \text{if } x \in F; \\ \infty & \text{if } x \in F^c. \end{cases}$$

Let $d(x, F)$ denote the distance of x to F. For $j \in \mathbb{N}$, define

$$f_j(x) = j(d(x, F) \wedge 1)$$

where $d(x, F)$ denotes the distance between the point x and the closed set F. For each $j \in \mathbb{N}$, define the function

$$f_j(x) = j\{d(x, F) \wedge 1\}.$$

Then, f_j is a bounded continuous function and $f_j \uparrow f$. Hence,

$$\begin{aligned} \frac{1}{n}\log P(X_n \in F) &= \frac{1}{n}\log E(e^{-nf(X_n)}) \\ &\leq \frac{1}{n}\log E(e^{-nf_j(X_n)}) \end{aligned}$$

and therefore,

$$\begin{aligned} \limsup_{n\to\infty} \frac{1}{n}\log P(X_n \in F) &\leq \lim_{n\to\infty} \frac{1}{n}\log E(e^{-nf_j(X_n)}) \\ &= -\inf_x\{f_j(x) + I(x)\}. \end{aligned}$$

It remains to show that

$$\lim_{j\to\infty} \inf_x\{f_j(x) + I(x)\} = I(F).$$

First, since $f_j(x) \leq f(x)$, we obtain

$$\begin{aligned} \inf_x\{f_j(x) + I(x)\} &\leq \inf_x\{f(x) + I(x)\} \\ &= \inf_{x\in F} I(x) = I(F). \end{aligned}$$

Step 2 The proof of the upper bound would be complete if we show the following reverse inequality:

$$\liminf_{j\to\infty} \inf_x \{f_j(x) + I(x)\} \geq I(F).$$

If $I(F) = 0$, this is automatic. Therefore, let us consider the case when $I(F)$ is positive but finite. Since

$$\begin{aligned}\inf_x \{f_j(x) + I(x)\} &= \min\left(\inf_{x\in F}\{f_j(x) + I(x)\}, \inf_{x\in F^c}\{f_j(x) + I(x)\}\right)\\ &= \min\left(I(F), \inf_{x\in F^c}\{f_j(x) + I(x)\}\right),\end{aligned}$$

it suffices to show that

$$\liminf_{j\to\infty} \inf_{x\in F^c} \left\{f_j(x) + I(x)\right\} \geq I(F). \tag{12.2.6}$$

We will prove this by contradiction. Assume that

$$\liminf_{j\to\infty} \inf_{x\in F^c} \{f_j(x) + I(x)\} < I(F).$$

Choose an $\epsilon > 0$ such that

$$\liminf_{j\to\infty} \inf_{x\in F^c} \{f_j(x) + I(x)\} \leq I(F) - 2\epsilon.$$

Then there exists a subsequence of $\{j\}$, which we denote by $\{j\}$ itself for the sake of notational simplicity, such that

$$\inf_{x\in F^c} \{f_j(x) + I(x)\} \leq I(F) - \epsilon.$$

For each j, there exists an $x_j \in F^c$ such that

$$f_j(x_j) + I(x_j) \leq I(F) - \epsilon. \tag{12.2.7}$$

By the definition of f_j, the above inequality implies that $d(x_j, F) \to 0$ as $j \to \infty$. Therefore, there exists a $y_j \in F$ such that $d(x_j, y_j) \to 0$. On the other hand, from (12.2.7), it follows that

$$\sup_j I(x_j) \leq I(F) - \epsilon.$$

Since I has compact level sets, there exists a subsequence of $\{x_{j_n}\}$, and an x^* in the level set

$$\{x : I(x) \leq I(F) - \epsilon\}$$

such that $d(x_{j_n}, x^*) \to 0$. By the triangle inequality, it follows that the sequence $d(y_{j_n}, x^*) \to 0$ as $n \to \infty$. Hence, $x^* \in F$, since F is a closed set. Therefore,

$I(x^*) \geq I(F)$, which contradicts $I(x^*) \leq I(F) - \epsilon$. Thus, the proof of (12.2.6) is complete when $0 < I(F) < \infty$.

Step 3 If $I(F) = \infty$, then (12.2.6) would read as

$$\liminf_{j\to\infty} \inf_{x\in F^c} \{f_j(x) + I(x)\} = \infty.$$

To prove this by contradiction, we assume that there exists a finite $M > 0$ such that

$$\liminf_{j\to\infty} \inf_{x\in F^c} \{f_j(x) + I(x)\} < M.$$

We can repeat the arguments given above with M in the place of $I(F)$. By the arbitrariness of M, the proof is over.

Step 4 We will prove the lower bound in the statement of LDP:

$$-I(G) \leq \liminf_{n\to\infty} \frac{1}{n} \log P\{X_n \in G\}$$

for all open sets G.

If $I(G) = \infty$, there is nothing to prove. Therefore, let $I(G) < \infty$. Consider any $x \in G$ such that $I(x) < \infty$. Choose $M > I(x)$, and $\delta > 0$ such that the open δ-ball $B_\delta(x)$ centered around x is contained in G. Define the function

$$h(y) = M\left(\frac{d(y,x)}{\delta} \wedge 1\right).$$

Then, h is a bounded, continuous function such that

$$0 \leq h(z) \leq M,$$

$h(x) = 0$, and $h(y) = M$ if y is outside $B_\delta(x)$.

$$\begin{aligned} E\left(e^{-nh(X_n)}\right) &\leq e^{-nM} P\{X_n \in B_\delta(x)^c\} + P\{X_n \in B_\delta(x)\} \\ &\leq e^{-nM} + P\{X_n \in B_\delta(x)\} \\ &\leq 2\max\left(e^{-nM}, P\{X_n \in B_\delta(x)\}\right). \end{aligned}$$

Therefore, upon taking the logarithm,

$$\log E\left(e^{-nh(X_n)}\right) \leq \log 2 + \max\left(-nM, \log P\{X_n \in B_\delta(x)\}\right).$$

Dividing by n, and taking $\liminf_{n\to\infty}$, we obtain

$$\begin{aligned} \max\left(-M, \liminf_{n\to\infty} \log P\{X_n \in B_\delta(x)\}\right) &\geq \lim_{n\to\infty} \frac{1}{n} \log E\left(e^{-nh(X_n)}\right) \\ &= -\inf_{y\in E}\{h(y) + I(y)\} \\ &\geq -h(x) - I(x) \\ &= -I(x) \end{aligned}$$

by definition of h. Since $M > I(x)$ and $B_\delta(x) \subseteq G$, it follows that

$$\begin{aligned}\liminf_{n\to\infty} \frac{1}{n} \log P\{X_n \in G\} &\geq \liminf_{n\to\infty} \frac{1}{n} \log P\{X_n \in B_\delta(x)\} \\ &\geq -I(x) \\ &\geq -I(G),\end{aligned}$$

which completes the proof. ■

Next, we present a useful result by Bryc for identifying the rate function when the sequence of random variables satisfies a condition known as exponential tightness.

Definition 12.2.2 *A sequence $\{X_n\}$ is said to be* **exponentially tight** *if, for each $M \in (0, \infty)$, there exists a compact set K in the complete separable metric space E such that*

$$\limsup_{n\to\infty} \frac{1}{n} \log P\{X_n \in K^c\} \leq -M. \tag{12.2.8}$$

Theorem: (Bryc) Let $\{X_n\}$ be a sequence of E-valued random variables that is exponentially tight. Suppose that the limit

$$\lim_{n\to\infty} \frac{1}{n} \log Ee^{-nh(X_n)}$$

exists for all $h \in C_b(E)$, and is denoted by $\Lambda(h)$. Then,

$$I(x) = -\inf_{h\in C_b(E)} \{\Lambda(h) + h(x)\}$$

is a rate function. Besides,

$$\Lambda(h) = -\inf_{x\in E}\{I(x) + h(x)\}.$$

Since we do not use this result in the sequel, we omit the proof of this theorem. The interested reader can find a proof of it in Ellis and Dupuis. With the equivalence of LDP and the Laplace-Varadhan principle established, we turn our attention to an abstract variational representation which is useful for proving the Laplace-Varadhan principle.

Definition 12.2.3 *Let E be a complete, separable metric space with the Borel σ-field denoted by $\mathcal{E}$. Let $\Pi(E)$ be the class of probability measures defined on $(E, \mathcal{E})$. For any given $\mu \in \Pi(E)$, define the* **relative entropy function** *$R(\cdot \mid \mu)$ as the mapping from $\Pi(E)$ into the extended real numbers given by*

$$R(\nu \mid \mu) = \int_E \log \frac{d\nu}{d\mu}(x)\, \nu(dx) \tag{12.2.9}$$

whenever λ is absolutely continuous with respect to η and $\log \frac{d\lambda}{d\eta}(x)$ is λ-integrable. Otherwise, define $R(\lambda \mid \eta)$ to be infinity.

The following simple and elegant result gives an abstract variational representation using the relative entropy function.

Theorem 12.2.3 *Let h be a bounded, real-valued, measurable function on a measurable space $(Z, \mathcal{A})$. Let $\mu \in \Pi(Z)$. Then,*

(i)
$$-\log \int_Z e^{-h(z)} \mu(dz) = \inf_{\nu \in \Pi(Z)} \left\{ R(\nu||\mu) + \int_Z h(z)\, \nu(dz) \right\}. \tag{12.2.10}$$

(ii) Let ν_0 be the probability measure on Z with $\nu_0 << \mu$ and

$$\frac{d\nu_0}{d\mu} = \frac{e^{-f}}{\int_Z e^{-f} d\mu}. \tag{12.2.11}$$

Then the infimum in part (i) is attained uniquely at ν_0.

Proof To prove (i), it suffices to consider ν such that

$$R(\nu||\mu) < \infty.$$

Therefore, take $\nu << \mu$. Since $\mu << \nu_0$, we get $\nu << \nu_0$.

$$\begin{aligned} R(\nu||\mu) &= \int \left(\log \frac{d\nu}{d\mu}\right) d\nu \\ &= \int \left(\log \frac{d\nu}{d\nu_0}\right) d\nu + \int \left(\log \frac{d\nu_0}{d\mu}\right) d\nu \\ &= R\left(\nu||\nu_0\right) - \int f d\nu - \log \int e^{-f} d\mu. \end{aligned}$$

Thus,

$$R(\nu||\mu) + \int f d\nu = R\left(\nu||\nu_0\right) - \log \int e^{-f} d\mu. \tag{12.2.12}$$

Take infimum on both sides of Equation (12.2.12) with respect to ν. Since $R(\nu||\nu_0) \geq 0$ and $R(\nu_0||\nu_0) = 0$, both parts (i) and (ii) of the theorem are proved. ■

Next, we prove a representation for the relative entropy $R(\nu||\mu)$, known as the Donsker-Varadhan variational formula.

Theorem 12.2.4 (Donsker-Varadhan representation) *Let E be a Polish space, and μ and ν be two Borel probability measures on E. Then,*

$$
\begin{aligned}
R(\nu||\mu) &= \sup_{f\in C_b(E)} \left\{ \int_E f d\nu - \log \int_E e^f d\mu \right\} \\
&= \sup_{\phi\in B_b(E)} \left\{ \int_E \phi\, d\nu - \log \int_E e^\phi d\mu \right\} \qquad (12.2.13)
\end{aligned}
$$

where $C_b(E)$ denotes the space of all bounded, continuous, real-valued functions defined on E, and $B_b(E)$, the space of all bounded, Borel-measurable, real-valued functions defined on E.

Proof

Step 1 We will first show that the two suprema in (12.2.13) are equal. Given any $\epsilon > 0$, choose a compact set K such that $(\mu + \nu)(K^c) < \epsilon$. Such a compact set K exists since any given finite measure on E is tight.

Take any function $\phi \in B_b(E)$, and consider its restriction to K. By Lusin's theorem ([65], 55), there exists a continuous function g defined on K such that

$$(\mu + \nu)(\{x : \phi(x) \neq g(x)\}) < \epsilon/2$$

and $\|g\|_\infty \leq \|\phi\|_\infty$. Pick an open set G in E such that

$$\{x : \phi(x) \neq g(x)\} \subset G \cap K$$

and $(\mu + \nu)(G \cap K) < \epsilon$. Let F denote the closed set $G^c \cap K$. Then, $\phi = g$ on F. By Tietze's extension theorem ([64], 179), g can be extended to a continuous function f defined on the whole of E and $\|f\|_\infty \leq \|g\|_\infty$.

Thus, $\phi = f$ on F and $(\mu + \nu)(F^c) < 2\epsilon$. Therefore,

$$\int \phi\, d\nu - \log \int e^\phi d\mu \leq \int f d\nu - \log \int e^f d\mu + 6\|\phi\|_\infty \epsilon.$$

Taking supremum over $h \in C_b(E)$ and then letting $\epsilon \to 0$, we obtain

$$\int \phi\, d\nu - \log \int e^\phi d\mu \leq \sup_{f\in C_b(E)} \left\{ \int f d\nu - \log \int e^f d\mu \right\}.$$

The arbitrariness of ϕ allows us to conclude that

$$\sup_{\phi\in B_b(E)} \left\{ \int \phi\, d\nu - \log \int e^\phi d\mu \right\} \leq \sup_{f\in C_b(E)} \left\{ \int f d\nu - \log \int e^f d\mu \right\}.$$

The suprema are equal since the reverse inequality is obvious.

Step 2 Let $H(\nu||\mu)$ denote the common value of the suprema. Then, $H(\nu\,||\,\mu) \geq 0$, since with $f \equiv 0$, we have $\int f d\nu - \log \int e^f d\mu = 0$. By part (i) of Theorem 12.2.3,

$$R(\nu\,||\,\mu) \geq -\int \psi\, d\nu - \log \int e^{-\psi}\, d\mu$$

for all $\psi \in B_b(E)$. Denote $-\psi$ as ϕ, and take supremum over $\phi \in B_b(E)$ to obtain

$$R(\nu\,||\,\mu) \geq \sup_{\phi \in B_b(E)} \left\{ \int \phi\, d\nu - \log \int e^{\phi}\, d\mu \right\}; \tag{12.2.14}$$

that is, $R(\nu\,||\,\mu) \geq H(\nu, \mu)$.

Step 3 It remains to show the reverse inequality: $R(\nu\,||\,\mu) \leq H(\nu, \mu)$. We assume that $H(\nu, \mu) < \infty$, since there is nothing to prove if $H(\nu, \mu) = \infty$. Then we claim that $\nu << \mu$.

For, take any Borel set B such that $\mu(B) = 0$. For any constant $M > 0$, define the function $\phi = M\, 1_B$. It is obvious that $\phi \in B_b(E)$, and

$$\int \phi\, d\nu - \log \int e^{\phi}\, d\mu \leq H(\nu, \mu) \tag{12.2.15}$$

holds for any $\phi \in B_b(E)$. Thus, $M\nu(B) \leq H(\nu, \mu)$ for all $M > 0$. Letting $M \to \infty$ allows us to conclude that $\nu(B) = 0$.

Step 4 Let $p = \frac{d\nu}{d\mu}$. Suppose that p is bounded and uniformly positive; then $\phi := \log p$ is in $B_b(E)$. For this choice of ϕ, equation (12.2.15) yields

$$R(\nu\,||\,\mu) \leq H(\nu, \mu).$$

If p is uniformly positive but not bounded, set $p_n = p \wedge n$ and $\phi_n = \log p_n$. Using (12.2.15) for the function ϕ_n and letting $n \to \infty$, we conclude that $R(\nu\,||\,\mu) \leq H(\nu, \mu)$ by the monotone convergence theorem.

If ϕ is not uniformly positive, then define, for any $a \in [0, 1]$, the measure

$$\nu_a = a\mu + (1 - a)\nu.$$

Define $\phi_a := \frac{d\nu_a}{d\mu} = a + (1 - a)\phi$. Then ϕ_a is uniformly positive for each $a \in (0, 1]$, and hence

$$R(\nu_a\,||\,\mu) \leq H(\nu_a, \mu).$$

We need to establish this inequality in the limit as $a \to 0$.

Step 5 $\lim_{a\to 0} R(\nu_a \,||\, \mu) = R(\nu \,||\, \mu)$, and $\lim_{a\to 0} H(\nu_a, \mu) = H(\nu, \mu)$.

Toward establishing the first limit, first note that the function $x \log x$ is convex on $(0, \infty)$, and hence

$$\begin{aligned} R(\nu_a \,||\, \mu) &= \int \phi_a \log \phi_a \, d\mu \\ &\leq (1-a) \int \phi \log \phi \, d\mu \\ &= (1-a) R(\nu \,||\, \mu). \end{aligned} \tag{12.2.16}$$

Also, $\log x$ is non-decreasing on $(0, \infty)$ so that

$$\log \phi_a \geq \max\{\log a, (1-a)\log \phi\}.$$

Hence,

$$\begin{aligned} R(\nu_a \,||\, \mu) &= a \int \log \phi_a \, d\mu + (1-a) \int \phi \log \phi_a \, d\mu \\ &\geq a \log a + (1-a)^2 R(\nu \,||\, \mu). \end{aligned} \tag{12.2.17}$$

Equations (12.2.16) and (12.2.17) imply that

$$\lim_{a\to 0} R(\nu_a \,||\, \mu) = R(\nu \,||\, \mu).$$

Next, we need to show that $\lim_{a\to 0} H(\nu_a, \mu) = H(\nu, \mu)$. Note that $H(\mu, \mu) = 0$, and $H(\nu_a, \mu)$ as a function of $a \in [0, 1]$ is convex and lower semicontinuous. Therefore, for each $a \in [0, 1]$,

$$0 \leq H(\nu_a, \mu) \leq aH(\mu, \mu) + (1-a)H(\nu, \mu) \leq H(\nu, \mu) < \infty,$$

which allows us to conclude that $H(\nu_a, \mu)$ is a continuous function of $a \in [0, 1]$. Here, we have used the fact that a convex, bounded, lower semicontinuous function on $[0, 1]$ is continuous. ■

Next, we prove a limit theorem for integrals by requiring weak convergence of a sequence of probability measures and boundedness of their relative entropy with respect to a reference measure.

Proposition 12.2.5 *Consider the probability space $(E, \mathcal{E}, \mu)$, where E is a Polish space and $\mathcal{E}$ its Borel σ-field. Let f be a real-valued, bounded, Borel-measurable function on E. Suppose that $\{\lambda_n\}$ is a sequence in $P(E)$ such that there exists a positive constant C satisfying*

$$\sup_n R(\lambda_n \,||\, \mu) \leq C < \infty.$$

Assume that $\lambda_n \to \lambda$ weakly as $n \to \infty$. Then the following hold:

(i) $\lim_{n\to\infty} \int_E f d\lambda_n = \int_E f d\lambda$, *and*

(ii) *if $\{g_n\}$ is a sequence of uniformly bounded, Borel-measurable functions that converges to g μ-a.s., then*

$$\lim_{n\to\infty} \int_E g_n \, d\lambda_n = \int_E g \, d\lambda.$$

Proof

Step 1 The measure λ is absolutely continuous with respect to μ. For, $\lambda_n \Longrightarrow \lambda$ as $n \to \infty$, and $R(\cdot \,||\, \mu)$ is lower semicontinuous, so that

$$R(\lambda \,||\, \mu) \leq \liminf_{n\to\infty} R(\lambda_n \,||\, \mu) \leq C < \infty.$$

The finiteness of $R(\lambda \,||\, \mu)$ implies that $\lambda << \mu$.

We know that f can be approximated μ a.s. by a sequence $\{f_j\}$ of bounded continuous functions, and for all j, $\|f_j\|_\infty \leq \|f\|_\infty$. It is clear that, for each j,

$$\lim_{n\to\infty} \int f_j \, d\lambda_n = \int f_j \, d\lambda.$$

Since $\lambda << \mu, f_j \to f$ λ-a.s. as well. Hence, by the dominated convergence theorem, we obtain

$$\lim_{j\to\infty} \int f_j \, d\lambda \to \int f d\lambda,$$

and a similar conclusion holds if λ is replaced by λ_n.

Step 2 To prove part (i) of the theorem, it suffices to show that

$$\lim_{j\to\infty} \sup_n \int |f_j - f| d\lambda_n = 0. \tag{12.2.18}$$

Fix any $\epsilon > 0$. Let $M = \|f\|_\infty$. Then,

$$\begin{aligned} \int |f_j - f| \, d\lambda_n &\leq \int_{\{|f_j - f| > \epsilon\}} |f_j - f| \, d\lambda_n + \epsilon \\ &\leq 2M\lambda_n(|f_j - f| > \epsilon) + \epsilon. \end{aligned}$$

We need to show that $\sup_n \lambda_n(|f_j - f| > \epsilon) \to 0$ as $j \to \infty$. For any fixed constant $k \in (1, \infty)$, define the set

$$A_{jnk} = \{|f_j - f| > \epsilon\} \cap \left\{ \frac{d\lambda_n}{d\mu} \leq k \right\}.$$

We can write

$$\lambda_n(|f_j - f| > \epsilon) = \int_{\{|f_j - f| > \epsilon\}} \frac{d\lambda_n}{d\mu}\, d\mu \\ \leq \int_{A_{jnk}} \frac{d\lambda_n}{d\mu}\, d\mu + \int_{A^c_{jnk}} \frac{d\lambda_n}{d\mu}\, d\mu. \tag{12.2.19}$$

Since $\mu(\{|f_j - f| > \epsilon\}) \to 0$ as $j \to \infty$, the first term on the right side of Equation (12.2.19) goes to zero uniformly in n. In order to bound the second term uniformly in n, consider

$$\sup_n \int_{A^c_{jnk}} \frac{d\lambda_n}{d\mu}\, d\mu \leq \frac{1}{\log k} \sup_n \int_{\{\frac{d\lambda_n}{d\mu} > k\}} \frac{d\lambda_n}{d\mu} \log \frac{d\lambda_n}{d\mu}\, d\mu \\ \leq \frac{1}{\log k} \sup_n \int \left(e^{-1} + \frac{d\lambda_n}{d\mu} \log \frac{d\lambda_n}{d\mu} \right) d\mu$$

since $k > 1$. Hence,

$$\sup_n \int_{A^c_{jnk}} \frac{d\lambda_n}{d\mu}\, d\mu \leq \frac{e^{-1} + C}{\log k},$$

which tends to zero as $k \to \infty$. The proof of part (i) of the proposition is thus completed.

Step 3 To prove part (ii) of the proposition, write

$$\int g_n\, d\lambda_n = \int g\, d\lambda_n + \int (g_n - g)\, d\lambda_n.$$

The first term converges to $\int g\, d\lambda$ by part (i). The second term converges to zero by following the arguments given in Step 2 to prove Equation (12.2.18) after replacing f_j and f by g_n and g, respectively, and letting n to ∞ (instead of taking supremum over n). Note that the continuity of f_j is not used in Step 2. ■

12.3 A Variational Representation Theorem

We will use the variational representation given in Equation (12.2.10) as the starting point for calculating $\log E[e^{-h(X)}]$ for any bounded, Borel-measurable function h, and X, a strong solution of a stochastic differential equation driven by a d-dimensional Wiener process W. We have seen that such a random variable X is a Borel-measurable function of W. Therefore, $h(X)$ is a bounded, Borel-measurable function of W, say, $f(W)$. For this reason, we derive, in this section, a variational representation for $\log E[e^{-f(W)}]$.

We will work on the Wiener space $(\Omega, \mathcal{F}, \mu)$, where $\Omega = C([0,1] : \mathbb{R}^d)$, $\mathcal{F}$ is the Borel σ-field for Ω under the topology of uniform convergence, and μ is the

d-dimensional Wiener measure. Let W be the canonical d-dimensional Wiener process given by $W_t(\omega) = \omega(t)$ for all $t \in [0, 1]$. Define the filtration $(\mathcal{F}_t)$ as the filtration $\sigma(W_s : 0 \leq s \leq t)$ augmented by subsets of μ-null sets of $\mathcal{F}$. Then, W is a Wiener martingale with respect to $\mathcal{F}_t$.

Next, we build the necessary notation:

1. Let $\mathcal{A}$ denote the class of all d-dimensional $\mathcal{F}_t$-progressively measurable processes v such that

$$E\left[\int_0^1 |v_s|^2 \, ds\right] < \infty.$$

2. Let $\mathcal{A}_b$ denote the subset of bounded elements of $\mathcal{A}$ so that $v \in \mathcal{A}_b$ means that there exists a $K < \infty$ such that $|v_t| \leq K$ for all $t \in [0, 1]$ a.s..
3. A stochastic process $\{Z_t\}$ on $(\Omega, \mathcal{F})$ is called a bounded simple process if it can be written in the form

$$Z_t = H_0 1_{\{0\}}(t) + \sum_{i=0}^{n} H_i 1_{(t_i, t_{i+1}]}(t) \;\; \forall \;\; t \in [0, 1],$$

where $H_i \in \mathcal{F}_{t_i}$ for all i, and $\{t_i\}$ satisfies $0 = t_0 < t_1 \cdots, < t_n = 1$ for some $n \in \mathbb{N}$. Besides, there exists a finite constant C such that $|H_i| \leq C$ for all i. We denote the class of bounded simple processes by $\mathcal{A}_s$.

From the construction of stochastic integrals, we know that, if $Z := \{Z_t\}$ belongs to $\mathcal{A}_b$, then there exists a sequence $Z^n := \{Z^n_t\}$ of processes from $\mathcal{A}_s$ that are bounded uniformly in n by the bound for Z, and

$$\lim_{n \to \infty} E \int_0^1 |Z^n_s - Z_s|^2 \, ds = 0.$$

It is also useful to recall the basic fact that a real-valued function f on a probability space $(E, \mathcal{E}, \eta)$, where E is a Polish space, can be approximated by a sequence of continuous functions $\{f_n\}$ in the almost sure sense. If $|f| \leq K$, then, for all n, we can take $|f_n| \leq K$.

The main result of this section is due to Boué and Dupuis.

Theorem 12.3.1 *Let f be a bounded, real-valued, Borel-measurable function defined on* $C([0, 1] : \mathbb{R}^d)$. *Then,*

$$-\log E\left[e^{-f(W)}\right] = \inf_{v \in \mathcal{A}} E\left\{\frac{1}{2}\int_0^1 |v(s)|^2 \, ds + f\left(W + \int_0^{\cdot} v(s) \, ds\right)\right\}. \tag{12.3.1}$$

Proof We will first prove that $-\log E[e^{-f(W)}]$ is bounded above by the expression on the right side of Equation (12.3.1). This is what we call the upper bound for the above variational formula.

Step 1 Consider any $v \in A_b$. Then $\int_0^t v_j(s)\, dW_j(s)$ is a square-integrable martingale for all $1 \leq j \leq d$. Therefore,

$$M(t) = \exp\left\{\sum_{j=1}^{d} \int_0^t v_j(s)\, dW_j(s) - \frac{1}{2}\int_0^t |v(s)|^2\, ds\right\}$$

is a martingale with respect to the Gaussian measure μ. Define the measure ν_v by

$$\nu_v(A) = \int_A M(1)\, d\mu \quad \forall\ A \in \mathcal{F}_1.$$

By the Girsanov theorem, $B(t) = W(t) - \int_0^t v(s)\, ds$ is a d-dimensional Wiener process under ν_v. In what follows, we will denote expectation with respect to ν_v by E_v. Consider

$$\begin{aligned} R(\nu_v||\mu) &= \int \left(\log \frac{d\nu_v}{d\mu}\right) d\nu_v \\ &= \int \left(\sum_{j=1}^{d} \int_0^1 v_j(s)\, dW_j(s) - \frac{1}{2}\int_0^1 |v(s)|^2\, ds\right) d\nu_v \\ &= E_v \left\{\sum_{j=1}^{d} \int_0^1 v_j(s)\, dB_j(s) + \frac{1}{2}\int_0^1 |v(s)|^2\, ds\right\} \\ &= E_v\left[\frac{1}{2}\int_0^1 |v(s)|^2\, ds\right] \end{aligned} \tag{12.3.2}$$

Therefore, one obtains

$$R(\nu_v||\mu) + \int f d\nu_v = E_v\left\{\frac{1}{2}\int_0^1 |v(s)|^2\, ds + f(B(s) + \int_0^{\cdot} v(s)\, ds)\right\}.$$

Therefore, by Theorem 12.2.3, we obtain

$$-\log E\left[e^{-f(W)}\right] \leq \inf_{v \in A_b} E_v\left\{\frac{1}{2}\int_0^1 |v(s)|^2\, ds + f(B(s) + \int_0^{\cdot} v(s)\, ds)\right\}. \tag{12.3.3}$$

Using the above inequality, we will establish that

$$-\log E\left[e^{-f(W)}\right] \leq E\left\{\frac{1}{2}\int_0^1 |v(s)|^2\, ds + f(W(s) + \int_0^{\cdot} v(s)\, ds)\right\} \tag{12.3.4}$$

for all $v \in \mathcal{A}$. Here, it should be noted that expectation on both sides is with respect to μ. To prove the inequality (12.3.4), we will first take a simple $v \in \mathcal{A}_b$; and there exists a $u \in \mathcal{A}_b$ such that the joint distribution of (W, v) under μ is equal to that of (B, u) under ν_u.

Step 2 Let v be a given simple bounded process of the form

$$v(t)(\phi) = X_0(\phi)1_{\{0\}}(t) + \sum_{j=0}^{k-1} X_j(\phi)1_{(t_j,t_{j+1}]}(t)$$

for all $t \in [0,1]$ and $\phi \in C([0,1];\mathbb{R}^d)$, with $X_j \in \mathcal{F}_{t_j}$ for all j, and $t_0 = 0$.

Define a map T_u by

$$T_u(\phi) = \phi - \int_0^{\cdot} u(s)(\phi)\, ds;$$

then,

$$\nu_u(T_u^{-1}(A)) = \mu(A) \quad \text{for all Borel sets } A \text{ in } C([0,1];\mathbb{R}^d).$$

Hence, if we can find a u such that

$$u(\phi) = v(T_u(\phi)) \quad \text{for all } \phi \in C([0,1];\mathbb{R}^d),$$

then the distribution of u under ν_u is equal to that of v under μ.

To find such a u, define a new set of random variables as follows:

$$Y_0(\phi) = X_0(\phi),$$

and for $j = 1, \ldots, k-1$,

$$Y_j(\phi) = X_j(\eta),$$

where η is any function that coincides with $\phi - \sum_{i=0}^{j-1} Y_i(\phi)(t_{i+1} - t_i)$ up to time t_j. Then, Y_j is $\mathcal{F}_{t_j}$-measurable for all j. Define

$$u(\phi) = Y_0(\phi)1_{\{0\}}(t) + \sum_{j=0}^{k-1} Y_j(\phi)1_{(t_j,t_{j+1}]}(t)$$

for all $0 \leq t \leq 1$. Then $u(\phi) = v(T_u(\phi))$ with probability one.

This implies that, for $B(\phi) = W(\phi) - \int_0^{\cdot} u_s(\phi)\, ds$ and $A \in \mathcal{B}(C([0,1];\mathbb{R}^d))$, $C \in \mathcal{B}(L^2([0,1];\mathbb{R}^d))$

$$\begin{aligned}
\nu_u(B \in A, u \in C) &= \nu_u\left(\left\{\phi : \phi - \int_0^{\cdot} u_s(\phi)\, ds \in A, u(\phi) \in C\right\}\right) \\
&= \nu_u(\{\phi : T_u(\phi) \in A, v(T_u(\phi)) \in C\}) \\
&= \mu(\{\psi : \psi \in A, v(\psi) \in C\}) \\
&= \mu(W \in A, v \in C),
\end{aligned} \tag{12.3.5}$$

which shows that the distribution of (B, u) under the measure ν_u is the same as the distribution of (W, v) under μ. Using this equivalence and (12.3.3), we obtain

$$\begin{aligned} -\log E^{-f(W)} &\le E_u \left\{ \frac{1}{2} \int_0^1 |u_s|^2 \, ds + f\left(B + \int_0^\cdot u_s \, ds \right) \right\} \\ &= E \left\{ \frac{1}{2} \int_0^1 |v_s|^2 \, ds + f\left(W + \int_0^\cdot v_s \, ds \right) \right\}, \end{aligned} \tag{12.3.6}$$

which implies (12.3.4) for all $v \in \mathcal{A}_s$. Let $L_\mu(W + \int_0^\cdot v_s \, ds)$ denote the measure on $C([0,1] : \mathbb{R}^d)$ that is induced by $W + \int_0^\cdot v_s \, ds$ under μ. Using (12.3.2), the equality on the right side of (12.3.6) implies that

$$R\left(L_\mu \left(W + \int_0^\cdot v_s \, ds \right) || \mu \right) = E \left\{ \frac{1}{2} \int_0^1 |v_s|^2 \, ds \right\} \tag{12.3.7}$$

for all $v \in \mathcal{A}_s$.

Step 3 Let $v \in \mathcal{A}_b$, so that $|v_s(\omega)| \le M < \infty$ for $0 \le s \le 1$, $\omega \in \Omega$. There exists a sequence of simple processes $\{v^n, n \in N\}$ such that $|v^n_s(\omega)| \le M < \infty$ for all $0 \le s \le 1, \omega \in \Omega$, and

$$\lim_{n \to \infty} E \int_0^1 |v^n_s - v_s|^2 \, ds = 0, \tag{12.3.8}$$

and thus $(W, \int_0^\cdot v^n_s \, ds)$ converges in distribution to $(W, \int_0^\cdot v_s \, ds)$ in $(C([0,1]; \mathbb{R}^d))^2$.

By virtue of Step 2, for each $n \in N$,

$$-\log E e^{-f(W)} \le E \left\{ \frac{1}{2} \int_0^1 |v^n_s|^2 \, ds + f\left(W + \int_0^\cdot v^n_s \, ds \right) \right\}. \tag{12.3.9}$$

It remains to show that the inequality above continues to hold in the limit as $n \to \infty$. Let $\mu_n = L_\mu(W + \int_0^\cdot v^n_s \, ds)$; then (12.3.7) implies that

$$\sup_{n \in N} R(\mu_n || \mu) = \sup_{n \in N} E \left\{ \frac{1}{2} \int_0^1 |v^n_s|^2 \, ds \right\} \le \frac{M^2}{2} < \infty. \tag{12.3.10}$$

Hence we can apply Proposition 12.2.5 (i) to obtain

$$\lim_{n \to \infty} Ef\left(W + \int_0^\cdot v^n_s \, ds \right) = Ef\left(W + \int_0^\cdot v_s \, ds \right). \tag{12.3.11}$$

Letting $n \to \infty$ in (12.3.9), we conclude that (12.3.2) is still valid for the limit process v, and thus, for any $v \in \mathcal{A}_b$.

Step 4 General $v \in \mathcal{A}$. We define

$$v^n_s(\phi) = v_s(\phi) 1_{\{|v_s(\phi)| \le n\}}, \quad 0 \le s \le 1, \quad \phi \in C([0,1]; \mathbb{R}^d). \tag{12.3.12}$$

Then, v^n is bounded for every $n \in N$, and thus Step 2 guarantees that (12.3.9) holds for each v^n. Let $\mu_n = L_\mu(W + \int_0^\cdot v_s^n\, ds)$; then (12.3.11) implies that

$$\sup_{n\in N} R(\mu_n||\mu) = \sup_{n\in N} E\left\{\frac{1}{2}\int_0^1 |v_s^n|^2\, ds\right\} \leq E\left\{\frac{1}{2}\int_0^1 |v_s|^2\, ds\right\} < \infty. \tag{12.3.13}$$

As in Step 3, Proposition 12.2.5 and the dominated convergence theorem yield (12.3.2) for any $v \in \mathcal{A}$. The desired upper bound has thus been established.

Now, we will give a proof of the lower bound in the variational representation formula; that is,

$$-\log \mathbb{E}e^{-f(W)} \geq \inf_{v\in\mathcal{A}} \mathbb{E}\left\{\frac{1}{2}\int_0^1 |v_s|^2\, ds + f(W + \int_0^\cdot v_s\, ds)\right\}.$$

Step 5 Let $(\Omega, \mathcal{F})$ be the Wiener space and f be a bounded, measurable function mapping Ω into $\mathbb{R}$. Let μ be the Wiener measure on Ω and $\Pi(\Omega)$ be the set of probabilities on Ω. Consider the measure η_0 where infimum is attained in the variational formula

$$-\log \int_\Omega e^{-f(x)}\, d\mu = \inf_{\eta\in\Pi(\Omega)} \left\{R(\eta||\mu) + \int_\Omega f(x)d\eta\right\}. \tag{12.3.14}$$

Then, η_0 is not only absolutely continuous with respect to μ, but it is in fact equivalent to μ on $\mathcal{F}$. It follows that, for each $t \in [0, 1]$, the restriction of η_0 to $\mathcal{F}_t$ is a probability measure that is equivalent to the restriction of μ to $\mathcal{F}_t$. Let R_t be the corresponding Radon-Nikodym derivative

$$R_t = \mathbb{E}\left[\frac{d\eta_0}{d\mu}|\mathcal{F}_t\right] = \mathbb{E}\left[\frac{e^{-f(x)}}{\int_\Omega e^{-f(x)}\mu(dx)}|\mathcal{F}_t\right]. \tag{12.3.15}$$

Then, $\{R_t; 0 \leq t \leq 1\}$ forms a μ-martingale that is bounded from below and above μ-a.s., respectively, by constants $\exp(-2||f||_\infty)$ and $\exp(2||f||_\infty)$. Moreover, since R_t is a martingale with respect to the augmentation under μ of the filtration generated by a Brownian motion, it can be represented as a stochastic integral $R_t = 1 + \int_0^t u_s\, dW_s$, where u_s is progressively measurable.

Since R_t is bounded from below, we can define $v_t = u_t/R_t$ and write

$$R_t = 1 + \int_0^t v_s R_s\, dW_s. \tag{12.3.16}$$

The random variable R_1 is bounded by a constant, and hence $\mathbb{E}(R_1^2) < \infty$. This observation and Equation (12.3.16) yield $\mathbb{E}\int_0^1 |v_s|^2 R_s^2\, ds < \infty$. Since R_t is bounded below by a constant, we have $\mathbb{E}\int_0^1 |v_s|^2\, ds < \infty$. Also, $d\eta_0/d\mu$ is bounded, so that one obtains

$$\int_{C([0,1]:\mathbb{R}^d)} \int_0^1 |v_s|^2 ds\, d\eta_0 < \infty. \tag{12.3.17}$$

These bounds and Equation (12.3.16) allow us to write

$$R_t = \exp\left[\int_0^t v_s\,dW_s - \frac{1}{2}\int_0^t |v_s|^2\,ds\right]. \tag{12.3.18}$$

Since R_t is a martingale, the Girsanov theorem identifies η_0 as the measure under which the process $\tilde{W} := W - \int_0^{\cdot} v_s\,ds$ is a Brownian motion. As in the proof of the upper bound to evaluate $R(\eta_0||\mu)$, we obtain

$$-\log \mathbb{E}e^{-f(W)} = \mathbb{E}^{\eta_0}\left\{\frac{1}{2}\int_0^1 |v_s|^2\,ds + f(\tilde{W} + \int_0^{\cdot} v_s)\right\} \tag{12.3.19}$$

Step 6 Let us first assume that f is continuous. Since progressively measurable processes can be approximated by bounded, simple processes in the L^2-sense, given $\epsilon > 0$, there exists a process v^* that is a bounded, simple process such that

$$\mathbb{E}^{\eta_0}\left\{\int_0^1 |v_s^* - v_s|^2\,ds\right\} < \frac{\epsilon}{2} \tag{12.3.20}$$

Let us write the process v^* in the form

$$v_t^*(\omega) = \xi_0(\omega)1_{\{0\}}(t) + \sum_{i=0}^{l-1} \xi_i(\omega)1_{(t_i,t_{i+1}]}(t), \qquad 0 \le t \le 1,\ \ \omega \in \Omega \tag{12.3.21}$$

where $0 = t_0 < t_1 < \cdots < t_l = 1$ and ξ_i is $\mathcal{F}_{t_i}$-measurable for each $i = 0, \ldots, l-1$. Each ξ_i can be approximated in $L^2(\mu)$ (and hence equivalently in $L^2(\eta_0)$ as well) by a smooth cylindrical functional with compact support, namely, $g_i(\omega_{s_1}, \ldots, \omega_{s_n})$, where $s_1 < s_2 < \cdots s_n \le t_i$. Replacing each ξ_i by g_i, and then using polygonalization in the time variable s, we can find a smooth progressively measurable functional z with continuous sample paths which approximates v^* in the sense that

$$\mathbb{E}^{\eta_0}\left\{\int_0^1 |z_s - v_s^*|^2\,ds\right\} < \frac{\epsilon}{2}.$$

It follows that, given $\epsilon > 0$, we can choose a progressively measurable process z as constructed above such that

$$\mathbb{E}^{\eta_0}\left\{\frac{1}{2}\int_0^1 |v_s|^2\,ds + f(\tilde{W} + \int_0^{\cdot} v_s\,ds)\right\} \ge \mathbb{E}^{\eta_0}\left\{\frac{1}{2}\int_0^1 |z_s|^2\,ds + f(\tilde{W} + \int_0^{\cdot} z_s\,ds)\right\} - \epsilon. \tag{12.3.22}$$

Consider the probability space $(\Omega, \mathcal{F}, (\mathcal{F}_t), \eta_0)$. Under the measure η_0, we have that $\tilde{W}(\omega) = \omega - \int_0^{\cdot} v_s(\omega)\,ds$ is a Brownian motion. Define

$$X(\omega) := \tilde{W}(\omega) + \int_0^{\cdot} v_s(\omega)\,ds,$$

and note that $X(\omega) = \omega$. The process X_t solves

$$X_t = \tilde{W}_t + \int_0^t v_s(X)\, ds. \tag{12.3.23}$$

In general, we can only assert that Equation (12.3.23) has a weak solution which is unique in law.

Define a probability measure η_1 on $(\Omega, \mathcal{F})$ by

$$\frac{d\eta_1}{d\eta_0} = \exp\left\{\int_0^1 (z_s - v_s)\, d\tilde{W}_s - \frac{1}{2}\int_0^1 |z_s - v_s|^2\, ds\right\}.$$

Then, $\eta_1 \equiv \eta_0$ and η_1 a.s., we can write, for all $0 \le t \le 1$,

$$\tilde{W}_t = \hat{W}_t + \int_0^t (z_s - v_s)\, ds,$$

where $\hat{W}$ is a Brownian motion on $(\Omega, \mathcal{F}, (\mathcal{F}_t), \eta_1)$. Thus, η_1 a.s., we have

$$W_t = \hat{W}_t + \int_0^t z_s\, ds. \tag{12.3.24}$$

We can rewrite Equation (12.3.24) as

$$Y_t = \hat{W}_t + \int_0^t z_s(Y)\, ds \tag{12.3.25}$$

Equation (12.3.24) has a strong pathwise unique solution by the choice of z. Therefore, $Y = h(\hat{W})$ for some Borel-measurable function h.

Note that η_1 depends on ϵ. Taking $\epsilon = 1/n$, let us denote the corresponding sequence of probability measures by $\eta^{(n)}$. Then, $\eta^{(n)}$ is the law of the solution of the equation

$$Y_t^{(n)} = \hat{W}_t^{(n)} + \int_0^t z_s^{(n)}(Y^{(n)})\, ds,$$

where $\hat{W}^{(n)}$ is a Wiener process with respect to $\eta^{(n)}$. Also, $\eta^{(n)} \to \eta_0$ weakly as $n \to \infty$. Thus, for any fixed constant $K > 0$ and any given $\epsilon > 0$, there exists an n such that

$$\mathbb{E}^{\eta_0}\left\{\frac{1}{2}\left[K \wedge \int_0^1 |v_s(X)|^2\, ds\right] + f(X)\right\} \ge \mathbb{E}^{\eta^{(n)}}\left\{\frac{1}{2}\left[K \wedge \int_0^1 |z_s^{(n)}(Y_n)|^2\, ds\right] + f(Y)\right\} - \epsilon. \tag{12.3.26}$$

Recalling that $Y_n = h_n(\hat{W}^{(n)})$, let us introduce the following notation:

$$\mathcal{L}_n(\cdot) := \frac{1}{2}\left[K \wedge \int_0^1 |z_s^{(n)}(h_n(\cdot))|^2\, ds\right] + f(h_n(\cdot)).$$

Since $\hat{W}^{(n)}$ is a $\eta^{(n)}$-Brownian motion and W is a μ-Brownian motion, we have $\mathbb{E}^{\eta^{(n)}}(\mathcal{L}_n(\hat{W}^{(n)})) = E(\mathcal{L}_n(W))$. Then,

$$\begin{aligned}
\mathbb{E}^{\eta_0}\left\{\frac{1}{2}\int_0^1 |v_s(X)|^2\, ds + f(X)\right\} &\geq \mathbb{E}^{\eta^{(n)}}(\mathcal{L}(\hat{W}^{(n)})) - \epsilon \\
&= \mathbb{E}(\mathcal{L}_n(W)) - \epsilon \\
&= \mathbb{E}\left\{\frac{1}{2}\left[K \wedge \int_0^1 |z_s^{(n)}(h_n(W))|^2\, ds\right] + f(h_n(W))\right\} - \epsilon \\
&= \mathbb{E}\left\{\frac{1}{2}\left[K \wedge \int_0^1 |z_s^*(W)|^2\, ds\right] + f(h_n(W))\right\} - \epsilon \\
&= \mathbb{E}\left\{\frac{1}{2}\left[K \wedge \int_0^1 |z_s^*|^2\, ds\right] + f\left(W + \int_0^{\cdot} z_s^*\, ds\right)\right\} - \epsilon,
\end{aligned} \tag{12.3.27}$$

where $z_s^* := z_s^{(n)} \circ h_n$ is progressively measurable. Now allow $K \to \infty$ using monotone convergence. The inequality (12.3.19) yields the lower bound for continuous f.

Step 7 If f is not continuous, let $\{f_j\}$ be a sequence of bounded continuous functions such that $||f_j||_\infty \leq ||f||_\infty < \infty$ and $\lim_{j\to\infty} f_j = f$, μ-a.s. The proceeding argument applied to each of the functions f_j implies that there exists a sequence of progressively measurable processes $\{z^{j*}, j \in \mathbb{N}\}$ that satisfies (12.3.27) for each j but with f replaced by f_j; that is

$$-\log \mathbb{E}e^{-f_j(W)} \geq \mathbb{E}\left\{\frac{1}{2}\int_0^1 |z_s^{j*}(W))|^2\, ds + f_j\left(W + \int_0^{\cdot} z_s^{j*}\, ds\right)\right\} - \epsilon. \tag{12.3.28}$$

Thanks to (12.3.10), we have

$$\sup_j R\left(L_\mu\left(W + \int_0^{\cdot} z_s^{j*}\, ds\right) || \mu\right) \leq \sup_j \mathbb{E}\left\{\frac{1}{2}\int_0^1 |z_s^{j*}|^2\, ds\right\} \leq ||f||_\infty. \tag{12.3.29}$$

It follows from this bound that the pair $(\int_0^{\cdot} z_s^{j*}, W)$ is tight, and hence there exists a subsequence such that $(\int_0^{\cdot} z_s^{j*}, W)$ converges in distribution to $(\int_0^{\cdot} z_s^*, W)$. It follows from (12.3.28), the dominated convergence theorem, and Proposition 12.2.5 that, for all sufficiently large j,

$$-\log \mathbb{E}e^{-f(W)} \geq \mathbb{E}\left\{\frac{1}{2}\int_0^1 |z_s^{j*}(W))|^2\, ds + f\left(W + \int_0^{\cdot} z_s^{j*}\, ds\right)\right\} - 2\epsilon, \tag{12.3.30}$$

which completes the proof of the lower bound. ∎

12.4 Sufficient Conditions for LDP

We will use the variational representation theorem proved in the last section to derive a set of sufficient conditions under which LDP holds. The sufficient conditions can be verified for a variety of stochastic differential systems that admit strong solutions. As mentioned in the previous section, it suffices to restrict our attention to bounded, Borel-measurable functions of a d-dimensional Wiener process, since unique strong solutions of stochastic differential equations started at a point $x_0 \in \mathbb{R}^d$ are Borel-measurable functions of the driving Wiener process.

Let $g_\epsilon : C([0,T], \mathbb{R}^d) \to E$ be a Borel-measurable function that depends on a parameter $\epsilon > 0$. Here, E is, as before, a complete separable metric space. For instance, one can take E to be $C([0,T], \mathbb{R}^d)$. Define the random variable $X_\epsilon = g_\epsilon(W(\cdot))$. The next theorem gives the conditions under which the Laplace-Varadhan principle holds for the family $\{X_\epsilon\}$ as $\epsilon \to 0$.

Theorem 12.4.1 *Suppose that there exists a function $g : C([0,T], \mathbb{R}^d) \to E$ such that the following conditions are satisfied:*

(i) The set $K_M = \{g(\int_0^\cdot v(s)\, ds : v \in \mathcal{A}_M\}$ is compact in E for all $M > 0$.

(ii) Let $\{v_\epsilon\}$ be any given family of $\mathcal{A}_M$-valued random variables such that, as $\epsilon \to 0$, v_ϵ converges in distribution to a limit, denoted by v. Then,

$$g_\epsilon\left(W(\cdot) + \frac{1}{\sqrt{\epsilon}} \int_0^\cdot v_\epsilon(s)\, ds\right) \to g\left(\int_0^\cdot v(s)\, ds\right). \tag{12.4.1}$$

Then the family $\{X_\epsilon\}$ satisfies the Laplace-Varadhan principle with a good rate function I given by

$$I(f) = \inf_{\{v \in L^2(0,T;\mathbb{R}^d): f = g\left(\int_0^\cdot v(s) ds\right)\}} \left\{\frac{1}{2}\int_0^T |v(s)|^2\, ds\right\} \tag{12.4.2}$$

for all measurable E-valued functions f, with the convention that the infimum of the empty set is infinity.

Proof

Step 1 First, we will show that for all $h \in C_b(E)$,

$$\inf_{x \in E}\{h(x) + I(x)\} \le \liminf_{\epsilon \to 0} \log \mathbb{E}\left[e^{-\frac{1}{\epsilon}h(X_\epsilon)}\right].$$

Toward this, define the function $f := \frac{1}{\epsilon} h \circ g_\epsilon$. Then, by Theorem 12.3.1

$$\begin{aligned} -\epsilon \log \mathbb{E}\left\{e^{-\frac{1}{\epsilon}h(X_\epsilon)}\right\} &= \inf_{v \in \mathcal{A}} \mathbb{E}\left(\frac{\epsilon}{2}\int_0^T |v(s)|^2\, ds + h \circ g_\epsilon(W + \int_0^\cdot v(s)\, ds)\right) \\ &= \inf_{v \in \mathcal{A}} \mathbb{E}\left(\frac{1}{2}\int_0^T |v(s)|^2\, ds + h \circ g_\epsilon(W + \frac{1}{\sqrt{\epsilon}}\int_0^\cdot v(s)\, ds)\right), \end{aligned}$$

which we will call $\inf_{v\in\mathcal{A}} J_v(\epsilon)$. By definition of infimum, given any $\delta > 0$, there exists a $v_\epsilon \in \mathcal{A}$ such that

$$\inf_{v\in\mathcal{A}} J_v(\epsilon) \geq J_{v_\epsilon}(\epsilon) - \delta. \tag{12.4.3}$$

It suffices to prove that

$$\liminf_{\epsilon\to 0} J_{v_\epsilon}(\epsilon) \geq \inf_{x\in E}\{h(x) + I(x)\}. \tag{12.4.4}$$

Step 2 To prove (12.4.4), let us first note that if M denotes $\|h\|_\infty$, then from (12.4.3), we can conclude that

$$J_{v_\epsilon}(\epsilon) \leq M + \delta$$

by taking $v \equiv 0$. Therefore,

$$\mathbb{E}\left(\frac{1}{2}\int_0^T |v_\epsilon(s)|^2\, ds\right) \leq 2M + \delta < \infty$$

for all $\epsilon > 0$. For any $N > 0$, define the stopping time

$$\tau_{\epsilon,N} = \inf\{t \in [0, T] : \int_0^t |v_\epsilon(s)|^2\, ds \geq N\} \wedge T.$$

Define $v_{\epsilon,N}(t) = v_\epsilon(t) 1_{[0,\tau_{\epsilon,N}]}(t)$. Then $v_{\epsilon,N} \in \mathcal{A}$.

$$P\{v_\epsilon \neq v_{\epsilon,N}\} \leq P\left\{\int_0^T |v_\epsilon|^2\, ds > N\right\} \leq \frac{2M + \delta}{N}.$$

Let A_N denote the set $\{v_\epsilon \neq v_{\epsilon,N}\}$, and

$$L(v) = \frac{1}{2}\int_0^T |v(s)|^2\, ds + h \circ g_\epsilon\left(W + \frac{1}{\epsilon}\int_0^\cdot v(s)\, ds\right)$$

for any $v \in \mathcal{A}$. Then

$$\begin{aligned} J_{v_\epsilon}(\epsilon) &= \int_{A_N} L(v_\epsilon)\, dP + \int_{A_N^c} L_{v_{\epsilon,N}}\, dP \\ &= \int_{A_N} L(v_\epsilon)\, dP - \int_{A_N} L_{v_{\epsilon,N}}\, dP + J_{v_{\epsilon,N}}(\epsilon) \\ &\geq -2M\frac{2M + \delta}{N} + J_{v_{\epsilon,N}}(\epsilon). \end{aligned}$$

Using this bound in Equation (12.4.3), we obtain

$$\inf_{v\in\mathcal{A}} J_v(\epsilon) \geq J_{v_{\epsilon,N}}(\epsilon) - 2M\frac{2M + \delta}{N} - \delta.$$

Therefore, it is sufficient to prove (12.4.4) for $v_{\epsilon,N}$ in the place of v_ϵ.

Step 3 Since $\int_0^T |v_{\epsilon,N}(s)|^2\, ds \le N$ for any $\epsilon > 0$, the family $\{v_{\epsilon,N}\}$ indexed by $\epsilon > 0$ is tight in the weak topology on the space $\mathcal{A}_N$. Hence, there exists a subsequence along which $v_{\epsilon,N}$ converge weakly to an $\mathcal{A}_N$ valued process, v. From Condition 1 of the theorem, we have

$$\begin{aligned}\liminf_{\epsilon\to 0} J_{v_{\epsilon,N}}(\epsilon) &\ge \mathbb{E}\left(\frac{1}{2}\int_0^T |v(s)|^2\, ds + h\circ g\left(\int_0^{\cdot} v(s)\, ds\right)\right)\\ &\ge \inf\left\{\frac{1}{2}\int_0^T |v(s)|^2\, ds + h(x)\right\}\end{aligned}$$

where infimum is taken over $\left\{(x,v)\in E\times L^2([0,T]:\mathbb{R}^d) : x = g\left(\int_0^{\cdot} v(s)\, ds\right)\right\}$. Thus,

$$\begin{aligned}\liminf_{\epsilon\to 0} J_{v_{\epsilon,N}}(\epsilon) &\ge \inf_{x\in E}\ \inf_{v\in L^2([0,T]:\mathbb{R}^d): x = g\left(\int_0^{\cdot} v(s)ds\right)} \left\{\frac{1}{2}\int_0^T |v(s)|^2\, ds + h(x)\right\} \qquad (12.4.5)\\ &\ge \inf_{x\in E}\{I(x) + h(x)\}.\end{aligned}$$

The proof of (12.4.4) is over.

Step 4 We will now prove the upper bound:

$$\inf_{x\in E}\{I(x)+h(x)\} \ge \liminf_{\epsilon\to 0}\log\mathbb{E}\left[e^{-\frac{1}{\epsilon}h(X_\epsilon)}\right]$$

for all $h\in C_b(E)$. Without loss of generality, let $\inf_{x\in E}\{I(x)+h(x)\} < \infty$. Call it α. By definition of infimum, given any $\delta > 0$, there exists a $x_0 \in E$ and $v_0 \in L^2([0,T];\mathbb{R}^d)$ such that

$$I(x_0) + h(x_0) \le \alpha + \delta/2 \quad \text{and} \quad \frac{1}{2}\int_0^T |v_0(s)|^2\, ds \le I(x_0) + \delta/2,$$

and $x_0 = g\left(\int_0^{\cdot} v_0(s)\, ds\right)$. From Theorem 12.3.1

$$\begin{aligned}&\limsup_{\epsilon\to 0} -\epsilon\log\mathbb{E}\left\{e^{-\frac{1}{\epsilon}h(X_\epsilon)}\right\}\\ &= \limsup_{\epsilon\to 0}\inf_{v\in\mathcal{A}} J_v(\epsilon)\\ &\le \limsup_{\epsilon\to 0} J_{v_0}(\epsilon)\\ &= \frac{1}{2}\int_0^T |v_0(s)|^2\, ds + \limsup_{\epsilon\to 0}\mathbb{E}\left(h\circ g_\epsilon\left(W_\cdot + \frac{1}{\sqrt{\epsilon}}\int_0^{\cdot} v_0(s)\, ds\right)\right)\\ &\le I(x_0) + \delta/2 + \limsup_{\epsilon\to 0}\mathbb{E}\left(h\circ g_\epsilon\left(W_\cdot + \frac{1}{\sqrt{\epsilon}}\int_0^{\cdot} v_0(s)\, ds\right)\right).\end{aligned}$$

(12.4.6)

By Condition 1, as $\epsilon \to 0$,

$$\mathbb{E}\left(h \circ g_\epsilon\left(W + \frac{1}{\sqrt{\epsilon}}\int_0^{\cdot} v_0(s)\, ds\right)\right)$$

converges to $h \circ g\left(\int_0^{\cdot} v_0(s)\, ds\right) = h(x_0)$. Thus the right side of Equation (12.4.6) is bounded above by $\inf_{x \in E}\{I(x) + h(x)\} + \delta$. Since δ is arbitrary, the proof is complete. ∎

Exercises

1. Derive the rate function I for each of the four distributions given in Example 12.1.2.
2. Prove that, for any $h \in C_b([0,1])$,

$$\lim_{n\to\infty} \frac{1}{n} \log \int_0^1 e^{-nh(x)}\, dx = -\min_{[0,1]} h(x).$$

3. Let μ and ν be any two probability measures on a measurable space. Show that the relative entropy $R(\nu \mid \mu)$ is nonnegative and that $R(\nu \mid \mu) = 0$ is equivalent to $\nu = \mu$.
4. Using the Donsker-Varadhan variational formula, prove that $R(\nu \mid \mu)$ is a convex and lower semicontinuous function of (ν, μ). Here, the space of probability measures is equipped with the topology of weak convergence.
5. Prove that $R(\cdot \mid \mu)$ has compact level sets.
6. Show that the LDP rate function is unique; that is, if I and J are two LDP rate functions for a sequence $\{X_n\}$, then $I(x) = J(x)$ for all x.
7. Let G and H be two complete separable metric spaces, and let $\phi : G \to H$, a continuous function. Suppose that $\{X_n\}$ is a sequence of G-valued random variables for which LDP holds with rate function I. Show that the sequence $\{\phi(X_n)\}$ satisfies LDP with rate function J where

$$J(y) = \inf\{I(x) : \phi(x) = y\}.$$

This result is known as the contraction principle.

Notes on Chapters

Chapter 1

The material contained in this chapter is standard. The Kolmogorov consistency theorem depends on the topology of the state space, and hence, is not an abstract measure-theoretic result. The reader can refer to [59] for a proof of the Kolmogorov consistency theorem when the state space of the random variables is a complete separable metric space. As a word of caution, notions such as realization and equivalence of a process have various meanings in the literature.

Chapter 2

A full account of Brownian motion can be found in [44]. The construction of Brownian motion by Ciesielski [6] simplifies that of Lévy. The KKL expansion of a Brownian motion is usually attributed to Karhunen and Loéve. It is worthwhile to note that Kosambi's work [46] appeared a bit earlier than the works of Karhunen [42] and Loéve [51]. We have omitted the proof of almost sure convergence of the quadratic variation of a Brownian motion when $\{P_n\}$ is a refining sequence of partitions with $||P_n|| \to 0$. The proof depends on backward martingales (see for eg., [30]).

Chapter 3

Martingale theory was developed systematically by Doob and Meyer. Excellent accounts of it can be found in the books by Doob [15], Meyer [53], and Dellacherie and Meyer [13]. The Doob-Meyer decomposition is the main result of this chapter. Protter [60] in a modern proof of the decomposition without using natural processes. In Section 3.5, one should probably refer to the Meyer process as the Doob-Meyer process. An example of a local martingale which is not a martingale is given in Chung and Williams [8].

Chapter 4

The Brownian semigroup and its generator are explained as in the works of Wentzell [75] and Itô [34]. The connection between pregenerators and martingales is given as a prelude to martingale problems studied in Chapter 7.

Chapter 5

The Itô integral is developed by a longer, constructive approach than by means of the extension of an isometry. The presentation of the Itô formula and Girsanov theorem follow as in [39]. The Burkhölder-Davis-Gundy inequality appears as an application of the Itô formula, and its proof is taken from [62]. The proof of the joint measurability of local time $L_t^a(X)$ (of a continuous semimartingale X) in t, a is rather long and is omitted. We refer the reader to [41] for a proof of this result. The Girsanov theorem is presented as in [39]. A good reference for stochastic integration is the book by Kuo.

Chapter 6

Stochastic differential equations form the central object of interest in this book. One can obtain existence and uniqueness of solutions to stochastic differential equations under continuity, coercivity, and monotonicity conditions on the drift and diffusion coefficients as in [47]. However, we have given the standard fare consisting of Lipschitz continuity and linear growth conditions on the coefficients. Our presentation on weak solutions follows along the lines of Ikeda and Watanabe [32] and Karatzas and Shreve [41]. Markov property of solutions is established, and under general hypotheses, solutions of stochastic differential equations are shown to be diffusion processes as in [39].

Chapter 7

The martingale problem invented by Stroock and Varadhan [71] is unique to stochastic analysis, with no counterpart in the theory of ordinary and partial differential equations. We present the basic theory as it appears in [70]. A study of martingale problems can be found in [24]. Though several simplifications and extensions of the theory have appeared, we have limited ourselves to the basic theory as in the original work of Stroock and Varadhan. We refer the reader to the book by Giga, [27] for a proof of the Calderon-Zygmund estimate for elliptic equations. The extension to parabolic equations is in the appendix of [70].

Chapter 8

The material in the first two sections is standard and well known. Wentzell [75] gives a beautiful account of the connection between stochastic differential equations and partial differential equations. The reader is also referred to the book by Bass and [41]. Our discussions of the Feynman-Kac formula and its application to finance theory are heuristic and brief since there are several excellent texts devoted on finance theory such as [4] and [68]. The Kolmogorov equations are derived in Section 8.6 following the work of Gihman and Skorohod [29] (also, see [25]).

Chapter 9

A full proof of Theorem 9.3.5 can be found in the book by Goberg and Krein [28]. Theorem 9.4.2 and the representation given by Equation 9.4.4 are taken from [40]. The representation was derived earlier by Hitsuda [31] using martingale techniques and the Girsanov theorem.

Chapter 10

Jump Markov processes are treated without introducing point processes and compensated Poisson random measures. For a more extensive treatment of jump Markov processes and weak convergence, one can refer to [36], [24] and [69]. Theorem 10.3.3 on non-explosion of jump Markov processes is due to Metivier [52]. The results in Section 10.4 on diffusion approximation is an adaptation of a result in [52]. The boundedness of the drift coefficient b seems necessary to obtain Theorem 10.4.3.

Chapter 11

An excellent development and description of ergodic theory for stochastic systems can be found in the books by Khasminskii [43] and Da Prato and Zabczyk [10]. An L^2-ergodic theorem for one-dimensional diffusions uses local times (see [72]) is presented in Section 11.2. A result of Varadhan [73] on existence and uniqueness of invariant measures for d-dimensional diffusions appears in Section 11.3. A more general result is due to Echeverria [21]. The latter is not presented since the proof, though beautiful, is more involved. Theorem 11.4.2 is the Krylov-Bogolyubov result on the existence of invariant measures and is widely applicable. A criterion to check uniqueness of invariant measures is given in Theorem 11.4.3 and is due to Chow and Khasminskii [7]. Theorem 11.5.4 is due to Khasminskii and is used to prove Theorem 11.5.5, a well-known result of Doob [15]. The Bismut-Elworthy theorem (see [2] and [23]) appears as

Theorem 11.5.6 which, by Doob's theorem, yields the ergodic behavior of d-dimensional stochastic systems.

Chapter 12

The chapter deals with Wentzell-Freidlin type large deviations result for small noise perturbations in stochastic systems (see [26]). The variational representation theorem in Section 12.3 and the sufficient conditions in Section 12.4 are due to Boué and Dupuis [3]. The weak convergence approach of Dupuis and Ellis [18] is well suited to prove the large deviations principle for diffusions.

References

[1] Bass, R. F. (1998) *Diffusions and Elliptic Operators*, Springer-Verlag, New York.

[2] Bismut, J. M. (1984) *Large Deviations and the Malliavin Calculus*. Birkhäuser, Berlin.

[3] Boué, M. and Dupuis, P. (1998) A variational representation for certain functionals of Brownian motion. *Ann. Prob.* **26**, 1641–1659.

[4] Björk, T. (2009) *Arbitrage Theory in Continuous time*. Third Ed., Oxford Univ. Press, Oxford.

[5] Chung, K. L. and Doob, J. L. (1965) Fields, optionality and measurability. *Amer. J. Math.*, **87**, 397–424.

[6] Ciesielski, Z. (1961) Hölder condition for realizations of Gaussian processes. *Trans. Amer. Math. Soc.*, **99**, 403–413.

[7] Chow, P. L. and Khasminskii, R. Z. (1997) Stationary solutions of nonlinear stochastic evolution equations. *Stoch. Anal. Appl.* **15**, 671–699.

[8] Chung, K. L. and Williams, R. J. (1990) *Introduction to Stochastic Integration*. Second Ed., Birkhäuser, Boston.

[9] Cramér, H. (1938) Sur un nouveau théorème-limite de la théorie des probabilités. In *Actualités Scientifiques et Industrielles*, no. 736 in Colloque consecré à la théorie des probabilités, pp 5–23, Hermann, Paris.

[10] Da Prato G. and Zabczyk, J. (1996) *Ergodicity for Infinite Dimensional Systems*, Cambridge University Press, London.

[11] Dembo, A. and Zeitouni, O. (1998) *Large Deviations Techniques and Applications*. Second Ed., Springer, New York.

[12] Dellacherie, C. (1972) *Capacités et Processus Stochastiques*. Springer-Verlag, Berlin.

[13] Dellacherie, C. and Meyer, P. A. (1976) *Probabilités and Potentiel*. Hermann, Paris.

[14] Doob, J. L. (1953) *Stochastic Processes*. John Wiley, New York.

[15] Doob, J. L. (1948) Asymptotic properties of Markoff transition probabilities. *Trans. Am. Math. Soc.* **63**, 393–421.

[16] Doob, J. L. (1984) *Classical Potential Theory and Its Probabilistic Counterpart*. Springer-Verlag, Berlin.

[17] Donsker, M. D. and Varadhan, S. R. S. (1975, 1977, 1983) Asymptotic evaluation of certain Markov process expectations for large time. *Comm. Pure Appl. Math.* I, II, III, IV **28, 29, 36**, 1–47, 279–301, 389–461, 183–212.

[18] Dupuis, P. and Ellis, R. S. (1997) *A Weak Convergence Approach to the Theory of Large Deviations*. Wiley-Interscience, New York.

[19] Dvoretzky, A., Erdös, P., and Kakutani, S. (1961) Nonincrease everywhere of the Brownian motion process. *Proc. 4th Berkeley Symp. Math. Stat. and Probab.* **2**, 103–116.

[20] Dynkin, E. B. (1965) *Markov Processes.* Vol. I and II, Springer-Verlag, Berlin.

[21] Echeverria, P. E. (1982) A criterion for invariant measures of Markov processes. *Z. Wahrsch. verw. Gebiete* **61**, 1–16.

[22] Ellis, R. S. (1999) The theory of large deviations: from Boltzmann's 1877 calculation to equilibrium macrostates in 2-D turbulence. *Physica D*, 106–136.

[23] Elworthy, K. D. (1992) Stochastic flows on Riemannian manifolds. *Diffusion Problems and Related Problems in Analysis*, eds. Pinsky, M. A. and Wihstutz, V., Vol II, Birkhäuser, 33–72.

[24] Ethier, S. N. and Kurtz, T. G. (1986) *Markov Processes: Characterization and Convergence.* J. Wiley and Sons, New York.

[25] Friedman, A. (1975) *Stochastic Differential Equations and Applications.* Academic Press, New York.

[26] Freidlin, M. I. and Wentzell, A. D. (1984) *Random Perturbations of Dynamical Systems.* Springer-Verlag, New York.

[27] Giga, M.-H., Giga, Y., and Saal, J. (2010) *Nonlinear Partial Differential Equations.* Birkhäuser, Boston.

[28] Gohberg, I. C. and Krein, M. G. (1970) *Theory and Applications of Volterra Operators in Hilbert Space*, Trans. Math. Monographs, AMS, Providence.

[29] Gihman, I. I. and Skorohod, A. V. (1972) *Stochastic Differential Equations.* Springer-Verlag, New York.

[30] Hida, T. (1980) *Brownian Motion.* Springer-Verlag.

[31] Hitsuda, M. (1968) Representation of Gaussian processes equivalent to Wiener process. *Osaka J. Math.* **5**, 299–312.

[32] Ikeda, N. and Watanabe, S. (1981) *Stochastic Differential Equations and Diffusion Processes.*, North-Holland, Amsterdam.

[33] Itô, K. (1984) *Introduction to Probability Theory.* Cambridge Univ. Press, Cambridge.

[34] Itô, K. (2004) *Stochastic Processes.* Edited by Ole E. Barndorff-Nielson and Ken-iti Sato, Springer-Verlag, New York.

[35] Itô, K. and McKean, H. P., Jr. (1974) *Diffusion Proceses and their Sample Paths.* Springer-Verlag, Berlin.

[36] Jacod, J. and Shiryayev, A. N. (2003) *Limit Theorems for Stochastic Processes.* Second Ed., Springer-Verlag, New York.

[37] Kakutani, S. (1944) Two-dimensional Brownian motion and harmonic functions. *Proc. Acad. Japan* **20**, 706–714.

[38] Kallenberg, O. (1997) *Foundations of Modern Probability.* Springer, New York.

[39] Kallianpur, G. (1980) *Stochastic Filtering Theory.* Springer-Verlag, New York.

[40] Kallianpur, G. and Oodaira, H. (1973) Non-anticipative representations of equivalent Gaussian processes. *Ann. Probab.* **1**, 104–122.

[41] Karatzas, I. and Shreve, S. E. (1991) *Brownian Motion and Stochastic Calculus.* Second Ed., Springer, New York.

[42] Karhunen, K. (1947) Über linear methoden in der Wahrscheinlichkeitsrechnung. *Ann. Acad. Sci Fenn.* **37**.

[43] Khasminskii, R. Z. (2012) *Stochastic Stability of Differential Equations.* Second Ed., Springer-Verlag, Berlin.

[44] Knight, F. (1981) *Essentials of Brownian motion and Diffusion*. Math. Surv. **18**, AMS, Providence, Rhode Island.

[45] Kolmogorov, A. N. (1931) Über die analytischen Methoden in der Wahrscheinlichkeitsrechnung. *Math. Ann.* **104**, 415–458.

[46] Kosambi, D. D. (1943) Statistics in function space. *J. Indian Math. Soc. (N.S.)* **7**, 76–88.

[47] Krylov, N. V., and Rozovskii, B. L. (1981) Stochastic evolution equations. *J. Sov. Math.* **16**, 1233–1277.

[48] Kuo, H. - H. (1975) *Gaussian Measures in Banach Spaces*, Lect. Notes in Math. 463, Springer-Verlag, New York.

[49] Kuo, H. - H. (2006) *Introduction to Stochastic Integration*, Springer Universitext.

[50] Lévy, P. (1956) A special problem of Brownian motion and a general theory of Gaussian random functions. *Proc. Third Berkeley Symp. Math. Statist. Probab. II*, 133–175. Univ. of California Press, Berkeley and Los Angeles.

[51] Loéve, M. (1945) Fonctions aleatoires de secode ordre. *C. R. Acad. Sci.* **220**.

[52] Metivier, M. (1988) *Stochastic Partial Differential Equations in infinite Dimensional Spaces*. Quaderni, Scuola Normale Superiore, Pisa.

[53] Meyer, P. A. (1966) *Probability and Potentials*. Blaisdell Publishing Company, Waltham, Mass.

[54] Meyer, P. A. (1967) Intégrales stochastiques. *Lecture Notes in Mathematics* **39**, 72–162. Springer-Verlag, Berlin.

[55] Mogulskii, A. A. (1976) Large deviations for trajectories of multi-dimensional random walks. *Theory Probab. Appl.* **21** 300–315.

[56] Neveu, J. (1965) *Mathematical Foundations of the Calculus of Probability*. Holden-Day, Inc., San Francisco.

[57] Port, S. and Stone, C. (1978) *Brownian motion and Classical Potential Theory*. Academic Press, New York.

[58] Paley, R. E. A. C., Wiener, N., and Zygmund, A. (1933) Note on random functions. *Math. Z.*, **37**, 647–668.

[59] Parthasarathy, K. R. (1967) *Probability Measures on Metric Spaces*. Academic Press, New York.

[60] Protter, P. E. (2004) *Stochastic Integration and Differential Equations*. Springer-Verlag, Berlin.

[61] Reed, M. and Simon, B. (1975) *Methods of Modern Mathematical physics. Vol. 1: Functional Analysis*, Academic Press, New York.

[62] Revuz, D. and Yor, M. (1994) *Continuous Martingales and Brownian Motion*. Second Ed., Springer-Verlag, Berlin.

[63] Riesz, F. and Sz-Nagy, B. (1955) *Functional Analysis*, Frederick Ungar Publishing Company, New York.

[64] Royden, H. L. (1968) *Real analysis*. Second Ed., Macmillan Publ. Co., New York.

[65] Rudin, W. (1966) *Real and Complex Analysis*. McGraw-Hill, New York.

[66] Sanov, I. N. (1957) On the probability of large deviations of random variables. *Mat. Sbor.* **42**, Engl. Trans. in Selected Translations in Mathematical Statistics and Probability I (1961), pp 213–244.

[67] Schilder, M. (1966) Some asymptotic formulae for Wiener integrals. *Trans. Amer. Math. Soc.* **125**, 63–85.

[68] Sengupta, A. (2005) *Pricing Derivatives*. McGraw-Hill, New York.

[69] Skorokhod, A. V. (1965) *Studies in the Theory of Random Processes.* Addison-Wesley, Reading, Massachusetts.

[70] Stroock, D. W. and Varadhan, S. R. S. (1969) Diffusion processes with continuous coefficients, I and II. *Comm. Pure and Appl. Math.* **22**, 345–400 and 479–530.

[71] Stroock, D. W. and Varadhan, S. R. S. (1979) *Multidimensional Diffusion Processes.* Springer-Verlag, Berlin.

[72] Sundar, P. (1989) Ergodic solutions of stochastic differential equations, *Stoch. and Stoch. Reports* **28**, 65–83.

[73] Varadhan, S. R. S. (1977) *Lectures on Diffusion Problems and Partial Differential Equations,* TIFR, Bombay.

[74] Varadhan, S. R. S. (1984) *Large Deviations and Applications.* CBMS-NSF Conference Series, SIAM, Philadelphia.

[75] Wentzell, A. D. (1981) *A Course in the Theory of Stochastic Processes.* McGraw-Hill, New York.

Index

The manufacturer's authorised representative in the EU for product safety is Oxford University Press España S.A. of El Parque Empresarial San Fernando de Henares, Avenida de Castilla, 2 - 28830 Madrid (www.oup.es/en or product.safety@oup.com). OUP España S.A. also acts as importer into Spain of products made by the manufacturer.
Printed and bound by CPI Group (UK) Ltd, Croydon, CR0 4YY

06/07/2026

02157632-0008